The Death of Nature

Carolyn Merchant

THE **DEATH**
OF **NATURE**

WOMEN, ECOLOGY,
AND THE SCIENTIFIC REVOLUTION

HarperSanFrancisco

A Division of HarperCollins*Publishers*

Acknowledgment is made for the permission of the *Journal of the History of Philosophy* to include a revised version of the author's article "The Vitalism of Anne Conway" (July 1979) in Chapter 11; of *Ambix* to include parts of the author's article "The Vitalism of Francis Mercury Van Helmont" (November 1979) in Chapters 4 and 11; of Indiana University Press to reprint from *Metamorphoses* by Publius Ovid, translated by Rolfe Humphries (Bloomington: Indiana University Press, 1955); and of Cornell University Press to reprint tables on p. 312 from William E. Monter, *Witchcraft in France and Switzerland*. Copyright © 1976 by Cornell University.

FIRST HARPER & ROW PAPERBACK EDITION PUBLISHED IN 1983.

Designed by Paul Quin

Library of Congress Cataloging-in-Publication Data

Merchant, Carolyn.
  The death of nature.

  Originally published in 1980; with new preface.
  Includes bibliographical references.
  1. Women in science.   2. Philosophy of nature.
3. Human ecology.   I. Title.
Q130.M47 1989   304.2   89-24516
ISBN 0-06-250595-5

91 92 93 MCN 10 9 8 7 6 5 4

For David and John,
Elizabeth and Ann

Contents

List of Illustrations

Time Line

Events in the Period of the Scientific Revolution

1484 Death of Pico della Mirandola

1486 *Malleus Maleficarum* published

1489 Marsillio Ficino accused of practicing magic

1510 Henry Cornelius Agrippa's *De Occulta Philosophia*

1513 Niccolò Machiavelli completes *The Prince*

1515 First printed edition of Ptolemy's *Almagest*

1517 Martin Luther's *95 Theses* against the abuse of indulgences

1518 Lucas Cranach paints "The Nymph of the Spring"

1525 German Peasants War begins in Swabia

1528 Death of Albrecht Dürer

1531 Thomas Elyot's *The Boke Named the Governour*

1535 Henry VIII dissolves Catholic monasteries in England

1541 Death of Paracelsus

1543 Copernicus dies in the year his *On the Revolutions of the Celestial Spheres* is published

1543 Andreas Vesalius's *On the Fabric of the Human Body*

1543 Timber Preservation Act in England

1546 Michael Servetus completes book in which the "lesser circulation" of the blood (heart to lungs) is described

1551 Erasmus Reinhold publishes the Copernican based *Prutenic Tables* leading to reform of the calendar

1553 Mary Tudor ("Bloody Mary") restores Catholicism to England and persecutes Protestants

1556 Georg Agricola's *De Re Metallica*

1558 Accession of Elizabeth I to throne of England

1558 John Knox's *First Blast of the Trumpet Against the Monstrous Regiment of Women*

1558 Johann Battista Della Porta's *Natural Magic*

1564 Death of John Calvin

1565 Bernardino Telesio's *De Rerum Natura Iuxta Propria Principia*

1572 Tycho Brahe's "new star" (*super nova*) challenges the incorruptibility of the heavens

1576 Jean Bodin's *Six Books of the Republic*

1578 Queen Elizabeth requests Londoners not to use sea-coal in their industries

1584 Reginald Scot's *Discoverie of Witchcraft*

1587 Death of Mary Stuart, Queen of Scots

1589 Accession of Henry IV to throne of France

1590 Edmund Spenser's *The Faerie Queen*

ca. 1590 Henri Boguet's *An Examen of Witches*

1594 Richard Hooker's *On The Laws of Ecclesiastical Polity*

1599 Tommaso Campanella's revolution in Naples

1600 Giordano Bruno burned in Rome

1600 William Gilbert's *On the Magnet* describes his experiments with magnetic and electrical substances

1602 Tommaso Campanella writes *City of the Sun* in prison

1603 Accession of James I to throne of England

1603 Fabricius of Aquapendente announces his discovery of the valves in the veins

1607 Founding of Jamestown in Virginia

1609 Johannes Kepler's *New Astronomy* announces the law of planetary ellipses and the law of areas

1610 Galileo's *Sidereal Messenger* announces the satellites of Jupiter, craters on the moon, and the discreteness of the Milky Way

1610 Accession of Louis XIII to throne of France

1611 John Donne's "Anatomie of the World" written

1616 Death of William Shakespeare

1616 Godfrey Goodman's *The Fall of Man*

1618 Thirty Years' War begins

1619 Valentin Andreä's *Christianopolis* written

1619 Kepler announces the "harmonic law" in his *Harmony of the World*

1619 Descartes' "vision" of the mathematical method

1622 Indians massacre English settlers in colony of Virginia

1623 Urban VIII becomes Pope

1624 Cardinal Richelieu becomes chief minister of France under Louis XIII

1624 Marin Mersenne's *La Vérité des Sciences*

1625 Accession of Charles I to throne of England

1627 Francis Bacon's *New Atlantis*

1627 Kepler's *Rudolphine Tables*

1633 Galileo censured by the Inquisition

1634 Cornelius V~rmuyden, Dutch engineer, prepares plan to drain fens and straighten the Bedford River in England

1637 Death of Robert Fludd

1638 Galileo's *Discourses on Two New Sciences*

1642 Birth of Isaac Newton

1642 English Civil War begins

1642 Death of Richelieu

1642 Death of Galileo

1643 Death of Louis XIII

1643 Accession of Louis XIV (at age 5) to throne of France

1643 Cardinal Mazarin becomes chief minister of France

1643 Torricelli's barometer

1644 Descartes' *Principles of Philosophy*

1644 Death of Johann Baptista Van Helmont

1644 Matthew Hopkin's witch trials in England

1649 Charles I of England beheaded

1649 Gerrard Winstanley and Diggers take over George Hill in Surrey

1649 Pierre Gassendi's *Syntagma Philosophiae Epicuri*

1650 Thomas Vaughan's *Anima Magica Abscondita*

1651 William Harvey's *On Generation*

1653 Henry More's *Antidote Against Atheism*

1654 Otto von Guericke's "Magdeburg Experiment" demonstrates the force of atmospheric pressure

1658 Death of Oliver Cromwell

1660 English Restoration

1660 Founding of the Royal Society

1661 Death of Cardinal Mazarin, Chief Minister of France

1661 Robert Boyle's *Sceptical Chemist*

1661 John Evelyn's *Fumifugium*

1662 Boyle's law of reciprocal relation of gas pressure to volume

1665 Newton lays foundation for law of gravitation, integral calculus, and dispersion of light

1667 John Milton's *Paradise Lost*

1667 The Duchess of Newcastle visits the Royal Society

1674 Anton von Leeuwenhoek views animalcules in his microscope

1678 Ralph Cudworth's *True Intellectual System*

1679 Death of Anne Conway

1685 Accession of James II to throne of England

1686 Bernard de Fontenelle's *Conversations on the Plurality of Worlds*

1687 Sir Isaac Newton's *Principia Mathematica*

1687 Elizabeth Cellier proposes a corporation of midwives

1688 English (or Glorious) Revolution

1689 Accession of William and Mary to throne of England

1690 John Locke's *Two Treatises on Government*

1696 Francis Mercury Van Helmont visits Gottfried Wilhelm von Leibniz at Hanover

1702 Accession of Queen Anne to throne of England

1713 William Derham's *Physico-theology*

1716 Leibniz/Clarke correspondence in the year of Leibniz's death

1700

1650 -

Acknowledgments

It is a pleasure to record my gratitude to the persons and organizations who have supported and contributed to this book. My teachers have been the people, the period, and the place in which I have lived in recent years. The initial inspiration for a critical reassessment of the Scientific Revolution came from classes and numerous conversations with my friend and fellow historian of science David Kubrin, who in the early 1970s was teaching courses in the San Francisco Liberation School. His perspectives on Isaac Newton, the mechanistic analysis of nature, and the dialectic have been fundamental to a conceptual reorganization of the period's history. Sustained support, ideas, and critical evaluation have been offered by Charles Sellers, a colleague and close friend. Both have read and criticized the manuscript through its many stages.

Many others have contributed their scholarly knowledge and stimulating conversation. I have gained much from colleagues at the University of San Francisco, especially the "Human Prospect" faculty group (sponsored by a grant from the National Endowment for the Humanities, EP–25821–76–1126), and from my science and humanities students, in whose presence the themes of this book were first developed and tested. The founding mothers, fathers, and students of "Strawberry Creek College," of the University of California at Berkeley offered me opportunities to deepen my analysis. Many references and thematic suggestions have come from members of the University of California, Berkeley, History Department, especially John Heilbron, Roger Hahn, Natalie Davis (now at Princeton University), Thomas Laqueur, Gene Brucker, and Jan de Vries. Historians of science Walter Pagel, Robert Kargon, Donna Haraway, Betty Jo Teter Dobbs, and Margaret Osler have read and evaluated individual chapters. I am also grateful for the valuable suggestions and support of Theodore Roszak, Clarence Glacken, Everett Mendelsohn, Robert Cohen, Erwin Hiebert, Nina

Gelbart, Elisabeth Gleason, Shirley Cartwright, Peter Dale Scott, Karen Hermassi, and Hilda Smith.

The research for this book was supported by a grant from the National Science Foundation, SOC 7682650. A year of writing, reflection, and final synthesis was made possible by a fellowship from the American Council of Learned Societies in conjunction with a fellowship at the Center for Advanced Study in the Behavioral Sciences, Stanford, California. I am grateful to the National Endowment for the Humanities, the Rockefeller Foundation, and the Andrew W. Mellon Foundation for grants contributing to my year at the Center; to the fellows of the classes of 1978 and 1979, who provided intellectual challenge and critical comment—particularly Nannerl Keohane, Yaron Ezrahi, Charles Neider, Houston Baker, Alfonso Ortiz, and Bruce Kuklick; and to Gardner Lindzey, Preston Cutler, and the entire staff of the Center for creating an atmosphere of intellectual stimulation, cheer, and comfort. I am especially indebted to Christine Hoth, Bruce Harley, and Patricia Knoblock for research aid, to Heather MacLean, Joanne Fox, and Susan Thistle for typing the manuscript, and to John Shopp and Katharine Reigstad for editorial suggestions. I am indebted to John Lesch, Roger Hahn, and Mirko Grmek for aid in obtaining a photograph of the Barrias sculpture.

Although many have given their time and support to the writing of this book, the interpretation presented here remains my own. The analysis relies on many quotations from primary sources. To avoid an undue burden of notes, quotations have often been listed sequentially in a single note at the end of a paragraph containing several phrases or, in the case of an extended discussion, in a single note at the beginning of several paragraphs. All spelling has been modernized. My earlier publications have been in the name Carolyn Iltis.

C. M.
Berkeley, California
1979

Preface: 1990

In the decade since *The Death of Nature* first appeared, its major themes have been reinforced. Today, a global ecological crisis that goes beyond the environmental crisis of the 1970s threatens the health of the entire planet. Ozone depletion, carbon dioxide buildup, chloroflurocarbon emissions, and acid rain upset the respiration and clog the pores and lungs of the ancient Earth Mother, rechristened "Gaia," by atmospheric chemist James Lovelock. Toxic wastes, pesticides, and herbicides seep into ground water, marshes, bays, and oceans, polluting Gaia's circulatory system. Tropical rainforests and northern old-growth forests disappear at alarming rates as lumberers shear Gaia of her tresses. Entire species of plants and animals become extinct each day. A new partnership between humans and the earth is urgently needed.

Celebrating the twenty-fifth anniversary of Rachel Carson's *Silent Spring* (1962), a 1987 conference on "ecofeminist perspectives" called upon women to lead an ecological revolution to restore planetary ecology. During the past decade, women over the entire globe have emerged as ecological activists. In Sweden, they have protested the use of herbicides on forests by offering jam made from tainted berries to members of Parliament. In India, they joined the Chipco, or "tree hugging" movement, to preserve fuelwood for cooking in protest over market lumbering. In Kenya's Greenbelt movement, they planted millions of trees in an effort to reverse desertification. In England, they camped for many years at Greenham Common to protest the deployment of nuclear missiles that threatened the continuation of life on earth. German women helped to found the Greens Party as a platform for a green future for the country and the planet. Native American women protested uranium mining linked with an increased number of cancer cases on their reservations. At Love Canal near Niagara Falls, housewives demanded action from New York state offices over an outbreak of birth defects and miscarriages in a neighborhood built on the site of a former hazardous chemical dump.

Simultaneously, feminist scholars were producing an explosion of books on ancient goddesses that became the basis for a renewed earth-rooted spirituality. They revived interest in statues, images, poetry, and rituals surrounding prehistoric earth goddesses, the Mesopotamian Innana, the Egyptian Isis, the Greek goddesses Demeter and Gaia, the Roman Ceres, and European paganism, as well as Asian, Latin American, and African female symbols and myths. Concerts, street theater, solstice and equinoctial rituals, poetry, bookstores, and lecture series celebrated human resonance with the earth.

Yet these celebrations of the connection between women and nature contain an inherent contradiction. If women overtly identify with nature and both are devalued in modern Western culture, don't such efforts work against women's prospects for their own liberation? Is not the conflation of woman and nature a form of essentialism? Are not women admitting that by virtue of their own reproductive biology they are in fact closer to nature than men and that indeed their social role is that of caretaker? Such actions seem to cement existing forms of oppression against both women and nature, rather than liberating either.

But concepts of nature and women are historical and social constructions. There are no unchanging "essential" characteristics of sex, gender, or nature. Individuals form concepts about nature and their own relationships to it that draw on the ideas and norms of the society into which they are born, socialized, and educated. People living in a given period construct nature in ways that give meaning to their own lives as elites or ordinary people, men or women, Westerners or Easterners. The historian must ask, "How have people historically conceptualized nature?" "How have they behaved in relationship to that construction?" "What historical evidence supports a particular interpretation?"

Between the sixteenth and seventeenth centuries the image of an organic cosmos with a living female earth at its center gave way to a mechanistic world view in which nature was reconstructed as dead and passive, to be dominated and controlled by humans. *The Death of Nature* deals with the economic, cultural, and scientific changes through which this vast transformation came about. In seeking to understand how people conceptualized nature in the Scientific Revolution, I am asking not about unchanging essences, but about connections between social change and changing constructions of nature. Similarly, when women today attempt to change society's domination of nature, they are acting to overturn modern constructions of nature and women as culturally passive and subordinate.

When the historian raises questions about the way nature was viewed in another era, she is also asking questions meaningful to her own epoch. One reason the historical changes described in *The Death of Nature* are of interest is that we may be experiencing a similar revolution today. The machine image that has dominated Western culture for the past three hundred years seems to be giving way to something new. Some call the transformation a "new paradigm"; others call it "deep ecology"; still others call for a postmodern ecological world view.

Emerging over the past decade are a number of scientific proposals that challenge the Scientific Revolution's mechanistic view of nature. According to physicist David Bohm, a mechanistic science based on the assumption that matter is divisible into parts (such as atoms, electrons, or quarks) moved by external forces may be giving way to a new science based on the primacy of process. In the early twentieth century, he argues, relativity and quantum theory began to challenge mechanism. Relativity theory postulated that fields with varying strengths spread out in space. Strong, stable areas, much like whirlpools in a flowing stream, represent particles. They interact with and modify each other, but were still considered external to and separate from each other. Quantum mechanics mounted a greater challenge. Motion is not continuous, as in mechanistic science, but occurs in leaps. Particles, such as electrons, behave like waves, while waves, such as light waves, behave like particles depending on the experimental context. Context dependence, which is antithetical to mechanism and part of the organic world view, is a fundamental characteristic of matter.

Bohm's process physics challenges mechanism still futher. He argues that instead of starting with parts as primary and building up wholes as secondary phenomena, a physics is needed that starts with undivided, multidimensional wholeness (a flow of energy called the holomovement) and derives the three-dimensional world of classical mechanics as a secondary phenomenon. The explicate order of the classical world in which we live unfolds from the implicate order contained in the underlying flow of energy.

Another challenge to mechanism comes from the new thermodynamics of Ilya Prigogine. The equilibrium and near-equilibrium thermodynamics of nineteenth-century classical physics had beautifully described, closed, isolated systems such as steam engines and refrigerators. Prigogine's far-from-equilibrium thermodynamics allows for the possibility that higher levels of organization can spontaneously emerge out of disorder when a system breaks down. His approach applies to social and ecological systems, which are open rather than closed, and helps to account for biological and social evolution.

The recent emergence of chaos theory in mathematics suggests that deterministic, linear, predictive equations, which form the basis of mechanism, may apply to unusual rather than usual situations. Chaos, in which a small effect may lead to a large effect, may be the norm. Thus a butterfly flapping its wings in Iowa can result in a hurricane in Florida. Most environmental and biological systems, such as changing weather, population, noise, nonperiodic heart fibrillations, and ecological patterns, may be governed by nonlinear chaotic relationships. Chaos theory reveals patterns of complexity that lead to a great understanding of global behaviors, but militate against overreliance on the simple predictions of linear differential equations.

What all these developments point to is the possibility of a new world view that could guide twenty-first-century citizens in an ecologically sustainable way of life. The mechanistic framework that legitimated the industrial revolution with its side effects of resource depletion and pollution may be losing its efficacy as a framework. A nonmechanistic science and an ecological ethic, however, must support a new economic order grounded in the recycling of renewable resources, the conservation of nonrenewable resources, and the restoration of sustainable ecosystems that fulfill basic human physical and spiritual needs. Perhaps Gaia will then be healed.

Introduction

Women and Ecology

Women and nature have an age-old association—an affiliation that has persisted throughout culture, language, and history. Their ancient interconnections have been dramatized by the simultancity of two recent social movements—women's liberation, symbolized in its controversial infancy by Betty Friedan's *Feminine Mystique* (1963), and the ecology movement, which built up during the 1960s and finally captured national attention on Earth Day, 1970. Common to both is an egalitarian perspective. Women are struggling to free themselves from cultural and economic contraints that have kept them subordinate to men in American society. Environmentalists, warning us of the irreversible consequences of continuing environmental exploitation, are developing an ecological ethic emphasizing the interconnectedness between people and nature. Juxtaposing the goals of the two movements can suggest new values and social structures, based not on the domination of women and nature as resources but on the full expression of both male and female talent and on the maintenance of environmental integrity.

New social concerns generate new intellectual and historical problems. Conversely, new interpretations of the past provide perspectives on the present and hence the power to change it. Today's feminist and ecological consciousness can be used to examine the historical interconnections between women and nature that developed as the modern scientific and economic world took form in the sixteenth and seventeenth centuries—a transformation that shaped and pervades today's mainstream values and perceptions.

Feminist history in the broadest sense requires that we look at history with egalitarian eyes, seeing it anew from the viewpoint not only of women but also of social and racial groups and the natural environment, previously ignored as the underlying resources on which Western culture and its progress have been built. To write history from a feminist perspective is to turn it upside down—to see social structure from the bottom up and to flip-flop mainstream values. An egalitarian perspective accords both women and men their place in history and delineates their ideas and roles. The impact of sexual differences and sex-linked language on cultural ideology and the use of male, female, and androgynous imagery will have important places in the new history.

The ancient identity of nature as a nurturing mother links women's history with the history of the environment and ecological change. The female earth was central to the organic cosmology that was undermined by the Scientific Revolution and the rise of a market-oriented culture in early modern Europe. The ecology movement has reawakened interest in the values and concepts associated historically with the premodern organic world. The ecological model and its associated ethics make possible a fresh and critical interpretation of the rise of modern science in the crucial period when our cosmos ceased to be viewed as an organism and became instead a machine.

Both the women's movement and the ecology movement are sharply critical of the costs of competition, aggression, and domination arising from the market economy's *modus operandi* in nature and society. Ecology has been a subversive science in its criticism of the consequences of uncontrolled growth associated with capitalism, technology, and progress—concepts that over the last two hundred years have been treated with reverence in Western culture. The vision of the ecology movement has been to restore the balance of nature disrupted by industrialization and overpopulation. It has emphasized the need to live within the cycles of nature, as opposed to the exploitative, linear mentality of forward progress. It focuses on the costs of progress, the limits to growth, the deficiencies of technological decision making, and the urgency of the conservation and recycling of natural resources. Similarly, the women's movement has exposed the costs for all human beings of competition in the marketplace, the loss of meaningful productive economic roles for women in early capitalist society, and the view of both women and nature as psychological and recreational resources for the harried entrepreneur-husband.

It is not the purpose of this analysis to reinstate nature as the mother of humankind nor to advocate that women reassume the role of nurturer dictated by that historical identity. Both need to be liberated from the anthropomorphic and stereotypic labels that degrade the serious underlying issues. The weather forecaster who tells us what Mother Nature has in store for us this weekend and legal systems that treat a woman's sexuality as her husband's property are equally guilty of perpetuating a system repressive to both women and nature. Nor am I asserting the existence of female perceptions or receptive behavior. My intent is instead to examine the values associated with the images of women and nature as they relate to the formation of our modern world and their implications for our lives today.

In investigating the roots of our current environmental dilemma and its connections to science, technology, and the economy, we must reexamine the formation of a world view and a science that, by reconceptualizing reality as a machine rather than a living organism, sanctioned the domination of both nature and women. The contributions of such founding "fathers" of modern science as Francis Bacon, William Harvey, René Descartes, Thomas Hobbes, and Isaac Newton must be reevaluated. The fate of other options, alternative philosophies, and social groups shaped by the organic world view and resistant to the growing exploitative mentality needs reappraisal. To understand why one road rather than the other was taken requires a broad synthesis of both the natural and cultural environments of Western society at the historical turning point. This book elaborates an ecological perspective that includes both nature and humankind in explaining the developments that resulted in the death of nature as a living being and the accelerating exploitation of both human and natural resources in the name of culture and progress.

The central problem of this book is informed by the concerns of the present. Yet the perspectives of our own age do not dictate the account that results. Instead they help us to formulate questions and to reveal aspects of the Scientific Revolution that might otherwise escape us and that have validity for a history of the period. Several different revolutions took place in the various sciences of the sixteenth and seventeenth centuries, and I do not attempt a comprehensive integration of them here. Yet during the same age in which the much celebrated Copernican revolution was transforming people's image of the heavens above, a more subtle yet equally pervasive revolution was altering their concept of the earth underfoot—the ancient center of the organic cosmos.

By examining the transition from the organism to the machine as the dominant metaphor binding together the cosmos, society, and the self into a single cultural reality—a world view—I place less emphasis on the development of the internal content of science than on the social and intellectual factors involved in the transformation. Of course, such external factors do not cause intellectuals to invent a science or a metaphysics for the conscious purpose of fitting a social context. Rather, an array of ideas exists, available to a given age; some of these for unarticulated or even unconscious reasons seem plausible to individuals or social groups; others do not. Some ideas spread; others temporarily die out. But the direction and cumulation of social changes begin to differentiate among the spectrum of possibilities so that some ideas assume a more central role in the array, while others move to the periphery. Out of this differential appeal of ideas that seem most plausible under particular social conditions, cultural transformations develop.

Nor is the specific content of science determined by external factors. Instead social concerns serve consciously or unconsciously to justify a given research program and to set problems for a developing science to pursue. Cultural norms and social ideologies, along with religious and philosophical assumptions, form a less visible but nonetheless important component of the conceptual framework brought to the study of a scientific problem. Through dialectical interaction science and culture develop as an organic whole, fragmenting and reintegrating out of both social and intellectual tensions and tendencies.

Between 1500 and 1700, the Western world began to take on features that, in the dominant opinion of today, would make it modern and progressive. Now, ecology and the women's movement have begun to challenge the values on which that opinion is based. By critically reexamining history from these perspectives, we may begin to discover values associated with the premodern world that may be worthy of transformation and reintegration into today's and tomorrow's society.

NOTE ON TERMINOLOGY.   *Nature* in ancient and early modern times had a number of interrelated meanings. With respect to individuals, it referred to the properties, inherent characters, and vital powers of persons, animals, or things, or more generally to human nature. It also meant an inherent impulse to act and to sustain action; conversely, to "go against nature" was to disregard this innate impulse. With respect to the material world, it referred to a dynamic creative and regulatory principle that caused phenomena and their change and development. A distinction was commonly made between *natura naturans,* or nature creating, and *natura naturata,* the natural creation.

Nature was contrasted with art (*technê*) and with artificially created things. It was personified as a female-being, e.g., Dame Nature; she was alternately a prudent lady, an empress, a mother, etc. The course of nature and the laws of nature were the actualization of her force. The state of nature was the state of mankind prior to social organization and prior to the state of grace. Nature spirits, nature deities, virgin nymphs, and elementals were thought to reside in or be associated with natural objects.

In both Western and non-Western cultures, nature was traditionally feminine. In Latin and the romance languages of medieval and early modern Europe, nature was a feminine noun, and hence, like the virtues (temperance, wisdom, etc.), personified as female. (Latin: *natura, -ae*; German: *die Natur*; French: *la nature*; Italian: *la natura*; Spanish: *la natura*.) The Greek word *physis* was also feminine.

In the early modern period, the term *organic* usually referred to the bodily organs, structures, and organization of living beings, while *organicism* was the doctrine that organic structure was the result of an inherent, adaptive property in matter. The word *organical*, however, was also sometimes used to refer to a machine or an instrument. Thus a clock was sometimes called an "organical body," while some machines were said to operate by organical, rather than mechanical, action if the touch of a person was involved.

Mechanical referred to the machine and tool trades; the manual operations of the handicrafts; inanimate machines that lacked spontaneity, volition, and thought; and the mechanical sciences.[1]

CHAPTER 1

Nature as Female

The world we have lost was organic. From the obscure origins of our species, human beings have lived in daily, immediate, organic relation with the natural order for their sustenance. In 1500, the daily interaction with nature was still structured for most Europeans, as it was for other peoples, by close-knit, cooperative, organic communities.

Thus it is not surprising that for sixteenth-century Europeans the root metaphor binding together the self, society, and the cosmos was that of an organism. As a projection of the way people experienced daily life, organismic theory emphasized interdependence among the parts of the human body, subordination of individual to communal purposes in family, community, and state, and vital life permeating the cosmos to the lowliest stone.

The idea of nature as a living organism had philosophical antecedents in ancient systems of thought, variations of which formed the prevailing ideological framework of the sixteenth century. The organismic metaphor, however, was immensely flexible and adaptable to varying contexts, depending on which of its presuppositions was emphasized. A spectrum of philosophical and political possibilities existed, all of which could be subsumed under the general rubric of *organic*.

NATURE AS NURTURE: CONTROLLING IMAGERY.

Central to the organic theory was the identification of nature, especially the earth, with a nurturing mother: a kindly beneficent female who provided for the needs of mankind in an ordered, planned universe. But another opposing image of nature as female was also prevalent: wild and uncontrollable nature that could render violence, storms, droughts, and general chaos. Both were identified with the female sex and were projections of human perceptions onto the external world. The metaphor of the earth as a nurturing mother was gradually to vanish as a dominant image as the Scientific Revolution proceeded to mechanize and to rationalize the world view. The second image, nature as disorder, called forth an important modern idea, that of power over nature. Two new ideas, those of mechanism and of the domination and mastery of nature, became core concepts of the modern world. An organically oriented mentality in which female principles played an important role was undermined and replaced by a mechanically oriented mentality that either eliminated or used female principles in an exploitative manner. As Western culture became increasingly mechanized in the 1600s, the female earth and virgin earth spirit were subdued by the machine.[1]

The change in controlling imagery was directly related to changes in human attitudes and behavior toward the earth. Whereas the nurturing earth image can be viewed as a cultural constraint restricting the types of socially and morally sanctioned human actions allowable with respect to the earth, the new images of mastery and domination functioned as cultural sanctions for the denudation of nature. Society needed these new images as it continued the processes of commercialism and industrialization, which depended on activities directly altering the earth—mining, drainage, deforestation, and assarting (grubbing up stumps to clear fields). The new activities utilized new technologies—lift and force pumps, cranes, windmills, geared wheels, flap valves, chains, pistons, treadmills, under- and overshot watermills, fulling mills, flywheels, bellows, excavators, bucket chains, rollers, geared and wheeled bridges, cranks, elaborate block and tackle systems, worm, spur, crown, and lantern gears, cams and eccentrics, ratchets, wrenches, presses, and screws in magnificent variation and combination.

These technological and commercial changes did not take place quickly; they developed gradually over the ancient and medieval eras, as did the accompanying environmental deterioration. Slowly over many centuries early Mediterranean and Greek civilization had mined and quarried the mountainsides, altered the forested landscape, and overgrazed the hills. Nevertheless, technologies were low level, people considered themselves parts of a finite cosmos, and animism and fertility cults that treated nature as sacred were numerous. Roman civilization was more pragmatic, secular, and commercial and its environmental impact more intense. Yet Roman writers such as Ovid, Seneca, Pliny, and the Stoic philosophers openly deplored mining as an abuse of their mother, the earth. With the disintegration of feudalism and the expansion of Europeans into new worlds and markets, commercial society began to have an accelerated impact on the natural environment. By the sixteenth and seventeenth centuries, the tension between technological development in the world of action and the controlling organic images in the world of the mind had become too great. The old structures were incompatible with the new activities.

Both the nurturing and domination metaphors had existed in philosophy, religion, and literature. The idea of dominion over the earth existed in Greek philosophy and Christian religion; that of the nurturing earth, in Greek and other pagan philosophies. But, as the economy became modernized and the Scientific Revolution proceeded, the dominion metaphor spread beyond the religious sphere and assumed ascendancy in the social and political spheres as well. These two competing images and their normative associations can be found in sixteenth-century literature, art, philosophy, and science.

The image of the earth as a living organism and nurturing mother had served as a cultural constraint restricting the actions of human beings. One does not readily slay a mother, dig into her entrails for gold or mutilate her body, although commercial mining would soon require that. As long as the earth was considered to be alive and sensitive, it could be considered a breach of human ethical behavior to carry out destructive acts against it. For most traditional cultures, minerals and metals ripened in the uterus of the Earth Mother, mines were compared to her vagina, and metallurgy was the human hastening of the birth of the living metal in the artificial womb of the furnace—an abortion of the metal's natural growth cycle before its time. Miners offered propitiation to the deities of the soil and subterranean world, performed ceremonial sacrifices, and observed strict cleanliness, sexual abstinence, and fasting before violating the sacredness of the living earth by sinking a mine. Smiths assumed an awesome responsibility in precipitating the metal's birth through smelting, fusing, and beating it with hammer and anvil; they were often accorded the status of shaman in tribal rituals and their tools were thought to hold special powers.

The Renaissance image of the nurturing earth still carried with it subtle ethical controls and restraints. Such imagery found in a culture's literature can play a normative role within the culture. Controlling images operate as ethical restraints or as ethical sanctions—as subtle "oughts" or "ought-nots." Thus as the descriptive metaphors and images of nature change, a behavioral restraint can be changed into a sanction. Such a change in the image and description of nature was occurring during the course of the Scientific Revolution.

It is important to recognize the normative import of descriptive statements about nature. Contemporary philosophers of language have critically reassessed the earlier positivist distinction between the "is" of science and the "ought" of society, arguing that descriptions and norms are not opposed to one another by linguistic separation into separate "is" and "ought" statements, but are contained within each other. Descriptive statements about the world can presuppose the normative; they are then ethic-laden. A statement's normative function lies in the use itself as description. The norms may be tacit assumptions hidden within the descriptions in such a way as to act as invisible restraints or moral ought-nots. The writer or culture may not be conscious of the ethical import yet may act in accordance with its dictates. The hidden norms may become conscious or explicit when an alternative or contradiction presents itself. Because language contains a culture within itself, when language changes, a culture is also changing in important ways. By examining changes in descriptions of nature, we can then perceive something of the changes in cultural values. To be aware of the in-

4

terconnectedness of descriptive and normative statements is to be able to evaluate changes in the latter by observing changes in the former.[2]

Not only did the image of nature as a nurturing mother contain ethical implications but the organic framework itself, as a conceptual system, also carried with it an associated value system. Contemporary philosophers have argued that a given normative theory is linked with certain conceptual frameworks and not with others. The framework contains within itself certain dimensions of structural and normative variation, while denying others belonging to an alternative or rival framework.

We cannot accept a framework of explanation and yet reject its associated value judgments, because the connections to the values associated with the structure are not fortuitous. New commercial and technological innovations, however, can upset and undermine an established conceptual structure. New human and social needs can threaten associated normative constraints, thereby demanding new ones.

While the organic framework was for many centuries sufficiently integrative to override commercial development and technological innovation, the acceleration of such changes throughout western Europe during the sixteenth and seventeenth centuries began to undermine the organic unity of the cosmos and society. Because the needs and purposes of society as a whole were changing with the commercial revolution, the values associated with the organic view of nature were no longer applicable; hence the plausibility of the conceptual framework itself was slowly, but continuously, being threatened.

In order to make this interpretation of cultural change convincing, it will be advantageous to examine the variations of the organic framework, focusing on its associated female imagery and pointing out the values linked to each of the variants. It will then be possible to show how, in the context of commercial and technological change, the elements of the organic framework—its assumptions and values about nature—could be either absorbed into the emerging mechanical framework or rejected as irrelevant.

The Renaissance view of nature and society was based on the organic analogy between the human body, or microcosm, and the larger world, or macrocosm. Within this larger framework, how-

5

ever, a number of variants on the organic theme were possibie. The primary view of nature was the idea that a designed hierarchical order existed in the cosmos and society corresponding to the organic integration of the parts of the body—a projection of the human being onto the cosmos. The term nature comprehended both the innate character and disposition of people and animals and the inherent creative power operating within material objects and phenomena. A second image was based on nature as an active unity of opposites in a dialectical tension. A third was the Arcadian image of nature as benevolent, peaceful, and rustic, deriving from Arcadia, the pastoral interior of the Greek Peloponnesus. Each of these interpretations had different social implications: the first image could be used as a justification for maintaining the existing social order, the second for changing society toward a new ideal, the third for escaping from the emerging problems of urban life. Drawing on the work of literary critics and historians of science and art, we can construct a spectrum of images of nature and delineate their associated value systems.

LITERARY IMAGES.   The Chaucerian and typically Elizabethan view of nature was that of a kindly and caring motherly provider, a manifestation of the God who had imprinted a designed, planned order on the world.[3] This order imposed ethical norms of behavior on the human being, the central feature of which was behavioral self-restraint in conformity with the pattern of the natural order. Each organic creature was responsible for maintaining its own place and expressing itself within the natural order and was a necessary part of the whole, but was not the whole itself. The Elizabethan first had to understand his or her own place dictated by the cosmic and social order and then to act in accordance with the traditional reason and restraint that would maintain the balance and harmony of the whole. This reverence for nature's law was expressed by Richard Hooker (1593): "See we not plainly that obedience of creatures unto the law of nature is the stay of the whole world." Nature operated "without capacity or knowledge," solely on the basis of "her dexterity and skill," as the instrument of God's expression in the mundane world. Whatever was known of God was taught by nature, "God being the author of Nature, her voice is but his instrument. By her from him we receive whatsoever in sort we learn."[4] Here nature is God's involuntary agent, a benevolent teacher of the hidden pattern and values God employed in creating the visible cosmos (*natura naturata,* the natural creation). A somewhat less orthodox view saw her as a creative force *(natura naturans)*—a soul with a will to generate mundane forms.

In Shakespeare's tragedy *King Lear,* the king, as human nature, represented Renaissance man, whose worldly existence was part of a larger patterned whole and of the contemporary hierarchical social order. His human nature symbolized the medieval-Renaissance cosmos whose patterns must not be violated. Lear's nature was a composite structure of the qualities of benevolence, comfort, and generosity, dictating honor and charity in his own ethical behavior as father of the household and reverence for his authority and wisdom on the part of his daughters.

Lear's daughter Cordelia represented utopian nature, or nature as the ideal unity of the opposites, a second image of nature within the larger spectrum of the organic framework. She was strength and gentleness hewn as one: "passion and order, innocence and maturity, defenselessness and strength, daughter and mother, maid and wife."[5] She represented simplicity in unity and the balance of the contraries. She was the mature integration of society and of nature as ideally reflected in the human being. She stood for unity, wholeness, and virtue—the utopian expression of human nature. Cordelia's nature was the human symbol of the new Jerusalem of the millenarian movements of the Middle Ages, the utopias of the Renaissance, and the religious sects of the subsequent civil war period in England, which attempted to improve society. The dialectical image of nature, here symbolized by a woman, represented the impetus to move society forward toward a new ideal.

Pastoral poetry and art prevalent in the Renaissance presented another image of nature as female—an escape backward into the motherly benevolence of the past. Here nature was a refuge from the ills and anxieties of urban life through a return to an unblemished Golden Age. Depicted as a garden, a rural landscape, or a peaceful fertile scene, nature was a calm, kindly female, giving of her bounty. Against an idyllic backdrop, sheep grazed contently, birds sang melodies, and trees bore fruit. Wild animals, thorns, snakes, and vultures were nowhere to be found. Human beings meditated on the beauties of nature far removed from the violence of the city.[6]

The pastoral tradition had its roots in nostalgia for the Homeric Golden Age, for the uncorrupted Garden of Eden, and escape from the ills of the city. It echoed the poetic tradition of Virgil (70–19 B.C.) and Juvenal (A.D. 60–140). Virgil wrote of spending old age "amid familiar streams and holy springs";[7] Juvenal yearned for the rural town in which

> Springs bubbling up from the grass, no need
>   for windlass or bucket,
> Plenty to water your flowers, if they need
>   it without any trouble.[8]

The Arcadia theme, eulogized in the pastoral poetry of Philip Sidney (1554–1586; *Arcadia,* 1590) and Edmund Spenser (1552–1599; *The Shepheard's Calendar,* 1579), appeared in many poetic and artistic settings in which nature was idealized as a benevolent nurturer, mother, and provider. The sixteenth-century French painting "St. Genevieve with Her Flock" depicts the virgin surrounded by a flock of sheep within a protective stone circle on a hillside of trees and blooming flowers, well outside the city in the background. Here the female image of nature and the virgin, symbol of the earth spirit, are fused with the circular symbolism of order and protective encasement. In Lucas Cranach's painting "The Nymph of the Spring" (1518), the female earth nymph rests in a bed of flowers while doves, symbols of peace, feed near the edge of a trickling stream and deer water on its farther bank (Fig. 1). The "Birth of Venus" (1482) and the "Primavera" (1477–8) of Sandro Botticelli portray the virgin in conjunction with the earth mother, who is covered with a gown and wreath of flowers, both symbols of female fertility. In the seventeenth century, Nicolas Poussin and other landscape painters illustrated the transitory nature of the Arcadian experience by sometimes inserting a death's head into their works of art.

But while the pastoral tradition symbolized nature as a benevolent female, it contained the implication that nature when plowed and cultivated could be used as a commodity and manipulated as a resource. Nature, tamed and subdued, could be transformed into a garden to provide both material and spiritual food to enhance the

Figure 1. *The Nymph of the Spring,* by Lucas Cranach (1518, Germany). This pastoral representation of a nymph of the woods and meadows implies the passive role of nature: her quiver of arrows, borrowed from the ancient huntress-goddess Diana, is laid aside, and she herself reclines invitingly on the ground.

comfort and soothe the anxieties of men distraught by the demands of the urban world and the stresses of the marketplace. It depended on a masculine perception of nature as a mother and bride whose primary function was to comfort, nurture, and provide for the well-being of the male. In pastoral imagery, both nature and women are subordinate and essentially passive. They nurture but do not control or exhibit disruptive passion. The pastoral mode, although it viewed nature as benevolent, was a model created as an antidote to the pressures of urbanization and mechanization. It represented a fulfillment of human needs for nurture, but by conceiving of nature as passive, it nevertheless allowed for the possibility of its use and manipulation. Unlike the dialectical image of nature as the active unity of opposites in tension, the Arcadian image rendered nature passive and manageable.

PHILOSOPHICAL FRAMEWORKS. In his *Timaeus,* Plato endowed the whole world with life and likened it to an animal. The deity "framed one visible animal comprehending within itself all other animals of a kindred nature." Its shape was round, since it had no need for eyes, ears, or appendages. Its soul was female, "in origin and excellence prior to and older than the body," and made "to be ruler and mistress, of whom the body was to be the subject." The soul permeated the corporeal body of the universe, enveloping it and "turning herself within herself." The earth "which is our nurse" was placed at the immovable center of the cosmos.[9]

For Plato, this female world soul was the source of motion in the universe, the bridge between the unchanging eternal forms and the changing, sensible, temporal lower world of nature. The Neoplatonism of Plotinus (A.D. 204–270), which synthesized Christian philosophy with Platonism, divided the female soul into two components. The higher portion fashioned souls from the divine ideas; the lower portion, *natura,* generated the phenomenal world. The twelfth-century Christian Cathedral School of Chartres, which interpreted the Bible in conjunction with the *Timaeus,* personified Natura as a goddess and limited the power attributed to her in pagan philosophies by emphasizing her subservience to God. Nature was compared to a midwife who translated Ideas into material things; the Ideas were likened to a father, the matter to a mother, and the generated species to a child. In Platonic and Neoplatonic symbolism, therefore, both nature and matter were feminine, while the Ideas were masculine. But nature, as God's agent, in her role as creator and producer of the material world, was superior to human artists both in creativity and in ease of production. She was more powerful than humans, but still subordinate to God.[10]

An allegory (1160) by Alain of Lille, of the School of Chartres, portrays Natura, God's powerful but humble servant, as stricken with grief at the failure of man (in contrast to other species) to obey her laws. Owing to faulty supervision by Venus, human beings engage in adulterous sensual love. In aggressively penetrating the secrets of heaven, they tear Natura's undergarments, exposing her to the view of the vulgar. She complains that "by the unlawful assaults of man alone the garments of my modesty suffer disgrace and division."

Natura was a replica of the cosmos. Set in her crown as jewels were the signs of the zodiac and the planets; decorating her robe, mantle, tunic, and undergarments were birds, water creatures, earth animals, herbs, and trees; on her shoes were flowers. Her torn tunic, indicative of the need to protect nature's secrets from misuse, exemplifies the rhetorical role of imagery. The specific moral of Alain's allegory is that nature has no power to enforce her own laws. After the disobedience that resulted in the fall of Adam and Eve, unity in the created world can only be maintained by moral choices; human reason must control human lust.[11]

Renaissance Neoplatonism illustrates the image of the macrocosm enlivened by the female soul. The Neoplatonic alchemist Robert Fludd (1574–1637) pictured the world soul as a woman connected by her right hand to God, represented by the Hebrew tetragrammaton—the four consonants JHVH—transmitted by a golden chain to the terrestrial world below (Fig. 2). Natura was also pictured as the mother goddess Isis, her flowing hair drawn through a sphere representing the world, her mantle decorated at the top by stars and the bottom by flowers and her womb by a half-moon whose rays fertilize the earth. On opposite ears are the sun and moon, representing the male and female principles in the natural world. In her left hand she holds a pail, symbol of the flooding Nile, which irrigated the earth, and with her right she shakes the sistrum, a rattle representing the unceasing constant motion and internal vigor of nature (Fig. 3).

The ancient philosophy of Aristotle, recovered in the twelfth century, was also based on the organic theory of the primacy of internal growth and development within nature. In his *Metaphysics,* Aristotle defined nature (*physis*) as "the source of movement of natural objects, being present in them either potentially or in complete reality."[12] Each individual object could be explained by specifying four "causes": material (the matter out of which it was made), formal (its shape or the arrangement of its parts), efficient (the moving force), and final (the purpose or ultimate end for which the object existed). In natural objects as opposed to artificially created products, the material and efficient causes were unified such that the material substratum had its principle of motion or change within it. The material of a tree or a child caused its growth, whereas a table had to be produced by a builder. Motion,

11

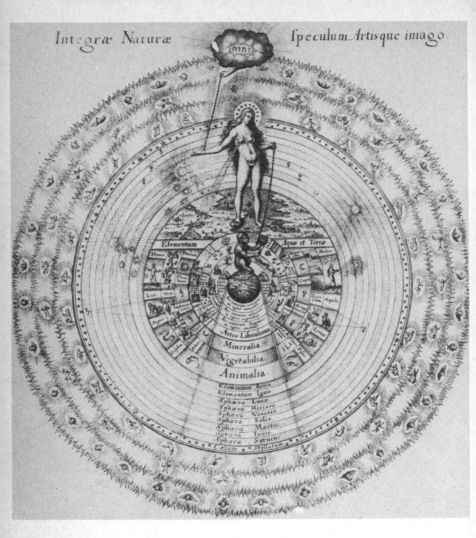

Figure 2. The Female Soul of the World, engraved by Johann Theodore de Bry and conceived by Robert Fludd in his *Utriusque Cosmi Maioris Scilicet et Minoris Metaphysica* (1617–21). Alchemist Robert Fludd's representation of the world soul as a woman shows the Western identification of the right hand with masculinity, God the Father, the sun and dominance; the left hand is traditionally associated with the feminine and subordinate moon, earth, and water.

12

like growth, was the internal development of the potential toward the final form or actual existence of an individual being—a girl becomes a woman, a chicken develops into a hen. In nature the end toward which an object developed was its form or shape; in artificial products, such as a house, the form was distinct from its purpose as a dwelling.

Aristotle differed from the Presocratics, who defined nature in terms of a material substratum such as water (Thales, ca. 585 B.C.), air (Anaximenes, ca. 525 B.C.), fire (Heraclitus, ca. 478 B.C.), but he concurred with them in stressing the primacy of growth, change, and process in the natural world. He opposed Plato's dualism between imperfect appearances and pure perfect forms by asserting that the form existed in the individual object rather than as a separate transcendent level of reality.

Aristotelian philosophy, while unifying matter and form in each individual being, associated activity with maleness and passivity with femaleness. Form reigned superior over dead, passive matter. Socially, Aristotle found the basis for male rule over the household in the analogy that, as the soul ruled the body, so reason and deliberation, characteristic of men, should rule the appetites supposedly predominant in women.

Aristotle's biological theory viewed the female of the species as an incomplete or mutilated male, since the coldness of the female body would not allow the menstrual blood to perfect itself as semen. In the generation of offspring, the female contributed the matter or passive principle. This was the material on which the active male principle, the semen, worked in creating the embryo. The male was the real cause of the offspring. "The female, as female, is passive, and the male, as male, is active, and the principle of movement comes from him." Power and motion were contributed solely by the semen. The male supplied none of the matter from which the embryo developed: "Nor does he emit anything of such nature as to exist within that which is generated, as part of the material embryo, but that he only makes a living creature by the power which resides in the semen." [13]

The female supplied the nutriment—the catamenia, or menstrual blood—on which the qualities of the male could operate. The combination of semen with menstrual blood was like the curdling of cheese, just as rennet acts by coagulating milk. The male contribut-

Figure 3.   *Isis,* in Athanasius Kircher, *Oedipus Aegypticus* (1652). The representation of Isis employs the traditional Western identification of cyclical forces with the female: she stands with one foot in the Nile, which yearly overflows its banks, and uses the pail in her left hand to irrigate the earth. The moon's rays on her womb fertilize the earth, the snakes in her headdress symbolize renewal (snakes can shed their skin), and the grain represents the seasonal harvest.

ISIDIS
Magnæ Deorum Matris
APVLEIANA DESCRIPTIO.

Nomina varia
Iſidis.

Iſis
Minerua
Venus
Iuno
Proſerpina
Ceres
Diana
Rhea ſeu
  Tellus
Peſſinuncia
Rhramnuſia
Bellona
Hecate
Luna
Polymor-
  phus dæ-
  mon.

Γῖσις πανδεχής πο-
λύμορφ@ δαί-
μων.
Μυελόνυμ@ φύσις,
ὕλη.

Explicationes ſym-
bolorum Iſidis.

A Diuinitatem, mun-
dum, orbes cœleſtes
BB Iter Lunæ flexuo-
ſum, & vim fœcun-
datiuam notat.
CC Tutulus, vim Lu-
næ in herbas, &
plantas.
D Cereris ſymbolum,
Iſis enim ſpicas in-
uenit.
E Byſſina veſtis mul-
ticolor, multifor-
mem Lunæ faciem
F Inuentio frumenti.
G Dominium in omni-
nia vegetabilia.
H Radios lunares.
I Genius Nili malo-
rum auerruncus.
K Incrementa & de-
crementa Lunæ.
L Humectat. vis Lunę.
M Lunæ vis victrix, &
vis diuinandi.
N Dominium in hu-
mores & mare.
O Terræ ſymbolũ, &
Medicinæ inuentrix.
P Fœcunditas, quæ ſe,
quitur terram irri-
gatam.
Q Aſtrorum Domina.
R Omnium nutrix.
S ⎱ Terræ mariſque
M ⎰ Domina.

Ravello F. M.

Ἀκρα Θεῶν Μήτηρ ταύτη πολυώνμ@ ΙΣΙΣ.

15

ed the movement necessary for the embryo to develop, as well as the form it would take.

Both the efficient and formal causes were derived from the male principle and were the active cause of the offspring. Nature in her creative work used the semen as a tool just as the carpenter, through motion, imparted shape and form to the wood on which he worked. Since no material was ever transferred from the body of the carpenter to the wood, by analogy the male did not contribute any matter, but rather the force and power of the generation.

The Aristotelian theory of generation appeared at the basis of some alchemical theories. To change base metals to gold or silver, or to create the philosopher's stone—a substance thought to effect such a change—the qualities or form of the lowly metal must be removed, leaving the passive primary matter. The new more noble form, the active male principle, was then introduced and unified with the base female matter to produce the new metal.

Aristotelian ideas about human generation were also projected onto the cosmos. In the sixteenth century, the marriage and impregnation of the female earth by the higher celestial masculine heavens was a stock description of biological generation in nature. The movements of the celestial heavens produced semen, which fell in the form of dew and rain on the receptive female earth. A famous passage in *On the Revolutions of the Celestial Spheres* (1543) by Nicolaus Copernicus (1473–1543), reviver of the heliocentric hypothesis, draws on the marriage of the masculine heavens with the female earth: "Meanwhile, the earth conceives by the sun and becomes pregnant with annual offspring."[14] Such basic attitudes toward male-female roles in biological generation where the female and the earth are both passive receptors could easily become sanctions for exploitation as the organic context was transformed by the rise of commercial capitalism.

A radical alternative to the hierarchical view that the female was inferior appeared in the monistic form of ancient gnosticism, based (like Shakespeare's Cordelia) on the unity of the opposites and the equality of male-female principles. The transmission of gnostic ideas of androgyny provided an alternative view of generation and carried with it more positive implications and attitudes toward nature as female in the Renaissance. But these ideas lay outside the mainstream of Western Christian culture. Gnostic texts available in the Renaissance had been preserved through extensive quotation in the polemical writings of the early church fathers who attempted to refute the doctrine.

The gnostic tradition was a body of texts (written in the first three centuries A.D.) condemned as a heretical form of early Christianity. Centered in hellenistic Alexandria, it synthesized Christianity with the spiritual teachings of Babylonia and Persia to the east, and Greece and Rome to the west. Gnosticism (stemming from *knowledge*) maintained an original unity of male-female opposites in a transcendent God, but a dualism between God and the mundane world, between good and evil, between spirit and matter. By emanation God produced a female generative principle, which created angels and then the visible world; the light of God mingled with matter in the lower world. The religious goal of gnosticism was other-worldy salvation through knowledge. In contrast to orthodox Christians, some sects worshipped the serpent for inducing Eve to taste the fruit of knowledge, the beginning of gnosis on earth. The coiled serpent biting its own tail symbolized the unity of the opposites good and bad, and the cosmic metamorphic cycles.

Some of the early gnostic Christians who described God as androgynous prayed to both the father and the mother and interpreted the account of the creation in the first chapter of Genesis to mean that a male-female God created men and women in its image. In these interpretations God was a dyad of opposites existing in harmony in one being. The divine mother was named Wisdom, or *Sophia,* a Greek translation of the Hebrew *hokhmah.* Wisdom was the creative power, "self-generating, self-discovering its own mother, its own father, its own sister, its own son: father, mother, unity, root of all things."[15] Her wisdom was bestowed on men and women. Human nature, like God, consisted of a unity of equal male-female principles. Evidence for the appeal of gnostic androgyny to women is indicated by their attraction to these heretical groups during the period A.D. 150–200, when Christianity was struggling to gain its stature as a world religion. Here they could play important roles from which they had been excluded by orthodox churches, including healer, evangelist, priest, prophet, and teacher.

Gnostic philosophy, and its assignment of equal importance to male and female principles in generation, permeated many alchemical treatises. The gnostic trinity—father, mother, and son—first appeared in a work of the third or fourth century, *Chrysopeia* ("Gold-Making"), attributed to Cleopatra. The *Emerald Tablet* of Hermes

17

Trismegistus (a mythical alchemist to whom was attributed the gnostic *Corpus Hermeticum,* A.D. 100–300) emphasized the equality of the two great male and female principles in nature, the sun and moon: "The sun is its father, the moon is its mother. Wind is carried in its belly, the earth is its nurse."[16] In the gnostic tradition in alchemy many of the earliest alchemical treatises were either written by or attributed to women: Isis; Mary the Jewess, identified with Moses' sister Miriam; Cleopatra; and Theosobia, sister of Zosimus, an alchemist of the fourth century A.D.

> Noticeable also is the importance given to women by the Hermetic [alchemist]. In Majer's etching the virgin is, like her ancestor Eve, the instigator. And a woman is the alchemist's symbol of nature. He follows her tracks, which lead to perfection. It may be recalled that Magdalene and Sophia are the most important and active figures in the *Pistis Sophia* and that the earthly incarnation of the heavenly mother is the main feature in the dogma of Simon Magus. Flamel's transmutation took place when his wife was present; and in the *Liber Mutus,* an alchemical tract, it is recommended that before starting the operation the alchemist and his wife should kneel and pray before the oven. The union of the soul and spirit, of the male and female essence, has its counterpart in heaven: the sun is the father and the moon, the mother.[17]

In gnostic texts available in the sixteenth century, the Mother Sophia, or eightness, created the king or father, who in turn generated the seven heavens.[18] The mother produced four terrestrial elements after the pattern of the higher celestial elements. In some gnostic interpretations, the archetypal Adam was created androgynous and composed of eight parts.

For Paracelsus (1490–1541), the eight gnostic matrices were all mothers. Moreover, in contrast to Aristotle, he assigned equality to the male and female principles in sexual generation:

> When the seed is received in the womb, nature combines the seed of the man and the seed of the woman. Of the two seeds, the better and stronger will form the other, according to its nature. The seed from the man's brain and that from the woman's brain together make only one brain; but the child's brain is formed according to the one which is the stronger of the two, and it becomes like this seed but never completely like it.[19]

For some Paracelsians, the eight parts of the gnostic Adam were divided into four celestial fathers and four terrestial elements or mothers. The fathers provided astral semina that fertilized the female elemental matrices, which, activated by the central fire of the earth, rose up to receive the astral semina. For some alchemists, the philosopher's stone embodied the unity of the contraries resulting from the conjunction of the four mothers and the four fathers.

Scholar Ralph Cudworth noted in 1678 that the pagans "call God male and female together" and that "the Orphic theology calls the first principle hermaphroditic, or male and female together; thereby denoting that essence that is generative or productive of all things."[20]

In alchemy, permeated by male-female dualism, the hermaphrodite Mercurius symbolized the androgynous unity of the opposites. The unification of male and female principles, represented by the alchemical marriage of the sun and the moon and by the union of the male mineral agent mercury and the female *materia prima* (prime matter), resulted in the male-female hermaphrodite (Fig. 4). The tree symbolized the vegetative nurturing principle; the snake, the animated life-giving principle. The heat of the alchemical oven represented the male element; the egg-shaped retort, the female womb.[21]

Thomas Vaughan (1622–1666), alchemist and mystic, described the sun and the moon as two equal peers, the consummation of their marriage as a total unity resulting in the nurturing of their seed in the womb of the earth:

> There is in every star and in this elemental world a certain principle which is the "bride of the sun." These two in their coition do emit semen, which seed is carried in the womb of nature. But the ejection of it is performed invisibly and in a sacred silence, for this is the conjugal mystery of heaven and earth, their act of generation, a thing done in private between particular males and females; but how much more think you between the two universal natures. Know therefore that it is impossible for you to extract or receive any seed from the sun without this feminine principle which is the wife of the sun. . . . Know then for certain that the magician's sun and moon are two universal peers, male and female, a king and queen regents, always young and never old. These two are adequate to the whole world and coextended through the whole universe. The one is not without the other, God having united them in His work of creation in a solemn, sacramental union. It will then be a hard and difficult enterprise to rob the husband of his wife, to part those asunder who God has put together, for they sleep both in the same bed, and he that discovers the one must needs see the other [see Fig. 5].[22]

Figure 4. Hermaphrodite, from *Aurora consurgens* (late fourteenth century). The alchemists imagined the marriage of the sun and the moon (the male and female principles) as producing a hermaphrodite— an integrated view of nature as both male and female.

What fate did these differing images of nature meet as the Scientific Revolution began to mechanize the world view? Both pastoral art and Aristotelian philosophy saw the female sex as passive and receptive. Pastoral imagery could easily be incorporated into a mechanized industrialized world as an escape from the frustrations of the marketplace. The pastoral had been an antidote to the ills of urbanization in ancient times, and it continued to play that role in the commercial revolution. It became particularly significant as an image dictating the transformation and cultivation of the wilderness in American culture. Explorers of the new world, who described the scenery as a lovely garden fragrant with flowers, brightened by the sounds of marvelous new and beautiful birds, provided an impetus for settlement across the Atlantic. The cultivation of a bountiful mother earth helped to hasten the disruption and exploitation of new and "virgin" lands.

The Aristotelian and Platonic conception of the passivity of matter could also be incorporated into the new mechanical philosophy in the form of inert "dead" atoms, constituents of a new machine-like world in which change came about through external forces, a scheme that readily sanctioned the manipulation of nature. The Neoplatonic female world soul, the internal source of activity in nature, would disappear, to be replaced by a carefully contrived mechanism of subtle particles in motion.

The idea of the androgynous equality of male-female principles would eventually become more clearly articulated as the principle of the dialectic, either as the unity of opposites in idealism or as the struggle of the opposites in dialectical materialism. I will have occasion to note its progress in several social and philosophical contexts as the struggle for equality, against social hierarchy, continued.

THE GEOCOSM: THE EARTH AS A NURTURING MOTHER. Not only was nature in a generalized sense seen as female, but also the earth, or geocosm, was universally viewed as a nurturing mother,

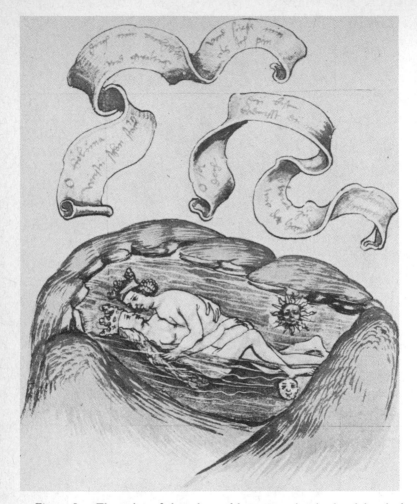

Figure 5. The union of the solar and lunar opposites in the alchemical work, from Arnold of Villanova, *Rosarium Philosophorum* (sixteenth century). The sacred marriage of male and female in alchemy is the mystical integration of co-creative opposites. The sun and air are associated with the male; water and moon, with the female.

sensitive, alive, and responsive to human action. The changes in imagery and attitudes relating to the earth were of enormous significance as the mechanization of nature proceeded. The nurturing earth would lose its function as a normative restraint as it changed to an inanimate dead physical system.

The macrocosm theory, as we have seen, likened the cosmos to the human body, soul, and spirit with male and female reproductive components. Similarly, the geocosm theory compared the earth to the living human body, with breath, blood, sweat, and elimination systems.

For the Stoics, who flourished in Athens during the third century B.C., after the death of Aristotle, and in Rome through the first century A.D., the world itself was an intelligent organism; God and matter were synonymous. Matter was dynamic, composed of two forces: expansion and condensation—the former directed outward, the latter inward. The tension between them was the inherent force generating all substances, properties, and living forms in the cosmos and the geocosm.

Zeno of Citium (ca. 304 B.C.) and M. Tullius Cicero (106–43 B.C.) held that the world reasons, has sensation, and generates living rational beings: "The world is a living and wise being, since it produces living and wise beings."[23] Every part of the universe and the earth was created for the benefit and support of another part. The earth generated and gave stability to plants, plants supported animals, and animals in turn served human beings; conversely, human skill helped to preserve these organisms. The universe itself was created for the sake of rational beings—gods and men—but God's foresight insured the safety and preservation of all things. Humankind was given hands to transform the earth's resources and was given dominion over them: timber was to be used for houses and ships, soil for crops, iron for plows, and gold and silver for ornaments. Each part and imperfection existed for the sake and ultimate perfection of the whole.

The living character of the world organism meant not only that the stars and planets were alive, but that the earth too was pervaded by a force giving life and motion to the living beings on it. Lucius Seneca (4 B.C.–A.D. 65), a Roman Stoic, stated that the earth's breath nourished both the growths on its surface and the heavenly bodies above by its daily exhalations:

> How could she nourish all the different roots that sink into the soil in one place and another, had she not an abundant supply of the breath of life? . . . all these [heavenly bodies] draw their nourishment from mate-

rials of earth . . . and are sustained . . . by nothing else than the breath of the earth. . . . Now the earth would be unable to nourish so many bodies . . . unless it were full of breath, which it exhales from every part of it day and night.[24]

The earth's springs were akin to the human blood system; its other various fluids were likened to the mucus, saliva, sweat, and other forms of lubrication in the human body, the earth being organized ". . . much after the plan of our bodies, in which there are both veins and arteries, the former blood vessels, the latter air vessels. . . . So exactly alike is the resemblance to our bodies in nature's formation of the earth, that our ancestors have spoken of veins [springs] of water." Just as the human body contained blood, marrow, mucus, saliva, tears, and lubricating fluids, so in the earth there were various fluids. Liquids that turned hard became metals, such as gold and silver; other fluids turned into stones, bitumens, and veins of sulfur. Like the human body, the earth gave forth sweat: "There is often a gathering of thin, scattered moisture like dew, which from many points flows into one spot. The dowsers call it *sweat,* because a kind of drop is either squeezed out by the pressure of the ground or raised by the heat."

Leonardo da Vinci (1452–1519) enlarged the Greek analogy between the waters of the earth and the ebb and flow of human blood through the veins and heart:

The water runs from the rivers to the sea and from the sea to the rivers, always making the same circuit. The water is thrust from the utmost depth of the sea to the high summits of the mountains, where, finding the veins cut, it precipitates itself and returns to the sea below, mounts once more by the branching veins and then falls back, thus going and coming between high and low, sometimes inside, sometimes outside. It acts like the blood of animals which is always moving, starting from the sea of the heart and mounting to the summit of the head.[25]

The earth's venous system was filled with metals and minerals. Its veins, veinlets, seams, and canals coursed through the entire earth, particularly in the mountains. Its humors flowed from the veinlets into the larger veins. The earth, like the human, even had its own elimination system. The tendency for both to break wind caused earthquakes in the case of the former and another type of quake in the latter:

The material cause of earthquakes . . . is no doubt great abundance of wind, or store of gross and dry vapors, and spirits, fast shut up, and as a man would say, emprisoned in the caves, and dungeons of the earth; which wind, or vapors, seeking to be set at liberty, and to get them home to their natural lodgings, in a great fume, violently rush out, and as it were, break prison, which forcible eruption, and strong breath, causeth an earthquake.[26]

Its bowels were full of channels, fire chambers, glory holes, and fissures through which fire and heat were emitted, some in the form of fiery volcanic exhalations, others as hot water springs. The most commonly used analogy, however, was between the female's reproductive and nurturing capacity and the mother earth's ability to give birth to stones and metals within its womb through its marriage with the sun.

In his *De Rerum Natura* of 1565, the Italian philosopher Bernardino Telesio referred to the marriage of the two great male and female powers: "We can see that the sky and the earth are not merely large parts of the world universe, but are of primary—even principal rank. . . . They are like mother and father to all the others."[27] The earth and the sun served as mother and father to the whole of creation: all things are "made of earth by the sun and that in the constitution of all things the earth and the sun enter respectively as mother and father." According to Giordano Bruno (1548–1600), every human being was "a citizen and servant of the world, a child of Father Sun and Mother Earth."[28]

A widely held alchemical belief was the growth of the baser metals into gold in womblike matrices in the earth. The appearance of silver in lead ores or gold in silvery assays was evidence that this transformation was under way. Just as the child grew in the warmth of the female womb, so the growth of metals was fostered through the agency of heat, some places within the earth's crust being hotter and therefore hastening the maturation process. "Given to gold, silver, and the other metals [was] the vegetative power whereby they could also reproduce themselves. For, since it was impossible for God to make anything that was not perfect, he gave to all created things, with their being, the power of multiplication."[29] The sun acting on the earth nurtured not only the plants and animals but also "the metals, the broken sulfuric, bituminous, or nitrogenous rocks; . . . as well as the plants and animals—if these are not made of earth by the sun, one cannot imagine of what else or by what other agent they could be made."[30]

Several theories accounted for the growth of metals and minerals. In the Aristotelian theory, the earth gave off exhalations under the influence of the sun's warmth. From the dry exhalations grew stones, from the wet metals, both being formed within the earth or on its surface. Sixteenth-century theories held that minerals grew from a combination of celestial influences, primarily the sun, and a formative power within the earth.[31]

In the liquid seed theory, the earth was a matrix or mother to the seeds of stones and minerals. The seeds of minerals and metals fermented water, transforming it into a mineral juice and then into the metal itself. Stones were generated from their own seminal principles or seeds, thereby preserving their species. Suitable nooks and crannies within the earth formed matrices or wombs for the nurturing and development of the infant seed. Neither the air nor the bodies of plants and animals were suitable mothers; only the rocks on the edges of ore veins, and special crannies in the earth's crust could act as matrices. Jerome Cardan (1550) and Bernard Palissy (1580) believed that metals reproduced by liquid seeds deposited by water; others set forth the thesis that mineral juices were formed into metals by degrees, expanding and taking on new matter until they grew into visible metals.[32]

A third theory, the lapidifying (stone-forming) juice theory, hypothesized a fluid or juice that circulated through the earth's body in the veins, and cracks, and pores beneath its surface just as blood ebbed and flowed within the human body. This *succus lapidescens,* combined with the principles of heat and cold, was the origin of stones and minerals. The mineral or stony matter was held in solution by the liquid *succus* until it evaporated out by the action of heat or deposited its matter through the action of cold.[33]

The earth's womb was the matrix or mother not only of metals but also of all living things. Paracelsus compared the earth to a female whose womb nurtured all life.

Woman is like the earth and all the elements and in this sense she may be considered a matrix; she is the tree which grows in the earth and the child is like the fruit born of the tree.... Woman is the image of the tree. Just as the earth, its fruits, and the elements are created for the sake of the tree and in order to sustain it, so the members of woman, all her qualities, and her whole nature exist for the sake of her matrix, her womb. . . .

And yet woman in her own way is also a field of the earth and not at all different from it. She replaces it, so to speak; she is the field and the garden mold in which the child is sown and planted.[34]

The earth in the Paracelsian philosophy was the mother or matrix giving birth to plants, animals, and men.

The image of the earth as a nurse, which had appeared in the ancient world in Plato's *Timeaus* and the *Emerald Tablet* of Hermes Trismegistus, was a popular Renaissance metaphor. According to sixteenth-century alchemist Basil Valentine, all things grew in the womb of the earth, which was alive, and vital, and the nurse of all life:

> The quickening power of the earth produces all things that grow forth from it, and he who says that the earth has no life makes a statement flatly contradicted by facts. What is dead cannot produce life and growth, seeing that it is devoid of the quickening spirit. . . . This spirit is the life and soul that dwell in the earth, and are nourished by heavenly and sidereal influences. . . . This spirit is itself fed by the stars and is thereby rendered capable of imparting nutriment to all things that grow and of nursing them as a mother does her child while it is yet in the womb. . . . If the earth were deserted by this spirit it would be dead. . . . [35]

Cambridge Platonist Henry More (1614–1687) referred to the sixteenth-century doctrine of the earth as a nurse and the possibility of its going dry and becoming sterile:

> For though we should admit, with Cardan and other naturalists, that the earth at first brought forth all manner of animals as well as plants, and that they might be fastened by the navel to their common mother the earth, as they are now to the female in the womb; yet we see she is grown sterile and barren, and her births of animals are now very inconsiderable. Wherefore what can it be but a providence, that while she did bear she sent out male and female, that when her own prolific virtue was wasted, yet she might be a dry nurse, or an officious grandmother, to thousands of generations?[36]

In general, the Renaissance view was that all things were permeated by life, there being no adequate method by which to designate the inanimate from the animate. It was difficult to differentiate between living and nonliving things, because of the resemblance in structures. Like plants and animals, minerals and gems were filled with small pores, tublets, cavities, and streaks, through which they seemed to nourish themselves. Crystalline salts were compared to plant forms, but criteria by which to differentiate the living from the nonliving could not successfully be formulated. This was due not only to the vitalistic framework of the period but to striking similarities between them. Minerals were thought to possess a lesser degree of the vegetative soul, because they had the capacity for medicinal action and often took the form of various parts of plants. By virtue of the vegetative soul, minerals and stones grew in the human body, in animal bodies, within trees, in the air and water, and on the earth's surface in the open country.[37]

Popular Renaissance literature was filled with hundreds of images associating nature, matter, and the earth with the female sex. The earth was alive and considered to be a beneficent, receptive, nurturing female. For most writers there was a mingling of traditions based on ancient sources. In general, the pervasive animism of nature created a relationship of immediacy with the human being. An I-thou relationship in which nature was considered to be a person-writ-large was sufficiently prevalent that the ancient tendency to treat it as another human still existed. Such vitalistic imagery was thus so widely accepted by the Renaissance mind that it could effectively function as a restraining ethic.

In much the same way, the cultural belief-systems of many American-Indian tribes had for centuries subtly guided group behavior toward nature. Smohalla of the Columbia Basin Tribes voiced the Indian objections to European attitudes in the mid-1800s:

> You ask me to plow the ground! Shall I take a knife and tear my mother's breast? Then when I die she will not take me to her bosom to rest.
> You ask me to dig for stone! Shall I dig under her skin for her bones? Then when I die I cannot enter her body to be born again.
> You ask me to cut grass and make hay and sell it, and be rich like white men! But how dare I cut off my mother's hair?[38]

In the 1960s, the Native-American became a symbol in the ecology movement's search for alternatives to Western exploitative attitudes. The Indian animistic belief-system and reverence for the earth as a mother were contrasted with the Judeo-Christian heritage of dominion over nature and with capitalist practices resulting in the "tragedy of the commons" (exploitation of resources available for any person's or nation's use). But as will be seen, European culture was more complex and varied than this judgment allows. It ignores the Renaissance philosophy of the nurturing earth as well as those philosophies and social movements resistant to mainstream economic change.

NORMATIVE CONSTRAINTS AGAINST THE MINING OF MOTHER EARTH. If sixteenth-century descriptive statements and imagery can function as an ethical constraint and if the earth was widely viewed as a nurturing mother, did such imagery actually function as a norm against improper use of the earth? Evidence that this was indeed the case can be drawn from theories of the origins of metals and the debates about mining prevalent during the sixteenth century.

The ancient Greek philosophers Anaxagoras (500–428 B.C.), Theophrastus (370–278 B.C.), and Dionysius of Periegetes (fl. A.D. 86–96) believed that metals were plants growing beneath the earth's surface and that veins of gold were like the roots and branches of trees. Metals were believed merely to be a lower form of life than vegetables and animals, reproducing themselves through small metallic seeds.

A popular Renaissance belief held about mining was the metaphor of the golden tree. The earth deep within its bowels produced and gave form to the metals, which then rose as mist up through the trunk, branches, and twigs of a great tree whose roots originated at the earth's center. The large branches contained the great veins of minerals, the smaller the metallic ores. Miners had

> . . . found by experience that the vein of gold is a living tree and that by all ways that it spreadeth and springeth from the root by the soft pores and passages of the earth putting forth branches even unto the uppermost part of the earth and ceaseth not until it discover itself unto the open air: at which time it showeth forth certain beautiful colors instead of flowers, round stones of golden earth instead of fruits and thin plates instead of leaves.[39]

Once the veins of ore were mined, it was believed that new ore grew again within a few years. Diamond mines in the East Indies were filled with new diamonds; sulfur mines were soon replenished. Iron ore continued to grow in the mines of Elba off the coast of Italy; the iron ore fields of Saga in Germany refilled within ten years. Silver twigs resembling plants grew in the shafts of abandoned mines in the Joachim Valley. In his *Art of Metals* (1640), Albaro Barba, director of the Potosi mines in the Spanish West Indies, wrote, "All of us know that in the rich hill at Potosi the stones, which divers years we have left behind us, thinking there was not plate enough in them to make it worth our labor, we now bring home and find abundant plate in them, which can be attributed to nothing but the perpetual generation of silver."[40] The image of Mother Earth and her generative role in the production of metals continued to be significant well into the eighteenth century.

What ethical ideas were held by ancient and early modern writers on the extraction of the metals from the bowels of the living earth? The Roman compiler Pliny (A.D. 23–79), in his *Natural History,* had specifically warned against mining the depths of Mother Earth, speculating that earthquakes were an expression of her indignation at being violated in this manner:

> We trace out all the veins of the earth, and yet . . . are astonished that it should occasionally cleave asunder or tremble: as though, forsooth, these signs could be any other than expressions of the indignation felt by our sacred parent! We penetrate into her entrails, and seek for treasures . . . as though each spot we tread upon were not sufficiently bounteous and fertile for us![41]

He went on to argue that the earth had concealed from view that which she did not wish to be disturbed, that her resources might not be exhausted by human avarice:

> For it is upon her surface, in fact, that she has presented us with these substances, equally with the cereals, bounteous and ever ready, as she is, in supplying us with all things for our benefit! It is what is concealed from our view, what is sunk far beneath her surface, objects, in fact, of no rapid formation, that urge us to our ruin, that send us to the very depths of hell. . . . when will be the end of thus exhausting the earth, and to what point will avarice finally penetrate!

Here, then, is a striking example of the restraining force of the beneficent mother image—the living earth in her wisdom has ordained against the mining of metals by concealing them in the depths of her womb. In addition the mining of gold contributed to human corruption and avarice. "The worst crime against mankind [was] committed by him who was the first to put a ring upon his fingers." In the age of barter, people were happy, but since then "man has learned how to challenge both nature and art to become the incitements to vice!"

While mining gold led to avarice, extracting iron was the source of human cruelty in the form of war, murder, and robbery. Its use should be limited to agriculture and those activities that contributed to the "honors of more civilized life":

> For by the aid of iron we lay open the ground, we plant trees, we prepare our vineyard trees, and we force our vines each year to resume their youthful state, by cutting away their decayed branches. It is by the aid of iron that we construct houses, cleave rocks, and perform so many other useful offices of life. But it is with iron also that wars, murders, and robberies are effected, . . . not only hand to hand, but . . . by the aid of missiles and winged weapons, now launched from engines, now hurled by the human arm, and now furnished with feathery wings. Let us therefore acquit nature of a charge that here belongs to man himself.

In past history, Pliny stated, there had been instances in which laws were passed to prohibit the retention of weapons and to ensure that iron was used solely for innocent purposes, such as the cultivation of fields.

In the *Metamorphoses* (A.D. 7), the Roman poet Ovid wrote of the violence done to the earth during the age of iron. In the preceding Golden Age, "people were unaggressive, and unanxious."

> And Earth, untroubled,
> Unharried by hoe or plowshare, brought forth all
> That men had need for, and those men were happy,
> Gathering berries from the mountain sides,
> Cherries, or black caps, and the edible acorns.
> Spring was forever, with a west wind blowing
> Softly across the flowers no man had planted,
> And Earth, unplowed, brought forth rich grain; the field,
> Unfallowed, whitened with wheat, and there were rivers
> Of milk, and rivers of honey, and golden nectar
> Dripped from the dark-green oak-trees.[42]

During the Iron Age, evil was let loose in the form of trickery, slyness, plotting, swindling, and violence, as men dug into the earth's entrails for iron and gold:

                                        The rich earth
Was asked for more; they dug into her vitals,
Pried out the wealth a kinder lord had hidden
In stygian shadow, all that precious metal,
The root of evil. They found the guilt of iron,
And gold, more guilty still. And War came forth.

The violation of Mother Earth resulted in new forms of monsters, born of the blood of her slaughter:

                                        Jove struck them down
With thunderbolts, and the bulk of those huge bodies
Lay on the earth, and bled, and Mother Earth,
Made pregnant by that blood, brought forth new bodies,
And gave them, to recall her older offspring,
The forms of men. And this new stock was also
Contemptuous of gods, and murder-hungry
And violent. You would know they were sons of blood.

Seneca also deplored the activity of mining, although, unlike Pliny and Ovid, he did not consider it a new vice, but one that had been handed down from ancient times. "What necessity caused man, whose head points to the stars, to stoop, below, burying him in mines and plunging him in the very bowels of innermost earth to root up gold?" Not only did mining remove the earth's treasures, but it created "a sight to make [the] hair stand on end—huge rivers and vast reservoirs of sluggish waters." The defiling of the earth's waters was even then a noteworthy consequence of the quest for metals.[43]

These ancient strictures against mining were still operative during the early years of the commercial revolution when mining activities, which had lapsed after the fall of Rome, were once again revived. Ultimately, such constraints would have to be defeated by proponents of the new mercantilist philosophy.[44]

An allegorical tale, reputedly sent to Paul Schneevogel, a professor at Leipzig about 1490–1495, expressed opposition to mining encroachments into the farmlands of Lichtenstat in Saxony, Germany, an area where the new mining activities were developing rapidly. Reminiscent of Alain of Lille's *Natura* and her torn gown and illustrative of the force of the ancient strictures against mining is the following allegorical vision of an old hermit of Lichtenstat. Mother Earth, dressed in a tattered green robe and seated on the right hand of Jupiter, is represented in a court case by "glib-tongued Mercury" who charges a miner with matricide. Testimony is presented by several of nature's deities:

> Bacchus complained that his vines were uprooted and fed to the flames and his most sacred places desecrated. Ceres stated that her fields were devastated; Pluto that the blows of the miners resound like thunder through the depths of the earth, so that he could hardly reside in his own kingdom; the Naiad, that the subterranean waters were diverted and her fountains dried up; Charon that the volume of the underground waters had been so diminshed that he was unable to float his boat on Acheron and carry the souls across to Pluto's realm, and the Fauns protested that the charcoal burners had destroyed whole forests to obtain fuel to smelt the miner's ores.[45]

In his defense, the miner argued that the earth was not a real mother, but a wicked stepmother who hides and conceals the metals in her inner parts instead of making them available for human use.

The final judgment, handed down by Fortune, stated that if men deign "to mine and dig in mountains, to tend the fields, to engage in trade, to injure the earth, to throw away knowledge, to disturb Pluto and finally to search for veins of metal in the sources of rivers, their bodies ought to be swallowed up by the earth, suffocated by its vapors . . . intoxicated by wine, . . . afflicted with hunger and remain ignorant of what is best. These and many other dangers are proper of men. Farewell."

In the old hermit's tale, we have a fascinating example of the relationship between images and values. The older view of nature as a kindly mother is challenged by the growing interests of the mining industry in Saxony, Bohemia, and the Harz Mountains, regions of newly found prosperity (Fig. 6). The miner, representing these newer commercial activities, transforms the image of the nurturing mother into that of a stepmother who wickedly conceals her bounty from the deserving and needy children. In the seventeenth century, the image will be seen to undergo yet another transformation, as natural philosopher Francis Bacon (1561–1626) sets forth the need for prying into nature's nooks and crannies in searching out her secrets for human improvement.

Henry Cornelius Agrippa's polemic *The Vanity of Arts and Sciences* (1530) reiterated some of the moral strictures against mining found in the ancient treatises, quoting the passage from Ovid por-

Figure 6. Miners diverting rivers and cutting down trees for the washing and refining of ores, drawing by Georg Agricola, in his *De Re Metallica* (1556). This scene shows an intensive use of wood and water resources in the mining industry that burgeoned in the late Middle Ages and the Renaissance. Nature is being thought of here as something to be exploited, rather than as a beneficent mother who nurtures her children.

traying miners digging into the bowels of the earth in order to extract gold and iron. "These men," he declared, "have made the very ground more hurtful and pestiferous, by how much they are more rash and venturous than they that hazard themselves in the deep to dive for pearls." Mining thus despoiled the earth's surface, infecting it, as it were, with an epidemic disease. Of all cultures that have mined for metals, "only the Scythians . . . condemned the use of gold and silver, resolving to keep themselves eternally free from public avarice. There was an ancient law among the Romans against the superfluity of gold. And indeed, it were to be wished that men would aspire with the same eagerness to heaven, that they descend into the bowels of the earth, allured with that vein of riches that are so far from making a man happy, that they repent too often of their time and labor so ill bestowed." [46]

If mining were to be freed of such strictures and sanctioned as a commercial activity, the ancient arguments would have to be refuted. This task was taken up by Georg Agricola (1494–1555), who wrote the first "modern" treatise on mining. His *De Re Metallica* ("On Metals," 1556) marshalled the arguments of the detractors of mining in order to refute them and thereby promote the activity itself.

According to Agricola, people who argued against the mining of the earth for metals did so on the basis that nature herself did not wish to be discovered what she herself had concealed:

> The earth does not conceal and remove from our eyes those things which are useful and necessary to mankind, but, on the contrary, like a beneficent and kindly mother she yields in large abundance from her bounty and brings into the light of day the herbs, vegetables, grains, and fruits, and trees. The minerals, on the other hand, she buries far beneath in the depth of the ground, therefore they should not be sought. [47]

This argument, taken directly from Pliny, reveals the normative force of the image of the earth as a nurturing mother.

A second argument of the detractors, reminiscent of Seneca and Agrippa, and based on Renaissance "ecological" concerns was the disruption of the natural environment and the pollutive effects of mining.

> But, besides this, the strongest argument of the detractors [of mining] is that the fields are devastated by mining operations, for which reason formerly Italians were warned by law that no one should dig the earth for metals and so injure their very fertile fields, their vineyards, and their olive groves. Also they argue that the woods and groves are cut down, for there is need of wood for timbers, machines, and the smelting of metals. And when the woods and groves are felled, then are exterminated the beasts and birds, many of which furnish a pleasant and agreeable food for man. Further, when the ores are washed, the water which has been used poisons the brooks and streams, and either destroys the fish or drives them away. Therefore the inhabitants of these regions, on account of the devastation of their fields, woods, groves, brooks, and rivers, find great difficulty in procuring the necessaries of life, and by reason of the destruction of the timber they are forced to greater expense in erecting buildings. Thus it is said, it is clear to all that there is greater detriment from mining than the value of the metals which the mining produces (Fig. 7).

Agricola may have been alluding to laws passed by the Florentines between 1420 and 1485, preventing people from dumping lime into rivers upstream from the city for the purpose of "poisoning or catching fish," as it caused severe problems for those living downstream. The laws were enacted both to preserve the trout, "a truly noble and impressive fish" and to provide Florence with "a copious and abundant supply of such fish." But these laws passed to safeguard the waters of the Arno and the areas of the Castino and Pistoian Apennines were not obeyed, and complaints mounted that "the rivers are empty of fish, and the fish are worse." By 1477, new laws protected the entire domain around Florence, including the Arno, Sieve, and Serchio rivers and their tributaries, from the diverting and damming of rivers and the poisoning of fish with lime, nutshells, or the reputedly toxic Aaron's Rod plant.[48]

Such ecological consciousness, however, suffered because of the failure of law enforcement, as well as because of the continuing progress of mining activities. Agricola, in his response to the detractors of mining, pointed out the congruences in the need to catch fish and to construct metal tools for the well-being of the human

Figure 7.   Veins of ore in hillsides, drawing by Georg Agricola, in his *De Re Metallica* (1556). Here the cut-off trees of the landscape imply their intensive use in the mining and refining of ores; the artist has used the term *vena,* "vein" (as we still do), to describe a lode of ore in the earth, thus likening the earth to a body, complete with veins.

race. His effort can be interpreted as an attempt to liberate the activity of mining from the constraints imposed by the organic framework and the nurturing earth image, so that new values could sanction and hasten its development and progress.

To the argument that, because the metals lie in the earth "enclosed and hidden from sight [and] should not be taken out," Agricola countered with the example of catching fish, which lie concealed in the depths of the waters.[49] "Nature has given the earth . . . to man that he might cultivate it and draw out of its caverns metals and other mineral products," without which the earth could not be cultivated, fish caught, sheep sheared, animals slaughtered, or food cooked. Without the metals, men would "return to the acorns and fruits and berries of the forest. They would feed upon the herbs and roots which they plucked up with their nails. They would dig out caves in which to lie down at night, a . . . condition . . . utterly unworthy of humanity, with its splendid and glorious natural endowment."

To the argument that the woods were cut down and the price of timber therefore raised, Agricola responded that most mines occurred in unproductive, gloomy areas. Where the trees were removed from more productive sites, fertile fields could be created, the profits from which would reimburse the local inhabitants for their losses in timber supplies. Where the birds and animals had been destroyed by mining operations, the profits could be used to purchase "birds without number" and "edible beasts and fish elsewhere" and refurbish the area.

The vices associated with the metals—anger, cruelty, discord, passion for power, avarice, and lust—should be attributed instead to human conduct: "It is not the metals which are to be blamed, but the evil passions of men which become inflamed and ignited; or it is due to the blind and impious desires of their minds." Agricola's arguments are a conscious attempt to separate the older normative constraints from the image of the metals themselves so that new values can then surround them.

Edmund Spenser's treatment of Mother Earth in the *Faerie Queen* (1595) was representative of the concurrent conflict of attitudes about mining the earth. Spenser entered fully into the sixteenth-century debates about the wisdom of mining, the two greatest sins against the earth being, according to him, avarice and lust. The arguments associating mining with avarice had appeared in the ancient texts of Pliny, Ovid, and Seneca, while during Spenser's lifetime the sermons of Johannes Mathesius, entitled *Bergpostilla, oder Sarepta* (1578), inveighed against the moral consequences of human greed for the wealth created by mining for metals.[50]

In Spenser's poem, Guyon presents the arguments against mining taken from Ovid and Agricola, while the description of Mammon's forge is drawn from the illustrations to the *De Re Metallica*. Gold and silver pollute the spirit and debase human values just as the mining operation itself pollutes the "purest streams" of the earth's womb:

> Then gan a cursed hand the quiet wombe
> Of his great Grandmother with steele to wound,

> And the hid treasures in her sacred tombe
> With Sacrilege to dig. Therein he found
> Fountaines of gold and silver to abound,
> Of which the matter of his huge desire
> And pompous pride eftsoones he did compound.[51]

The earth in Spenser's poem is passive and docile, allowing all manner of assault, violence, ill-treatment, rape by lust, and despoilment by greed. No longer a nurturer, she indiscriminately, as in Ovid's verse, supplies flesh to all life and lacking in judgment brings forth monsters and evil creatures. Her offspring fall and bite her in their own death throes. The new mining activities have altered the earth from a bountiful mother to a passive receptor of human rape.

John Milton's *Paradise Lost* (1667) continues the Ovidian image, as Mammon leads "bands of pioners with Spade and Pickaxe" in the wounding of the living female earth:

>   ... By him first
> Men also, and by his suggestion taught,
> Ransack'd the Center, and with impious hands
> Rifl'd the bowels of thir mother Earth
> For Treasures better hid. Soon had his crew
> Op'nd into the Hill a spacious wound
> And dig'd out ribs of Gold.[52]

Not only did mining encourage the moral sin of avarice, it was compared by Spenser to the second great sin, human lust. Digging into the matrices and pockets of earth for metals was like mining the female flesh for pleasure. The sixteenth- and seventeenth-century imagination perceived a direct correlation between mining and digging into the nooks and crannies of a woman's body. Both mining and sex represent for Spenser the return to animality and earthly slime. In the *Faerie Queen,* lust is the basest of all human sins. The spilling of human blood, in the rush to rape the earth of her gold, taints and muddies the once fertile fields.

The sonnets of the poet and divine John Donne (1573–1631) also played up the popular identity of mining with human lust. The poem "Love's Alchemie" begins with the sexual image, "Some that have deeper digged loves Myne than I,/say where his centrique happiness doth lie."[53] The Platonic lover, searching for the ideal or "centrique" experience of love, begins by digging for it within the female flesh, an act as debasing to the human being as the mining

39

of metals is to the female earth. Happiness is not to be obtained by avarice for gold and silver, nor can the alchemical elixir be produced from base metals. Nor does ideal love result from an ascent up the hierarchical ladder from base sexual love to the love of poetry, music, and art to the highest Platonic love of the good, virtue, and God.

The same equation appears in Elegie XVIII, "Love's Progress":

> Search every spheare
> And firmament, our Cupid is not there:
> He's an infernal god and under ground,
> With Pluto dwells, where gold and fire abound:
> Men to such Gods, their sacrificing Coles,
> Did not in Altars lay, but pits and holes.
> Although we see Celestial bodies move
> Above the earth, the earth we Till and love:
> So we her ayres contemplate, words and heart
> And virtues; but we love the Centrique part.[54]

Lust and love of the body do not lead to the celestial love of higher ideals; rather, physical love is associated with the pits and holes of the female body, just as the love of gold depends on the mining of Pluto's caverns within the female earth, "the earth we till and love." Love of the sexual "centrique" part of the female will not lead to the aery spiritual love of virtue. The fatal association of monetary revenue with human avarice, lust, and the female mine is driven home again in the last lines of the poem:

> Rich Nature hath in women wisely made
> Two purses, and their mouths aversely laid:
> They then, which to the lower tribute owe,
> That way which that Exchequer looks, must go.

Avarice and greed after money corrupted the soul, just as lust after female flesh corrupted the body.

The comparison of the female mine with the new American sources of gold, silver, and precious metals appears again in Elegie XIX, "Going to Bed." Here, however, Donne turns the image upside down and uses it to extol the virtues of the mistress:

> License my roaving hands, and let them go,
> Before, behind, between, above, below.
> O my America! my new-found-land,

> My kingdome, safelist when with one man man'd
> My Myne of precious stones, My Emperie,
> How blest am I in this discovering thee!

In these lines, the comparison functions as a sanction—the search for precious gems and metals, like the sexual exploration of nature or the female, can benefit a kingdom or a man.

Moral restraints were thus clearly affiliated with the Renaissance image of the female earth and were strengthened by associations with greed, avarice, and lust. But the analogies were double-edged. If the new values connected with mining were positive, and mining was viewed as a means to improve the human condition, as they were by Agricola, and later by Bacon, then the comparison could be turned upside down. Sanctioning mining sanctioned the rape or commercial exploration of the earth—a clear illustration of how constraints can change to sanctions through the demise of frameworks and their associated values as the needs, wants, and purposes of society change. The organic framework, in which the Mother Earth image was a moral restraint against mining, was literally undermined by the new commercial activity.

CHAPTER 2

Farm, Fen, and Forest

European Ecology in Transition

In seeking to understand how, between 1500 and 1700, the organic conception of the cosmos gave way to a mechanistic model, an ecological perspective is essential. Focusing on early modern Europe as an ecosystem means more than discovering that today's kind of environmental crisis has occurred in the past. It means a special sensitivity to the dialectical relationship between human behavior and institutions, on the one hand, and the natural environment, on the other.

An ecosystem model presents an earth's-eye view of history. By looking at history "from the ground up," factors having an impact on the earth's resources can be analyzed and a new and different interpretation of historical change developed, based on the assumption that the natural and human environments together form an interrelated system. Although the natural environment alone can be seen as a total ecosystem composed of interrelated physical and biological components, when human factors are included a more complete picture emerges. An ecosystem model of historical change looks at the relationships between the resources associated with a given natural ecosystem (a forest, marsh, ocean, stream, etc.) and the human factors affecting its stability or disruption over historical time periods. Historical change becomes ecological change, emphasizing human impact on the system as a whole. Conversely, ecological change is the history of ecosystem maintenance and disruption.

While such a perspective does full justice to the conventionally studied sources of change arising within culture—demographic, economic, political, technological, and ideological—it does not take the natural environment for granted, as does traditional history. Instead of dichotomizing nature and culture as a structural dualism, it sees natural and cultural subsystems in dynamic interaction.

Only an ecosystem approach to early modern Europe can deal adequately with the question of how changes arising within human culture affected and were affected by the natural environment. The important cultural changes in this case were the rise and fall of population, the conflict between landlord and peasant over control of natural resources, technological innovation, the spread of the capitalist market, and changing attitudes toward nature and the earth. Particularly important is the question of how environmental quality was affected by the transition from peasant control of natural resources for the purpose of subsistence to capitalist control for the purpose of profit.

The other side of the dialectical relationship raises the question of how the resulting changes in the natural environment influenced human culture and institutions and how people's experience of an increasingly manipulated nature undermined the organic model and made way for the mechanistic model. And this leads, in turn, to a final set of questions about how the mechanistic model reinforced and accelerated the exploitation of nature and human beings as resources and to a view of conservation as the restoration and management of ecosystems. A new interpretation of the history of early modern Europe that integrates nature and culture can be made by investigating the interactions and causal relationships among these factors.

MANORIAL FARM ECOLOGY.   Basic to the agrarian ecosystem of premodern Europe was the relationship between the peasant community and the land. Evolved over centuries of adaptation to the productive capabilities of the natural environment on the one hand and the state of agricultural technology on the other, the peasant community produced a level of subsistence by following traditional patterns of cooperation upheld by powerful cultural norms. In the early medieval period, these practices and norms tended to result in relatively high crop productivity combined with the maintenance of soil fertility. The use of plow agriculture integrated crop planting with the raising of cows, pigs, and horses. The number of animals kept corresponded to the number of acres to be manured to ensure fertile soil. In some areas, this interdependent animal and crop system incorporated the practice of carrying eroded soil back up the slopes to restore washed-out ground. The woodlands supplied a constant source of fuel, and the commons provided adequate fields for grazing and planting.[1]

The family and communal holdings constituted the fundamental economic units. The earliest agrarian communities, composed of persons living in the same geographical area, have been called "territorial communes." Whereas each family farmed its own individual plot, all those living in the community's territory shared its natural resources: forests, pastures, and water. The communal resources were regulated by officers elected or appointed by the members of the peasant community. The community grounds within the village and inside the edges of forests were subject to governance concerning the gathering of firewood and bedding litter for animals, common pasture, and hunting. Peasant jurisdiction and self-regulation of these common lands exemplified an interaction between individual needs and those of the group as a whole. Cooperation and interdependence maintained the health of the ecosystem. The village officials supervised rules for plowing, planting, harvesting, pasturing, and fencing.

Yet built into the premodern ecosystem was a source of instability. Through force and the need for military security, a hierarchical structure of landlord domination had imposed itself on the communal structure of agrarian society, extracting surplus value in the form of labor, services, rents, and taxes. The amount landlords exacted was regulated by long-established tradition, and medieval landlords did not, under stable conditions, strive to maximize their gains. But the built-in pressure for seigneurial privileges—the rights traditionally accorded a feudal lord over his domain—was in constant tension with the peasant pressure for community control of common rights and resources. When combined with other interacting destabilizing forces—population pressure and technological innovation—this tension could produce sharp conflict, altered relations between landlords and peasants, and significant changes in the ecosystem as a whole.

By the twelfth and thirteenth centuries, conflict was already evident over the use and control of technology for energy production. The energy of the preindustrial economy was drawn from renewable sources—wood, water, wind, and animal, including human, power. Watermills had been introduced into Europe slowly ever since Roman times, reaching Great Britain by the eighth century and Scandinavia by the twelfth, and were used for grinding grain, fulling (or shrinking and thickening) cloth, sawing timber, extracting oil from olives, and making paper. Windmills began to appear in the twelfth century and came into more general use as energy suppliers by the sixteenth and seventeenth, particularly for the draining of fens and marshes in England, France, and the Low Countries. In the thirteenth century, woodlands began to be more important as sources of energy, having been earlier used primarily for hunting, the grazing of swine, and cottage fuel. But the use of these energy sources had differing significance for the lives of peasants, artisans, landlords, entrepreneurial adventurists, and representatives of the state. Problems arose over the control of woodlands for the building of ships, the substitution of coal for wood in the trades as timber supplies became scarce, and the use of the lord-of-the-manor's watermill.[2] The impact of access to resources for differing interest groups can be illustrated by selected examples chosen from three historically changing ecosystems—the farm, the fen, and the forest.

The medieval manor was the basis of the European agrarian economy until its decline in the fourteenth and fifteenth centuries. Here conflicts occurred between the prince, or landlord, who controlled the resources of the manorlands, and the peasants, who were primarily unfree and lived in communal villages or in isolated cottages using the manorial commons for wood, grazing, and planting.

The use of the manor's watermill as energy for the grinding of grain was controlled by the lord, who raised revenue from its use by the villagers and tenant farmers. At harvest time, grain was hauled to the mill and, for a fee, ground into flour, a practice that encouraged peasants to keep hand-operated mills in their homes and therefore to resist the advent of wind and water as new energy sources. Lords were careful to see that competing mills were not erected on manor lands, but it was more difficult to prevent peasants from keeping hand-operated mills at home.

Handmills survived into the eighteenth and nineteenth centuries in areas devoid of sufficient waterways, as security against winter frost, summer drought, and siege. Hidden handmills thus competed with the manor mill, human muscle with the manor's centralized inanimate energy sources—water and wind.

Evidence of centuries of attempts by landlords to obliterate peasant handmills has been found in scattered documents. In twelfth-century England, for example, officials of the manor charged with executing the lord's right to enforce justice entered cottages and, buttressed by decrees dictating that "the men shall not be allowed to possess any handmills," smashed illegal mills. In Hertfordshire in 1274, artisans had refused to use not only the official watermill for grain grinding but also the lord's fulling mill for the beating of cloth. Mills and materials were seized and destroyed and court proceedings instituted. A second uprising by peasants in 1326 demanded cottage milling rights; eighty mills began operating in the homes of peasants and artisans, only to be confiscated five years later. Then in 1381 an open insurrection demanded a local charter allowing the use of handmills, only to be defeated this time by a royal statute denying domestic milling privileges. The issue resurfaced in 1547 when grinding rights were granted to the cottagers of Kingsthorpe manor. The English examples were repeated in other countries over most of Europe and testify, not to the resistance of peasants to technological innovation, but to their economic need to retain control over their own technology.

Yet the impact of these conflicts over energy technology was merely symptomatic, compared with the interaction between population and landlord-peasant relations in which, despite fluctuations in climate, disease, and marriage ages, a slow upward trend in population resulted; in turn, the larger manorial populations increased the power of the lords. During the twelfth and thirteenth centuries, population increases all over Europe resulted in an era of reclamation. Woodlands and marshes were converted to arable lands, and wastelands turned to pasture. Regulations over the use of meadows, pastures, and woodlands began to be tightened, and family holdings were divided and subdivided. The simultaneous growth of towns as centers for trade and crafts created new pressures to increase croplands. By the early fourteenth century, most wasteland was being used, and forests had shrunk dramatically over much of Europe.

As more and more pasture and wastelands were converted to arable, the earlier territorial commune was gradually replaced by the more tightly cooperative village commune, beginning in Germany and Poland and gradually spreading over much of western Europe and Russia. Groups of peasants voluntarily pooled their holdings into fields, which were cultivated on a rotation basis every two or three years. This open three-field system not only allowed for more intensive cultivation and higher crop productivity, but it also demanded greater cooperation and group regulation of water use, pasturing, and wood gathering, as well as the sharing of oxen and horses for plowing. The use of the heavy plow and of horses instead of oxen (resulting from the introduction of the horse collar) helped to increase yields and improve nutrition for a time.[3]

As population continued to increase, many village communes employing the open-field system instituted compulsory tillage regulations—all farming procedures were performed by everyone at the same time, and all grew the same crops. Everyone pastured animals on the same fallow land and plowed and harvested at times set by the village officials. Outsiders wishing to use the commons had to gain approval of the officials and were frequently excluded. This intensely cooperative system placed the good of the group above the individual, each family submitting to the group regulations administered by its own elected officials. But in areas such as Denmark and Norway, where individuals were more isolated and where pastoral husbandry took precedence over the cooperative three-field system, communal ties and organization remained weak. In many regions, therefore, population pressure resulted not in a "tragedy of the commons" produced by competitive self-interest, but in increased cooperative activity and in group regulation of the ecosystem.

But population growth, dealt with at the village level by increased cooperation, when combined with landlord taxation, led ultimately to the breakdown of the medieval agrarian economy and ecosystem. The landlords' practice of extracting from unfree peasants (those not subject to fixed rents) any income above subsistence meant that these peasants were unable to give back to the land what they took. They had insufficient reserves to reinvest in animals for plowing and manuring the soil. In many regions, soils became quickly exhausted and badly eroded. Here declining soil fertility led to the cultivation of unplowed land formerly used as pasture and of marginal soils. Continuously declining soil fertility, coupled with decreasing livestock, led to declining crop productivity, availability of food for subsistence, and the nutrition necessary for the maintenance of the population as a whole. Population growth exacerbated by landlord exactions meant that eventually there was not enough land per person to ward off famine in a poor harvest year.[4]

In 1315, following an unusually wet spring and summer, food shortages became acute over most of western Europe, and food prices inflated greatly. Those too poor to pay the higher prices died of starvation. The deaths mounted into 1317, despite a somewhat improved harvest the following year. Food shortages and malnutrition became common in the first half of the fourteenth century.

A general population decline was accelerated by the advent of the bubonic plague, to which a generally malnourished and weakened population had become especially susceptible. Outbreaks of the plague began in 1348, reappearing again in the years 1360–1361, 1369, and 1374. Populations all over Europe were drastically reduced, some by as much as 60 percent, the low points being reached between 1400 and 1450. But during the hundred years between 1350 and 1450, while the European population was at these new lows, the environment recovered. Ecosystem balances altered by preplague population pressures had been generally restored to their medieval state by 1500. Forests had grown back, renewing timber supplies, soils had recovered their fertility, and the cultivation of marginal lands had been abandoned. In France, the medieval ecosystem that had existed prior to the plague outbreaks of 1348 had been restored in both the south and the north by about 1550.[5]

Yet the Black Death, so terrible in scope for the population at large, not only reconstituted the ecological balance but raised the status of peasants. Labor was now scarce, and the surplus labor that had been paid by unfree peasants to the landlord was not as plentiful, resulting in a decrease in living standard for the lords.

The resulting change in the structure of the relationship between landlord and peasant had differing consequences in different European locales, depending on the exercise of either landlord or peasant power.[6] In areas where peasant collectivity and self-regulation were strong and landlords weak, principally in central Europe, ecological balances between people and land could be maintained. The tradition of cooperation and cohesion within the community provided unity and strength in the face of confrontation with the landlords. In western Germany, where cooperation in regulating use of the commons for grazing and for field rotation had been strong, peasants organized in village after village to ensure their rights of inheritance to the land, the payment of fixed rents, and the election of village officials. Outsiders wishing to use communal resources and grazing areas had to obtain the approval of the commune. The number of animals each person could graze, the amount of firewood to be cut, and the acreage to be tilled was regulated for the benefit of all. By the seventeenth century, in western Germany peasants controlled up to 90 percent of the land.

In East Prussia, open conflict broke out in the Peasant's War of 1525, as peasants rebelled against the power of the lords. Landlords had taken over the regulation of grazing, hunting, and fishing rights on the commons and had removed some commons lands for use by landless day laborers who had built cottages in the village. Supplies of firewood were short and expensive. Peasants rose in armed rebellion to demand the return of self-regulation of common resources and control over village officials.

In Switzerland, South Germany, Austria, and Scandinavia, common land, called the *Allmend*, which included not only forest and pasture but also vineyards, orchards, and gardens, was allotted to the members of the community, with any excess income being reserved for the poor. Collective ownership of land and resources was the practice in parts of Poland, Hungary, and Moldavia until the eighteenth and early nineteenth centuries. In many parts of inland Pomerania, Brandenburg, Silesia, and Saxony, an ecologically and communally healthy peasant society continued to flourish as late as 1945, retaining its traditional character and keeping the soil in a fertile condition.

Yet the resurgence of peasant status was not uniform, especially in areas where landlords remained powerful. Where princes, land-lords, or the state were strong enough to continue to levy high taxes and to control resources, soil ecology suffered because the peasant could not invest in sufficient animals to maintain soil fertility. In eastern Germany, where peasant organization was weak and princes strong, landlords reasserted their seigneurial rights, con-trolled rents and wages, and charged high fees for permission to move to new locales.

In France, despite a strong peasantry that maintained its rights to small landholdings, the state extracted surplus value in the form of taxes, thereby making adequate manuring of the land impossible. With population growth and increased demand for grain, land de-voted to the pasturing of animals was cut back. Lack of manure again led to cultivation of poorer soils and further depletion of pas-ture, which in turn reinforced the scarcity of animals and lack of manure. The seventeenth century was a period of zero population growth in rural areas, marked by large numbers of famines. A pro-longed series of peasant revolts from 1616 until 1647 resulted, di-rected not against the landlords but against the taxes levied by the state, while landlords, acting in their own self-interest, attempted to protect peasants from taxation. A series of plagues, famines, and bad harvests, beginning in 1627, drove the poor to cities where they competed for bread and where mobs of rioters, including armed women, seized bread and grain hoarded by the lords and millers.[7] French social structure, founded on a strong exploitative state, un-dermined ecological balance and resulted in a repetition of medi-eval demographic patterns, soil exhaustion, and famine.

Similar ecological and economic patterns over southern Europe, especially in the Mediterranean lands where topsoils were light and dry, hit the poor extremely hard. Whereas in northern Europe the Wars of Religion, to some extent, slowed population growth, in the Mediterranean countries a succession of almost daily social revolts and food riots broke out in the latter half of the sixteenth century. Naples (1561), Corsica (1564–1569), Palermo (1560), Mantua (1569), Urbino (1571), Genoa (1575), and the Turkish Empire (1590–1600), to mention only a few, were scenes of insurgence, up-rising, and armed warfare, testifying both to the spread of poverty and to oppression by landlords, nobility, and state tax collectors. Crime, violence, banditry, and vagabondry became commonplace among the lower orders of society.

THE MARKET AND CAPITALISM. In northern Europe, with its wetter heavier soils, events followed a somewhat different course primarily because of the emergence of a force that would presently shatter the premodern ecosystem. Arising in the city-states of Renaissance Italy and spreading to northern Europe was an inexorably expanding market economy, intensifying medieval tendencies toward capitalist relations of production and capitalist modes of economic behavior. As trade quickened throughout western Europe, stimulated by the European discovery and exploitation of the Americas, production for subsistence began to be replaced by more specialized production for the market. The spreading use of money provided not only a uniform medium of exchange but also a reliable store of value, facilitating open-ended accumulation. Inflation generated by the growth of population and the flood of American gold accelerated the transition from traditional economic modes to rationally maximizing modes of economic organization. Many landlords whose land rents had been fixed at preinflationary rates were now faced with declining income and rising expenses. In addition, the growth of cities as centers of trade and handicraft production created a new class of bourgeois entrepreneurs who supplied ambitious monarchs with the funds and expertise to build strong national states, undercutting the power of the regionally based landowning nobility.

The impact of the market on the agrarian ecosystem can be exemplified by the Netherlands and England.[8] Here preindustrial capitalism, through its division of labor, produced an initial restoration of the soil ecosystem. A growing class of prosperous farmers began to introduce new agricultural techniques that increased market yields and personal profits. Although the early effects of agricultural improvement were healthy for the soil, built into the emerging capitalist market economy was an inexorably accelerating force of expansion and accumulation, achieved, over the long term, at the expense of the environment and the village community—the natural and human resource bases. The same expanding market force that began in the Netherlands and England in the sixteenth and seventeenth centuries with the application of organic fertilizers, agricultural improvement, and specialization for market profits, ultimately culminates in today's agribusiness and the exportation of "green revolution" techniques to developing countries. Today, inorganic nitrate fertilizers and chemical pesticides, which leave long-lasting soil-depleting residues and have unanticipated side effects, the monoculture of high-yield grains subject to large-scale devastation by pests and disease, and the impetus to continually bring new "virgin" lands into cultivation for the market all disrupt established ecosystem balances. Secondly, the tendency toward growth, expansion, and accumulation inherent in capitalism results in the displacement of subsistence farmers from the land and the disruption of traditional patterns of human-land integration (as can be seen in such countries as Mexico, Java, and the Philippines).

The early stages of this process are to be found in the Dutch and English rural economies of the sixteenth and seventeenth centuries. But, while the ultimate rationale in both cases was an emerging agrarian market, the human-land relationship was disrupted less by the autonomous Dutch peasantry than it was in England where strong landlords displaced subsistence farmers from the farm, fen, and forest ecosystems.

Dutch rural society, especially in the northern Netherlands, had been based on a peasantry composed of freeholders owning *in toto* more land than the nobility, church, or urban bourgeoisie, with property ownership, transfer rights, and relatively little open-field common land. Here, where the nobility was weak, the peasantry was individualistic and autonomous, owning and leasing land, lending money to other peasants, and engaging in urban market trade. Some villages maintained a homogeneous, egalitarian character, while others had already developed more stratified social structures. Many poorer peasants made their living in the marshes digging and selling peat—the major energy source in regions devoid of forests; others supported themselves by fishing, fowling, and collecting reeds. Beginning in the early sixteenth century, production for urban markets, agricultural improvement, and increasing economic specialization began to alter class relationships.

Coupled with an expanding international Dutch trading economy, the rural economy began to specialize its household operations. Instead of dividing holdings, thereby reducing agricultural productivity, some Dutch peasants began to produce nonagricultural products, while others intensified and improved their farming operations, gradually forming a new class of prosperous farmers. Farmers increased the size of their livestock herds through a rigorous pasture-fertilization program. The spreading of manure on arable fields was extended to pastures, replacing the older practice of fertilizing pastures solely from the droppings of grazing animals. Liquid manure, peat ash, and clover improved pasture quality, increasing yields of dairy products. On arable fields, farmers also intensified fertilization, introduced new crops into a four-field (or more) rotation system, and began to produce new types of products for new markets—coleseed for cooking and lighting fuel, rapeseed for oil and fodder, and soil-improving legumes. Wheat, rye, and bean production was increased for urban markets.

The key to larger yields, use of new types of soil, and increased crop varieties was an organic fertilizer—manure. The European famines of the fourteenth and sixteenth centuries had resulted from ecological imbalances caused by population increases leading to increases in arable land at the expense of pasture, thereby reducing livestock for manuring the fields and maintaining soil fertility. In the Low Countries, this cycle was broken as Flemish farmers took the lead in introducing new sources of manure and a new specialized vocabulary describing the new forms of fertilizer. An extensive manure trade was pioneered by the city of Groningen, an area with rich peat layers covering sand. Human excrement, or night soil, was offered by the city to farmers attempting to cultivate the underlying sandy soils. Ships exporting peat to Holland returned with additional night soil. Sheep and pigeon dung were also exported to the tobacco district around Amersfoort. Urban industries contributed additional fertilizers to the rural regions—verval, a clay leftover from brick making, ash from soap boilers, and peat ash from other industries.

During the seventeenth century, Dutch horticulture also received much attention, as witnessed by a proliferation of new books on the production of vegetable crops and fruit trees. Onions, carrots, turnips, cabbages, cherries, and strawberries were grown and could be exported in exchange for boatloads of night soil. Dutch farming for the market was thus based on a total program of organic recycling. Moreover, due to rural specialization and the relative lack of common land, the process did not, as in England, create vast numbers of landless unemployed day laborers.

In the Low Countries, rural specialization for the market represented an early phase of capitalist development, turning subsistence peasants into market farmers. Despite the fact that the Dutch, acting on the basis of their peasant heritage, used the ecologically sound method of intensive manuring in order to increase yields, they nevertheless reaped substantial profits as a result. Under the flowering of capitalism, organic farming, an inherent component of peasant subsistence agriculture, was intensified, not only to grow crops with which to buy necessary tools and domestic goods in the village market but also to produce surplus profits to enhance the farmer's status. For the well-to-do Dutch market farmer, this meant not only a large dairy herd and several farm buildings, but also an excess inventory of bed linens, dishes, and utensils far above even the comfort level. But here the ecologically healthy practice of organic farming, coupled with a strong peasantry and the relative lack of common land for enclosure, dulled the human effects of capitalist farming as practiced simultaneously in the British Isles.

In contrast to the Dutch, English farmers increased their production for the market in a manner that displaced people in return for short-term profits. As in the Low Countries, the growth of a market economy and agricultural improvement increased the yields of English farmers during the expansive years of the sixteenth century. Here, however, landlords were strong and could increase their holdings by consolidating smaller plots, enclosure, cutting down forests for pasture and cultivation, and draining fens. They then rented the lands to tenant farmers, who paid wage laborers to work them. Despite peasant rebellions in the early sixteenth century and the upheavals of the Civil War in the mid-seventeenth century, landlords still retained control of approximately 70 percent of the land at the end of the seventeenth century.[9]

The English tenant farmers who, lacking initiative, were unable to meet rent payments and therefore to retain title might ultimately be relegated to the status of landless day laborers or the ever-increasing group of vagrants, "masterless" persons, and beggars. But farmers who, in the face of increasing demand for land and increasing prices, could raise production and make sufficient profit to buy the title to their land and perhaps that of several neighbors could gain enough wealth to rise in status from husbander (or small landholder) to yeoman (or larger freeholder), or eventually to gentleman. A middle class of well-to-do capitalist yeoman farmers producing crops for the market and wool for the growing export textile industry grew substantially during the Tudor and Stuart eras of the late sixteenth and early seventeenth centuries.

While most yeomen added to their land holdings by bits and pieces, some, along with gentlemen landlords, enclosed common lands, raised rents, and leased land on short-term loan for high profits. The profit motive caused the more industrious yeoman to work hard, live frugally, take risks, experiment with new techniques of husbandry, exploit new methods of increasing crop yields, and then, with the profits, to expand acreage or to enclose arable lands from the commons for sheep pasture. This reduced the proportion of common fields available to the peasant for subsistence farming. Severely criticized for these high-handed enclosure methods, the new landholder was portrayed as a land-devouring cormorant who did not hesitate "to hoard up corn with many a bitter ban, from widows, orphans, and the lab'ring man." [10] Complaints stated that "sheep and cattle drove out Christian laborers." [11] Thus, in contrast to the Dutch, enclosure represented for the English the most prevalent method of entering the market economy.

A program of agricultural improvement aided the ambitious yeoman and gentleman farmers in increasing their profits and status. Books on farming methods eulogized the role of the farmer in the economy as "the nerve and sinew which holdeth together all the joints of the monarchy." [12] Husbandry was praised as a noble occupation equal to or superior to that of the soldier or merchant. Sixteenth-century writers stressed rustic virtues, the healthy atmosphere of country life, and earth and soil as the very source of human existence. Agriculture was the suckling mother of mankind, and "the mother and nurse of all other arts. . . . The earth being the very womb that beares all, and the mother that must nourish and maintain all." [13]

An array of new "how-to" books on agricultural improvement appeared for the benefit of the yeoman farmer and gentleman landlord. They contained information on the breeding of animals, grafting of fruit trees, the art of planting, new ploughs and tilling methods, means of preserving fruits and vegetables, the building of fish ponds, and techniques of beekeeping. The title page of William Blith's *The English Improver Improved* (1652) promised that all lands could be improved, "some to be under a double- and treble-, others under a five- or sixfold, and many under a tenfold, yea, some under a twentyfold improvement." A variety of tools for trenching and plowing so as to increase yields were described. Methods of improvement included "floating and watering" land, "draining fen, reducing bog, and regaining sea lands," the planting of woodlots, composting and manuring of soils, planting new types of crops, vegetables, and fruit trees, warding off crows, destroying vermin, and by "such enclosures as prevents depopulation and advanceth all interests." Other works gave advice on preparing rich ground for the orchard, on planting and grafting fruit trees, and controlling predators. Calendars guided the seasonal activities of the yeoman in manuring fields and sowing peas in January, threshing barley at Candlemas in February, and so on. Some of these recommendations, if put into practice, would restore fertility, and conserve timber; others, however, were indicative of new attitudes sanctioning increased yields for personal profit through the manipulation of nature. In large measure, English agricultural improvement was oriented toward improvement of the farmer's own status. It initiated a tradition of progressive scientific agriculture focused on land management for increased yields.

FENS. The ecological effects of the expanding agrarian market economy are vividly illustrated by its encroachment on another ecosystem and its associated human subsistence economy—the fenlands of England. While marshlands all over Europe had been reclaimed for agricultural use since earliest times, in seventeenth-century Holland and England the process was accelerated by the continuing expansion of the rural market economy and the export of Dutch hydraulic technology.

The Dutch had achieved early renown as hydraulics experts by reclaiming land from the sea. Sluices built in the sea dikes drained water at ebb tide and were closed again during flood tide. The first windmills to be constructed as an energy source for drainage date back to 1408, but at that time drainage canals were not sufficiently extensive to carry the water pumped into them. By the early seventeenth century they were used on a large scale for pumping water out of inland lakes and as energy for early industries.[14]

Dutch engineers such as Jan Leeghwater (1575–1650), Gilles van den Houten, Cornelius Janszoon Meyer, and Cornelius Vermuyden (1590–1677) formulated comprehensive drainage plans and improved the construction of windmills, dikes, and sluices, technological expertise that could be exported to countries such as Italy, France, and England. Dutch capitalists who financed reclamation projects at home soon extended their financial backing to projects abroad. Whereas Dutch engineering at home was generally directed toward reclaiming new lands from the sea, when applied abroad it frequently displaced people whose form of life was integrally bound to the marsh. The Pontine marshes of Rome and the fens of England submitted to hydraulic improvement under protest by local inhabitants whose livelihoods depended upon them.

The draining of the English fens during the seventeenth century provides a striking example of the effects of early capitalist agriculture on ecology and the poor. The fen country of England, an extensive area north of London and Cambridge in the region of Peterborough, Ely, and Bedford, had been in the Middle Ages a region of marshes, meres, and meadows connected by open channels, affording pastures around its edges in the dry summers while almost entirely covered with water in the winters. In the seventeenth century, a period of drainage and reclamation of these marshes began that extended over the next three centuries, ultimately transforming the region into neatly planted fields of grain, sugar beets, and potatoes, separated by the geometric outlines of canals. When the fenland began to submit to dikes, sluices, pumps, and windmills, social struggles ensued. Progress, commerce, and technology permanently transformed the ecological balance of nature and the economic livelihoods of the fen inhabitants.[15]

Before the centuries of drainage, the fens were well stocked with fish and fowl that afforded food for the people who made their home in the region. As the drainage projects proceeded, the abundant wildlife sharply decreased, and birds became elusive. By the end of the seventeenth century it became necessary to trick them by decoys in netted channels, whereas once they had simply been driven into nets by fowlers and their assistants. Birds of the undrained fens included the golden-eye, stilt, puffin, wild goose, crane, snipe, curlew, osprey, cormorant, herne, sea-pye, and numerous others. "Coots, didappers, rails, water-hens, combined with eggs to fill [the] pots" of the fen dwellers, while godwits were fattened for markets in London.

A 1534 law restricting fowling forbade taking eggs of mallards, widgeon, teal, and wild geese from the fens and capturing the birds themselves by "nets and other engines" from June to August when the adults were moulting and could not fly from their captors. Because the latter restriction took its toll on the poor, who often lived by their cunning in surprising water birds, it was repealed sixteen years later.

The economy of the fenlands had been maintained by an ecological balance between human needs, animal grazing, crop yields, and soil fertility for hundreds of years of peasant tradition. Each peasant had free access to the common regardless of his or her wealth. The drainers destroyed this balance and threatened the means of support of the fen dwellers without supplying alternative employment. Although the gentry generally supported the draining, which would mean larger land areas to rent and increased income, an occasional lord saw that the draining "instead of helping the general poor . . . would undo them and make those that are already rich far more rich."[16]

A shortage of commons for grazing had already appeared in Holland Fen (in England) by the time of Queen Elizabeth, and had spread to other fen areas by the 1620s and 1630s. An increase in the number of cottagers using the commons, combined with the increasing exercise of the grazing rights of the lord, put pressure on the land. After the drainage operations, the problem was sharply compounded, as the common lands were cut to one-third their former size. The drier land near the village, which had in the past been reserved for meadow from which winter fodder was cut, was now used for summer grazing. Fewer animals could be sustained through the winter, with the result that less manure was produced for fertilizing grainfields the following year. Soil fertility and crop yields thus declined.

Lands leased after drainage benefited the landlord and those who had invested capital in the drainage operations. Most of the land recovered was now planted with coleseed, which produced oil for lamps, and oats, both important new products of the fenlands. Only the wealthier peasants could afford to lease the newly drained pasture. The fen drainage projects, according to a seventeenth-century pamphlet, had irrevocable effects on "the faces of thousands of poor people."[17]

Prior to the seventeenth century, lack of capital and adequate technology formed the major obstacles to drainage of the English fens. Although James I supported the project, several bills proposing draining were defeated. Charles I, who followed him to the throne, preferred to leave the project to the residents of the area, but most initial attempts ended in failure. Then in 1634, over protest by the fen dwellers who objected to the draining of their lands, the Dutch engineer Cornelius Vermuyden was invited to draw up a plan for straightening the natural course of the Bedford River. Vermuyden proposed a series of cuts, drains, and sluices to increase its gradient, thereby producing summer pasture and lessening the impact of winter flooding.

In 1631, a drainage project in the north, undertaken by Sir Anthony Thomas, met active opposition from dispossessed fen dwellers. After seven years of pasturing and planting by the intruders, the fen people rioted, took up arms against the drainers, "broke sluices, laid waste their lands, threw down their fences, spoiled their corn, demolished their houses, and forcibly retained possession of the land."[18] In another instance, the Earl of Lindsey, with accomplices, drained and enclosed fens, built houses, and cultivated the reclaimed pastures, only to have the commoners and fenmen, who had failed to obtain redress through petition, destroy the "drains and buildings" and the "crops then ready to be reaped."

Intellectual opposition to the fen projects was based on both ecological and social arguments. Some people maintained that it was wrong to interfere with the design of nature, because "Fens were made fens and must ever continue such." William Camden, noting the example of the ill-famed Pontine marshes of Italy, which had been beset by periodic erosion, flooding, and malaria, admonished that "Many think it the wisest and best course ... not to intermeddle at all with that which God hath ordained." (In Italy, as in England, marsh dwellers destroyed dikes and sluices.) Winter floods, it was stated, had served a useful function in making the summer meadows more fruitful: "Overflowing much enricheth those grounds, so that more draining would be very hurtful to them. These grounds ... cannot be spared or bettered by the industry of the undertakers." In some instances, draining the fens had made them too cold for pasture, and their value had actually been lowered.

The fens in their undrained state provided "great plenty and variety of fish and fowl, which here have their seminaries and nurseries; which will be destroyed on draining thereof." In addition, they furnished cottages with reeds, fodder, and hassocks, as well as "mattweede for churches, chambers, beds and many other fen commodities of great use both in town and country." A 1650 pamphlet argued that the fens, left as they were, grew fodder in the summer for feeding cattle in winter, which in turn provided manure to enrich the uplands and improve the growth of corn. In dry summers, the fens saved the lives of upland cattle, thus increasing the supplies of butter, cheese, beef, hides, and tallow.

Protests turned into further actions as the projects proceeded. Rioters and saboteurs filled in newly dug channels, demolished dikes, reversed the direction of drainage flow by reopening sluices during flood tide, and, with pitchforks and scythes, prevented overseers' cattle from grazing on fenland commons. Solidarity among the displaced people of the fens was maintained in the taverns, where drinking songs were composed for the occasion:

> Come Brethren of the water, and let us all assemble,
> To treat upon this matter, which makes us quake and tremble
> For we shall rue it, if't be true, that Fens be undertaken,
> And where we feed in Fen and Reed, they'll feed both Beef and Bacon.
>
> . . .
>
> The feather'd fowls have wings, to fly to other nations;
> But we have no such things, to help our transportations;
> We must give place (oh grievous case) to hornéd beasts and cattle
> Except that we can all agree to drive them out by battle.

Drainers replied to such arguments by asserting the superiority of domesticated sheep over wild fowl, well-fed cattle over even the largest of eels, and grain over sedge. Moreover, they argued, the fen air was dank and unhealthy, the water putrid, and earth boggy, while "swarms of stinging gnats and troublesome flies" pestered residents and travelers. Ultimately, in 1637, in accordance with a command of Charles I, the order was given to local officials to quell the rioters and "imprison some of the offenders." But in 1638 Oliver Cromwell, "Lord of the Fens" and leader of the parliamentary forces in the ensuing English Civil War, stepped forward to protest the injustices done to the fen people by lords and private entrepreneurial adventurers.

After the Civil War (1642–1646), the work was resumed and, despite setbacks, was continued over the next three centuries, in the name of progress, commercial growth, and national supremacy. The drained fens would now advance "the trade of clothing and spinning of wool," would "increase manufactures, commerce, and trading at home and abroad, [would] relieve the poor by setting them on work, and [would in] many other ways redound to the great advantage and strengthening of the nation."

As in the case of soil improvement, the draining of the English fens primarily benefited the landed and moneyed classes. The disruption of natural ecological balances also upset human ecology. Improvements in the productivity of the land were shared neither by the poor nor by the original occupants of the marshes—the fish, fowl, and marsh plants that over thousands of years had evolved a complex set of ecological interdependences.

The marsh—drained for pasture, blamed for disease, and exploited for its wildfowl—has today become a *cause célèbre* of ecology. The diversity and interdependence of its complex food chains, saline balances, and irreplaceable flora and fauna are important elements in the ecologists' argument that the preservation of diversity and genetic variability is essential to the survival of life on the planet. The Everglades of Florida, the Horicon Marsh of Wisconsin, and the Suisun Marsh of the San Francisco Bay are the homes of unique species continually being threatened by the same technological and market-oriented values that irrevocably altered the English fenlands.

FORESTS. As in the cases of the farm and fen ecosystems, early capitalist modes of production accentuated human impact on the forests over and above effects attributable solely to population pressure. Technological improvements associated with an expanding economy based on inorganic, nonrenewable metallic wealth increased the potential for human exploitation of the woodlands. While population pressure had taken its toll on the forest ecosystem before the demographic collapse of the fourteenth century, after ecosystem recovery mercantile capitalism hastened the dramatic decline of timber resources in the sixteenth century.

In England, by the early modern era, the heavy forests of pre-Roman times had given way to groves and woodlands, as pasture for sheep and fields for grain were created by burning or cutting timber and grubbing up the stumps—a process known as *assarting*. In the earlier periods, the clearing of land for fields and houses had enhanced the formation of communities and the bonding together of inhabitants in regions once heavily wooded.[19]

Yet at the time of the Norman Conquest (1066), "a great forest with wooded glades . . . lairs of wild beasts, deer both red and fallow, [and] wild boars and bulls" still existed just outside London.[20] In 1253, "the forest of Leicester was so great and thick and wide that a man could scarce go by the wood-ways for the multitude of dead wood and wind-fallen boughs."[21] In France and England, wooded areas were initially preserved as monastery lands and as hunting grounds for the aristocracy, the legal term *forest* being used to designate lands where game was reserved for kings and the nobility. Wooded lands were also cultivated and settled by monasteries, which then sold the wood for tithes. Thirteenth-century ordinances bestowing forest and water rights on the king and persons of high status attempted to regulate against abuses.

In the twelfth and thirteenth centuries, when rapid increases in population all over Europe had initiated an era of land reclamation, woodlands were converted to arable fields, and marshes were drained for pastures for draught animals. Forested areas in England, France, Germany, and Italy were markedly diminished. The iron industry of the English Weald, which depended on oak to produce charcoal for smelting, had already begun to take its toll of timber by the thirteenth century. Around populated areas, shortages of firewood caused an inflation in price by as much as 780 percent, compared to general inflation of 291 percent, producing a crisis in energy sources. By the late thirteenth century in London, it was becoming necessary to import sea coal from Newcastle, a soft coal with a high sulfur content, which when burned polluted the air with black soot and irritating, choking smoke.[22]

During the period of demographic collapse, when marginal arable lands were abandoned, forests regenerated and soils regained their fertility, facilitating restoration of the medieval ecosystem.

Slow recovery during the fifteenth and early sixteenth centuries brought the English population back to 2.3 million by 1525, and to 2.8 million by the mid-sixteenth century. Between 1550 and 1600, growth was more rapid, bringing the population to 4 million by 1600 and to 6 million by 1700.[23]

In England, people who became landless through the enclosure movement or the inability to pay rents to the landlord frequently settled in the forest. A 1578 report on the Crown Forest of Inglewood noted 178 encroachments by poor squatters of an acre or less, which by 1619 had increased to 757. In 1569, new enclosures in the Forest of Westwood were pulled down in an uprising by 300–400 poor laborers and servants who needed the forest for grazing their animals and who, by combining earnings, sharing crops, and using the forest, had been able to make a scant living. They were quickly brought to trial, however, and the enclosures replaced.[24]

Although increased population pressures contributed to encroachments on the forests for pasture and firewood, the most significant cause of their demise was the growth of early industry. The shipbuilding, soap, glass, iron, and copper-refining industries depended on the forests for their source of energy. Although the forest reserves of the early 1500s had been amplified in England by the dissolution of the monasteries in 1535, by the late sixteenth century, shortages of wood had become acute and the substitute fuel coal, again, increasingly important. During the next hundred years, England was forced to change from wood to coal as its chief form of industrial energy. No other European country except Holland developed its coal industry on such a wide scale.[25]

Whereas the medieval economy had been based on organic and renewable energy sources—wood, water, and wind—the emerging capitalist economy taking shape over most of western Europe was based not only on the nonrenewable energy source—coal—but on an inorganic economic core—metals: iron, copper, silver, gold, tin, and mercury—the refining and processing of which ultimately depended on and further depleted the forests. Over the course of the sixteenth century, mining operations quadrupled as the trading of metals expanded. Large commercial banking organizations such as the Fugger Company of Augsburg owned and operated silver mines in Tyrol, gold mines in Silesia, mercury in Spain, copper in Hungary. The Fuggers had twenty-five different offices and made profits of up to 50 percent on their investments, ten times greater than the Medicis, who in 1460, just a century before, had operated eight offices.[26]

The commercial bankers represented a high degree of capitalist organization and were extremely powerful. But, based on the mining industry, they ultimately depended on the forests for charcoal used in the smelting of ores. The metallurgical industries, which processed iron, lead, tin, and copper, required enormous stretches of timber just to make one ton of charcoal. The prevalence of trees and wood influenced the location of the smelting and refining operations; it was cheaper to move the ore to the forest than vice versa. Active iron industries developed in England in the regions of Sussex and Kent, where there was heavy forestation. The iron was smelted in a blast furnace, and the molten metal was run into molds to form pig iron. Then in forges it was heated and hammered into wrought iron. After that it went through a slitting mill, where it was drawn out into thin rods. The malleable wrought iron was used in the making of horseshoes, farming implements, locks, bolts, and tools. Cast iron, not treated in this way, was brittle and used for pots and pans, cannons, ballistics, and guns. The furnaces, forges, and slitting mills were all at different locations, because in each case a source of wood for fuel was needed. Beginning around 1540, large-scale mills and foundries producing paper, gunpowder, cannons, copper, brass, sugar, and saltpeter began to supplement production by domestic workshops. They employed 60–200 and, in rare cases, up to 600 persons. These new industrial establishments used pipes, kettles, pots, tubs, and cisterns, as well as the furnaces, boilers, and buildings, and therefore required large investments of capital, in contrast to the looms, grinding wheels, and forges of the home workshop.[27]

Furnaces used in the smelting of iron ore were much larger after the mid-sixteenth century, the new ones being often 30 feet high, with walls 6 feet thick. Large 20-foot bellows operated by a water-wheel helped to increase a furnace's iron production to between 100 and 500 tons per year. The growth of the mining industries during the reigns of Elizabeth, James I, and Charles I demanded the construction of mining shafts and pumps driven by horse power, as ores were increasingly raised from depths of 20 to 50 fathoms, in contrast to the earlier extractions of only a few fathoms. By the time of the Civil War, thousands of men and women were employed as wage workers in the new mining and metallurgical industries.

In addition to the metal industries, other uses of wood contributed to the drain on timber supplies. Oak was needed for the beams, rafters, and floors of houses, public buildings, and fortifications, and for building the docks, bridges, locks, and barges necessary to commercial transportation on navigable waterways. The brewing industry required oak casks, while the soap and glass industries needed charcoal and wood ash in order to produce soap and the important new product, glass.

But the industry most dependent on wood and most critical to sixteenth-century commercial expansion and national supremacy was shipbuilding. Mature oak of 80 to 120 years was required for the hulls of ships, firs were used in the masts, and pitches and tars for sealing. Famous oak reserves in England included the royal forests of Dean, New, and Alice Holt, acreages eventually set aside after the Restoration of 1660 exclusively for naval timber.[28]

The shipbuilding industries of Italy, England, and the Netherlands were organized along capitalist lines, requiring a variety of wage workers, large capital outlays, and immense cranes and saw mills for the construction of vessels. The largest shipyard in Europe, the Venetian Arsenal, covered sixty acres and was owned by the state. It employed artisans such as carpenters, caulkers, sail makers, pulley makers, and iron smiths. The state not only owned the industry but also controlled the capital and fixed the wages and hours. Supplies of oak for the frames of ships decreased alarmingly during the fifteenth century as oaks sent down the rivers from the forests of Trevisana and Friuli became depleted. Although supplies of fir and larch for the masts and internal planking remained adequate through the sixteenth century, progressive depletion of oak forced the barge-building industry to move inland from Venice. By the end of the sixteenth century, the Mediterranean, with the exception of Ragusa (present-day Dubrovnik, Yugoslavia), had been largely divested of its oak supplies.

A conservation consciousness was first developed by the Venetian Arsenal beginning about 1470. New regulations provided that the Arsenal, rather than local officials, would determine the cutting of oaks. Private timber and shipping interests were excluded after 1520, although enforcement through penalties was fraught with bureaucratic difficulty. Regular surveys of oaks were instituted in 1568, followed by a policy of marking trees reserved for the Arsenal with its own seal, while those allocated to private shipbuilders were branded with a double seal. Ditches, stone pillars, and hedges designed to keep out wagons and cattle were built around groves of mature timber. Special "captains" were chosen by the Arsenal from among its carpenters to live in the woodlots, to seize wood-gathering peasants and timber poachers, and to oversee cutting by licensed state officials. An unauthorized person carrying an axe into one of the Arsenal's timber reserves could be whipped and fined. In the region surrounding Montello, one of the more important forests, all furnaces of charcoal burners were destroyed, and peasants were prohibited from receiving alms from a monastery within its confines lest they poach wood during the visit.

In England, concern for shipbuilding lumber during the reign of Queen Elizabeth (1558–1603) resulted in a law preventing the cutting of oaks larger than a foot in diameter in areas near the coast, where the navy could easily harvest and transport them. Despite timber conservation for naval use, lumbering continued unabated during the reign of Elizabeth, due to the granting of extensive cutting rights.[29]

As a consequence of English timber exploitation, coal mining rose exponentially. Between 1540 and 1640, outputs from coal mines increased from a few hundred tons a year, mined by part-time tenant farmers and husbanders, to 10,000–15,000 tons, extracted by hundreds of full-time miners. Coal shipped out of Newcastle-Upon-Tyne to other parts of England rose sixteenfold during this same period. As London needed more and more coal to replace its dwindling supplies of wood, imports increased twenty- to twenty-fivefold.[30]

Sea coal left heavy black soot and clouds of choking smoke wherever it was burned. It was said that Queen Elizabeth was "so grieved and annoyed with the taste of the smoke of sea coals" that in 1578 she asked the brewers of London and other industries not to use any coal in their operations, but to rely only on wood. Since wood was not generally available except at high prices, this demand could not be met. Complaints over sea coal rose dramatically during the seventeenth century. A 1627 petition against alum workers stated that the smoke of sea coal tainted the pastures and poisoned the fish of the Thames. London women complained about "the smell of this city's sea coal smoke."[31]

Despite the timber crisis of the sixteenth century, the Stuarts, James I and Charles I, extended the cutting grants within the royal reserves begun by Elizabeth. A survey of the royal timber supplies conducted in 1608 by James I, showing a total of 784,748 mature timber trees suitable for naval use, merely resulted in the release of more royal timbers to raise revenue. By the time of the 1660 Restoration, only 68 hunting grounds preserved by the nobility remained, the largest tracts being seldom more than 20 square miles in area. The situation had thus become critical and the country never fully recovered from its wasteful practices and policies.[32]

The disruption of the forest ecosystem by the rise of early modern industry, coupled with the careless use and mismanagement of resources, bears striking parallels to current environmental issues and is illustrative of the fact that today's environmental crisis is not new in kind, only in degree. As in the sixteenth century, we must look to alternative sources of energy. One of these is coal, with its concomitant air pollution, the same alternative available to our sixteenth-century predecessors. Another, alien to an age wedded to the ideology of progress, is to return to the renewable resource economy of the preindustrial period. Water, wind, sun, wood, and intermediate labor-intensive technologies could help to restore the ecological balances disrupted by modern industrialization. But this recognition does not mean that technological innovation must be sacrificed or that we must return to the medieval watermill. Rather, it means that we need to develop technologies that harmonize with natural cycles rather than exploit resources.

But viewing historical change as ecological change goes beyond the recognition that our current environmental situation has a history. The integration of ecosystem history with human culture reveals both the limits imposed on demographic and national expansion as a result of the disruption of soil and forest ecology, and the alteration of human insitutions and laws to restore ecosystem balance through conservation. Further, it introduces into historical interpretation additional complexity. An ecosystem model reveals the limits of demographic, economic, or political factors as single underlying explanations in history. Without consideration of how the resource of soil fertility affects population growth and decline, peasant-landlord conflict, and market expansion, the history of the change from feudalism to early capitalism is inadequate. The rise of both democracy and capitalist economic institutions in Europe and America were directly dependent on the exploitation of natural resources—metals, soils, grasses, timber, furs, etc. The disruption of associated ecosystems (forests, prairies, marshes, lakes, oceans) and their human components affects the course of history in the form of social uprisings, wars, laws, and technological innovations, and has an important impact on human health, nutrition, and welfare.

Conversely, psychological adaptation to altered environments helps to explain the rise of intellectual movements, conceptual structures, and new human behaviors. As European cities grew and forested areas became more remote, as fens were drained and geometric patterns of channels imposed on the landscape, as large powerful waterwheels, furnaces, forges, cranes, and treadmills began increasingly to dominate the work environment, more and more people began to experience nature as altered and manipulated by machine technology. A slow but unidirectional alienation from the immediate daily organic relationship that had formed the basis of human experience from earliest times was occurring. Accompanying these changes were alterations in both the theories and experiential bases of social organization which had formed an integral part of the organic cosmos.

CHAPTER 3

Organic Society
and Utopia

Theories about nature and theories about society have a history of interconnections. A view of nature can be seen as a projection of human perceptions of self and society onto the cosmos. Conversely, theories about nature have historically been interpreted as containing implications about the way individuals or social groups behave or ought to behave. The power of the organic metaphor derived from the unifying structure it imposed on social and cosmic reality. But as the ecology and economy of farm, forest, and fen were altered by new forms of human interaction with nature, traditional models of organic society and modes of social organization were likewise being undermined and transformed.

Three variations of the organic theory of society are important to the transition from organicism to mechanism. The first was based on medieval society as a hierarchy. The body politic was metaphorically modeled on the organic unity of the human body, and, as a hierarchy of status groups, represented a conservative view of the social order consistent with the experience of feudal lords and the medieval church. The second variation tended toward a leveling of the hierarchies and was based on the actual experience of the village community. The third, a revolutionary form of the organic theory, advocated the complete overthrow of social hierarchies. It symbolized the utopian aspirations of medieval and early modern groups for an end to the established order and its replacement by an egalitarian community in which people returned to a Golden Age of harmony with nature. The three variants of the organic theory—hierarchical, communal, and revolutionary—served either as ideologies for actually existing social structures or as ideals for the transformation of those structures.

ORGANIC SOCIETY. The hierarchical model of organic society had been articulated in 1159 by John of Salisbury in his *Policraticus*, based on a lost work attributed to Plutarch (A.D. 350–430) and written by Salisbury before the thirteenth century recovery of Aristotle's *Politics*.[1] Reflecting the cultural perspective of feudal lords and territorial princes, it conceived of the commonwealth as a "person-writ-large." Its body was endowed with life and ruled by reason in the form of the prince, who, together with the clergy, functioned as its soul. Judges and governors, who communicated its dictates to the provinces, represented its sense organs—the eyes, ears, and tongue. The good of the commonwealth was invested in its senate, which occupied the position of the heart. Of the hands, one was armed and protected the citizenry from outside attack, while the other, unarmed, disciplined them from within. Both were restrained by the reason and justice of the prince. Keepers of the state's finances were confined to the stomach and intestines, who for the body's health must avoid anal retention and congestion of their holdings. The feet were farmers, craftspeople, and menial workers, so numerous as to cause the organism to resemble a centipede, rather than a human.[2]

Each part of this vast society was indispensable to its unified purpose and function. Thus, "take away the support of the feet . . . and it cannot move forward by its own power, but must creep painfully and shamefully on its hands, or else be moved by means of brute animals." According to Salisbury, the citizens formed a corporate community who wished to preserve the prince because the prince in his role protected and gave affection to his subjects. A "faithful and firm cohesion" existed only in conjunction with "an enduring union of wills and . . . a cementing together of souls." Tyranny by the prince or magistrates would effectively add such weight to the head as to render the body incapable of support or result in a suicidal slashing of the throat. "The happiness of no body politic will be lasting unless the head is preserved in safety and vigor." Due to the interconnectedness and mutual dependence of its parts, "an injury to the head . . . is brought home to all the members, and . . . a wound unjustly inflicted on any member tends to the injury of the head."

The medieval theory of society thus stressed the whole before the parts, while emphasizing the inherent value of each particular part. The unity of the one was of higher value than the objectives of the many. The connection between the parts was integrated through a universal harmony pervading the whole. This organic cement bound together the macrocosm, the community, and the parts of each individual being or microcosm. John of Salisbury illustrated these harmonic bonds in the commonwealth by the comparison of harmony in the community to harmony in a musical instrument. Just as musicians tune their strings skillfully and gently to create harmony out of discord, rather than breaking them through too much tension, so the prince as soul of the commonwealth must achieve a fine tuning between strictness and mercy so that the minds of all his subjects can be brought together in one perfect harmony. And, just as the world soul held together the cosmic sphere, so the human soul represented unity among bodily parts. In the cosmic hierarchy, each subordinate whole was itself an integrated interconnected system of parts, bound, in turn, to a larger, more comprehensive whole above it.

The organic society was to be modeled on nature's prime examples of communal colonies—bees and ants. Civil life should imitate nature, as exemplified by the political constitution of the bee. All workers must join together in common to produce welfare for the whole and sweetness in the honey while the queen superintends. Care must be taken that the workers continue to produce diligently and do not indulge in luxury lest the citizens become "effeminate and bow their necks to manly valor." In articulating the analogy between a commonwealth and a person, Salisbury's theory of society stressed the political and psychological interdependence and integration of the citizens within the whole, as well as the hierarchical relationship among the parts of the human body and the body politic.

The translation of Aristotle's *Politics* into Latin in 1260, and its incorporation into the systematic philosophy of St. Thomas Aquinas (1225?–1274) amplified Salisbury's organic concept of the state. Aquinas presented an integrated system of nature and society based, like Salisbury's, on hierarchical gradations. Each part had its own place, rights, duties, and value, which together contributed to the perfection of the whole universal community. Both nature and society were composed of parts so that the purpose or end of the lower was to serve the higher, while that of the higher was to guide the lower toward the common moral good. Each part sought the perfection of its own particular nature, growing and developing from within.[3]

On this model, monarchy was the form of political organization most natural to the cosmic and divine order. Monarchy derived its authority by divine right from God and distributed that authority to the various social estates through civil law. The medieval sovereign represented God in the temporal realm; the church his will in the spiritual realm. For the philosopher Nicolas of Cusa (1401–1464), the church was the body of God, the clergy its soul, and God its all-pervading, unifying spirit. The clergy held the corporeal body of the church together, each part of the body having its own portion of the soul in a hierarchical ordering. The papacy represented the soul in the brain, the patriarchate the soul in the ears and eyes, the archiepiscopate the soul in the arms, the episcopate the soul in the fingers and so on.

In sixteenth-century England, where the organic metaphor was pervasive, every conceivable analogy between the parts of the body and the body politic was explored. Lungs, liver, gall bladder and kidneys all had their correspondences. The belly in Shakespeare's play *Coriolanus* (first produced in 1607) discourses at length about its importance in storing, digesting, and distributing food to all parts of the body, just as the Roman senators digest matters affecting the commonwealth for the public benefit. In *Henry V* (first produced in 1598) the Archbishop of Canterbury compares civil society to a beehive from which social order can be learned. For most

Elizabethans democracy was an "ugly hydra, a "many-headed monster," in which the feet of the mob had attacked the head.[4]

The hierarchical form of the organic model was consistent with a political ideology that emphasized order and stability, and can be placed at the conservative or right end of the political spectrum. The interest of the state assumed central importance in comparison to the individual parts. The whole was greater than the sum of the parts, but could be ontologically different from them; the common interest and happiness of the state could be different from the interests and happinesses of the individual members or their sum.

The hierarchical variant depended on inequalities among the members and did not concern itself with equal rights or economic standards, since different parts of the social organism had different economic functions of differing importance to the whole. Each status group or rank performed its own function and received an appropriate reward without expectation of changing its power or position as a result. The labor force, comprising the feet of the organism, was tied directly to the land, while land itself was transferred within families of the nobility and gentry usually on the basis of primogeniture—inheritance by the eldest son (Fig. 8). The organic bonds within a landowning family lay not in familial affection or emotional satisfaction but solely in the economics of the kinship bond. Genetic relationships held together large networks of kin whose mutual interests lay in maintaining the long-term economic and social status of the family group. Power within the family was hierarchical, with the eldest son frequently in the position of simply waiting for his father to die before assuming control, while the youngest faced little prospect of inheriting the estate.[5]

But sixteenth-century social changes had already begun to shatter this hierarchical model. Although mobility was not absent from medieval society, the growth of a market economy in both wage labor and property produced more upwardly as well as downwardly mobile individuals, undermining the integrative ideology of the hierarchical model. An increase in the father's power within the nuclear family accompanied a decline in kinship economic functions in the wider family structure. The rise of politically powerful nation-states magnified the role of the head of state, distorting the ideal set forward by Salisbury.[6]

Jean Bodin (1530–1596), French political theorist, and King

Figure 8. *The Classes of Men,* woodcut by Hans Weiditz (ca. 1530). Organic hierarchical society is represented as a tree, with peasants as its roots. On the second level are artisans, journeymen, and merchants; on the third level, bishops, cardinals, nobles, and princes; on the fourth, king, pope and emperor; and, at the top, again two peasants, on whom the organic society depends for sustenance.

James I of England (1566–1625; acc. 1603) both modified Salisbury's model in justifying the new authority of the head in the emerging nations. Bodin enlarged the absolute and perpetual power of the sovereign over citizens and subjects. Nobles and merchants qualified for citizenship; artisans and retailers did not. But equality, even among the citizens, was absurd. Women were subject to the authority of the father and were unfit for public affairs. Female rulers like Mary, Queen of Scots, and Elizabeth I of England had overturned the natural order. Succession to the throne should descend through the male line, since nature had deprived women of "strength, foresight, pugnacity, [and] authority." Bodin's treatise was a conservative defense of the authority of absolute monarchy in response to the disruptive social tendencies of the French Wars of Religion.[7]

James I of England also enlarged the patriarchal authority of the head of the kingdom. "The king towards his people is rightly compared to a father of children, and to a head of a body composed of diverse members. . . . As the discourse and direction flows from the head . . . so it is betwixt a wise prince and his people." If disease in one part of the body affected the other organs, the head might be forced to "cut off some rotten members . . . to keep the rest of the body in integrity." In ironic anticipation of the beheading of his successor Charles I, James queried, "But what state the body can be in, if the head . . . be cut off I leave to the reader's judgment."[8]

The magnification of the role and power of the head undercut the integrative bonds of the organic analogy and was symptomatic of social transformations that would hasten its dissolution as a primary metaphor. By the mid-seventeenth century, political theorist Thomas Hobbes (1588–1679) would use it to describe, as a person, a state that was in reality a cleverly contrived machine (see Chapter 8). While organismic social thought did not die out completely, after the seventeenth-century, atomistic and mechanical metaphors

75

began to describe the emergence of a pluralism of interest groups in democratic societies.

Twentieth-century ecology emerged from the intellectual framework historically associated with an organic approach to nature and society. Employing the hierarchical variant of the organic model, ecologists in the early decades of this century used concepts such as "organic community," "mutual interdependence," and "evolution toward higher forms" on a hierarchical scale to provide an understanding not only of the organization of bacterial colonies, grassland climax vegetation, and bee and ant communities, but also of human tribal societies and the world economy. They stressed an evolution toward greater cooperation on a worldwide basis and argued that nature could provide the model for an ethic of human sharing, integration, and unity. But the emergence of fascist tyranny based on a centralized organismic model glorifying the father as absolute dictator undermined the evolutionary hierarchical component of their argument, and ecology turned in a mathematically reductionistic direction.[9]

O RGANIC COMMUNITY.   Whereas the medieval hierarchical model emphasized distinctions in status and estate, a second variation of the organic theory, growing out of peasant experience and village culture, was based on the leveling of differences and stressed, instead, the primacy of community, the collective will of the people, and the idea of internal self-regulation and consent. Here the communal whole was still greater and more important than the sum of the parts, but the parts were of equal or nearly equal value. Ideas that stressed the common consent of the people and mutual will of the community represented a formulation of the organic theory at the communal or socialist end of the political spectrum. Despite numerous variations in local rural social stratification and economic patterns, the peasant society of much of western Europe can accurately be described in terms of cohesive community responsibilities and common traditions and exemplifies the communal variant of the organic model.

In the communal variant, both the law of God and nature dictated an original equality among the parts of the village community, cooperative land use, and communal sharing of tools and goods. Moreover, the body of the people had the right to choose its own head by elective right. Consent of the community must validate the actions of the village officials who remained parts of the whole and subordinate to it. The principles of people's sovereignty, internal self-regulation, and decision making formed the political rationale for the peasant rebellions that challenged the reassertion of authority by landlords when traditional balances between land and peasant were upset.[10]

Although a spectrum of actual communal forms existed in the thousands of European rural village communities, the unifying principle common to all was the good of the group over that of the individual. The early territorial communes, composed of families living on individual holdings in spatial proximity, shared natural resources that were managed by local elected officials. The village communes that gradually replaced them further reinforced cooperation and group unity by adding land and plow animals to the shared resources and by adding the regulation of pasture and plowing to the duties of the corporate officials. The tradition of village cohesiveness overrode divisiveness and quarreling among individual inhabitants.

Agrarian communism in such villages reached a new level of cooperation with the introduction of compulsory tillage in response to population increases and land shortages: all persons in the village plowed, planted, and harvested at the same time in order to increase productivity. In many central European communities, as family groups split, land was periodically redistributed, to equalize the fertility, productivity, and accessibility of plots. These variations on the model of the organic community were established widely over most of medieval and early modern Europe, surviving in much of central and eastern Europe well into the eighteenth and nineteenth centuries.

Between the two poles of organic hierarchy and communal consent lay a number of political possibilities, exemplified in church, state, and intellectual doctrine. One variant, for example, deemphasized rule by the head and held that the community or town was composed only of qualified citizens, excluding the undifferentiated masses. An assembly of representatives made decisions for the larger community and was responsible to it. Reformation sects also drew on the organic theory in formulating the ideal of corporate Christian community.

Like the hierarchical variant, the communal model was undermined by the growth of a market economy based on property rights and exchanges in land and money. Advances in agricultural improvement and the growth of rural industry broke down communal farming practices and the communal control of resources.[11]

By the seventeenth century, John Locke (1632–1704), in his theory of property rights could challenge the idea that communal ownership of all parts of the earth was given forever by the laws of nature and scripture. The acquisition of land, worked by an individual's own labor, became that person's individual property. "As much land as a man tills, plants, improves, cultivates, and can use the product of, so much is his property. He by his labor does, as it were, enclose it from the common." By mixing human labor with the acorns, apples, beasts, and earth that "Nature, the common mother of all," had given, people removed these "spontaneous products" from the common and made them and the land their own property. In creating the world, God had commanded "man . . . to subdue the earth," and in so doing, he "gave authority . . . to appropriate" private possessions. "And hence subduing or cultivating the earth and having dominion, we see, are joined together. The one gave title to the other."[12]

Were it not for money—"gold, silver, diamonds . . . that fancy or agreement hath put value on"—any surplus products would perish and land itself would remain unimproved. Exchange of gold and silver for land and its products was preferable to simple barter, because the metals "may be hoarded up without injury to any one, these metals not spoiling or decaying in the hands of the possessor."

While land planted with useful products—wheat, barley, tobacco, and sugar—was of greater value than unimproved land from the common, a market for the products was also essential. "For I ask, what would a man value ten thousand or a hundred thousand acres of excellent land, ready cultivated and well stocked, too, with cattle, in the middle of the inland parts of America, where he had no hopes of commerce with other parts of the world, to draw money to him by the sale of the product?" Without a market it would not be worth the time and labor to grow anything over the subsistence level. A market economy based on money exchanges, property rights, agricultural improvement, and the domination of the earth would thus undercut the theory as well as the practice of organic community.

ORGANIC UTOPIAS. A third variation of the organic model expressed the needs of peasants and artisans for social revolution. The utopian millenarian tradition called for the complete overthrow of the established social order and its replacement by an egalitarian communal society and state of nature like that anticipated during the millennium—a thousand-year period when Christ would reign on earth and Satan would be banished. The millenarian movements had a historical continuity dating from the medieval crusades of the poor, Joachim of Fiore, the Brethren of the Free Spirit, the Amaurians, Thomas Munster, and the Anabaptists, to the religious sects of the English Civil War—Seekers, Ranters, Levelers, Antinomians, and Muggeltonians. These groups, along with the intellectuals who identified with them, shared a belief in the emergence of a new age of liberty and love in which God would appear from within and there would be equal sharing of food, clothing, and property among all people. At various times and places, leaders and groups emerged who tried to hasten the arrival of the millenarian age and to create communities based on an egalitarian state of nature and communal ownership of wealth. The reasons for failure ranged from military defeat by the forces of the church or state to betrayal by the leaders of rebellion.[13]

Millenarianism represented a preindustrial form of social revolution. It differed from the movements of the industrial revolution by preparing people, through revelation, to accept revolutionary change, as opposed to politicizing the working class. Signs in the heavens, prophets, and saviors would appear predicting the arrival of the millennium. For example, persisting since the Middle Ages was the prophecy that Frederick I (Barbaroso), King of Germany, would be resurrected; he had died in 1190 on the third crusade and was idealized as a savior of the poor who would bring with him a communal state. He would banish the Pope as Antichrist, and destroy his cohorts—the clergy, the wicked, and the rich, well-fed laity. Throughout the Middle Ages, prophets and written manifestos sustained revolutionary influences and the real hopes of the poor for a new social order.

In the early seventeenth century, two utopian plans, Tommaso Campanella's *City of the Sun* (1602) and Johann Valentin Andreä's *Christianopolis* (1619), articulated a philosophy of commu-

nal sharing that responded to the interests of artisans and the poor
for a more egalitarian distribution of wealth based on an original
harmony between people and nature.[14] They contrast markedly with
a third utopia, *The New Atlantis* of Francis Bacon (1627), which
undermined and transformed the concept of an organic utopian
community. Yet historians have largely emphasized the similarity
of the three works for the emergence of modern science and educational theory. As we shall see later, Bacon's ideas were rooted in an
emerging market economy that tended to widen the gap between
upper and lower social classes by concentrating more wealth in the
hands of merchants, clothiers, entrepreneurial adventurers, and
yeomen farmers through the exploitation and alteration of nature
for the sake of progress. Andreä's and Campanella's utopian communities postulated a more egalitarian view of woman and man,
artisan and master, than Bacon's more hierarchical and patriarchal
community. But Bacon's inductive methodology, which helped to
establish a precedent by which all persons could verify the truth for
themselves, was also fundamental to the growth of egalitarianism.

From the perspective of today, there are both positive and negative aspects to Campanella's and Andreä's utopias. Some of their
ideas are basic to subsequent "back to the land" utopian movements that have rejected the division of labor and the alienation of
people from productive work brought about by capitalist modes of
economic organization. Yet Campanella advocated a program of
eugenics considered repressive in the wake of Hitler's genetically
based holocaust, while Andreä's ideal society was based on a rigid
Calvinist moralism.

Tommaso Campanella (1568–1639) followed the Renaissance
naturalist philosopher, Bernardino Telesio (1509–1588), in his belief that God was immanent within nature and that all matter was
alive. Change was explained by the dialectical warring of the opposites, heat and cold, as source of the dynamic motion of matter (see
Chapter 4). In his attempt to establish an ideal society based on holistic presuppostions about nature, Campanella, in 1599, led a revolutionary movement to overthrow Spanish rule in Naples and the
entire province of Calabria, in southern Italy. Portents in the heavens predicting the advent of the millennium indicated to him the
necessity of immediate political overthrow and the fusion of all religious sects as the world returned to the golden age of simple, natu-

ral, primitive faith. The new society would liberate the people from the tyrannical slavery of political and religious usurpers and establish communal property and rule by brotherhood. A new, unique God, immanent in nature, would be revealed, coupled with a true spiritual religion, which, rather than debasing nature, would exalt it as a divine artisan. Under a new social rule, body and spirit would function in harmony.[15]

Signs in the heavens indicated that the millennium was near, and it was important to prepare the people of Calabria for a new city of God in which all of humanity would live in health and sainthood. In anticipation, the community had to be freed from Spanish domination. Inspired by Campanella's words, people and nobles would unite in armed rebellion.

Conditions in Naples were ripe for revolt. All classes of society were fraught with internal dissension. The people were angry about inequalities in social conditions and fiscal measures. The lords were unhappy with the government of the viceroy. Papacy and bishops were overwhelmed with juridical questions. Of the 70,000 people in Naples, 10,000 to 15,000 laborers were worn and driven by toil.

In preparation for the revolt, Campanella began preaching to large audiences and describing in eloquent terms the meaning of the signs in the heavens. His objective was to convert the people to a true spiritual religion which would lie at the foundation of his ideal republic. Aided by the voice of his friend Dionisio Ponzio and the energy of the nobleman Maurizio de Rinalis, the conspirators prepared to deliver the country from the tyranny of Spain by an uprising in Catanzaro, near the tip of Italy, and an armed attack on the Chateau d'Arena.

By the end of August 1599, everything was ready for the revolt. In a fiery speech to his companions on the mountain of Stilo, Campanella proclaimed victory to be near and described the happiness that would belong to all in the perfect city built by their communal effort. But the plans were shattered by the arrival of the troops of the Spanish viceroy, who arrested Campanella and his coconspirators on September 6, 1599. Campanella was captured and taken to Naples in chains, charged with heresy and conspiracy, and thrown into prison. He was retained there for twenty-seven years, during which time he was tortured four times.

Most of his voluminous works, including the utopian *City of the*

Sun, were written while in prison without access to books and under deplorable physical conditions. In the early years of his life, Campanella was a revolutionary who developed concrete plans for a better society and actively attempted to carry them out. After years of imprisonment by the Spanish Inquisition, he was released for a short time in 1626 but within a month was arrested and held prisoner in the Vatican until 1629. He spent the end of his life in France (1633–1639), where he received protection from Cardinal Richelieu, in whom he had by then placed his hope for a unified world. His final allegiance to Pope Urban VIII (pope from 1623 until 1644) and the French monarchy, in which he exchanged rebellion for protection and the freedom to publish and lecture, represented a conservative end to his revolutionary career.

Valentin Andreä (1586–1650) must be categorized as a reformer rather than a revolutionary, yet his reforms, like those of Campanella, were directed toward the creation of an egalitarian society. His *Christianopolis* (1619) described a utopian society in which the labor of all people was equalized and educational training existed for both sexes.

Like Campanella, Andreä was serious about making his utopian *Christianopolis* a social reality. In 1620, he moved to Calw, in the Black Forest, where as a Lutheran pastor he formed a social system that expanded from a congregational core into a larger community. At the basis of his society was a union of textile and dye workers in the Calw textile industry, consisting initially of thirteen families. The families, known as the Färberstift, all contributed money toward the education of the young, especially for orphans and the children of the poor, the operation of a library, care of the sick, elderly, and widowed, and support for the community church and its services. The emphasis on educational advancement by reformers such as Andreä and the Czech Jan Amos Comenius (1592–1670) soon spread to England, where it influenced the circle known as the Invisible College which included such seventeenth-century scientists and educators as Samuel Hartlib, John Dury, Theodore Haak, and Robert Boyle, preparing the way for England's first scientific society, The Royal Society of London, founded in 1660.[16]

The social context within which Campanella and Andreä initially proposed their utopian communities suggests that these societies were responses to the real needs of seventeenth-century people for change and that both authors put forward these ideals as serious alternatives to existing social conditions. These communal societies are representative of an organic philosophy that placed people within rather than above nature. They represent ideals in many ways as inspiring to us today—as we search for antidotes to the problems of urbanization and industrialism—as they were to sixteenth-century reformers trying to alter the power relations of hierarchical society.

HOLISM. The people of Campanella's City of the Sun were to dwell in an organic holistic cosmos. Their earth was like an animal, drawing its source of motion from within, and was alive with blood from its bowels. As in the Hermetic *Emerald Tablet*, the earth was considered to be a mother and the sun a father:

> They assert two principles of the physics of things below, namely that the sun is the father and the earth the mother; the air is an impure part of the heavens; all fire is derived from the sun. The sea is the sweat of the earth, or the fluid of earth combusted, and fused within its bowels, but is the bond of union between air and earth, as the blood is of the spirit and flesh of animals. The world is a great animal, and we live within it as worms live within us.[17]

Recognized today as keys to viable ecosystems in nature are the interrelationships and organic unity among a system's parts, and the maintenance of ecological diversity. In native and traditional cultures which sustain equilibrium with their environments, rules, rituals, and religious practices have the latent function of preventing overuse of resources and overpopulation. In the City of the Sun, such principles subtly guided community norms and practices. Nature was an organic whole in which both natural and human cycles were integrated. Agricultural practices and animal breeding were performed in harmony with the seasonal cycles—a marked contrast, as we shall see, with Bacon's *New Atlantis*. Only land necessary for survival was cultivated; groves and woods were retained for wild animals. The breeding of oxen, sheep, hens, ducks, and geese took place in harmony with natural cycles of the winds and the seasonal rising of certain propitious stars and constellations. The holistic principles on which the community based its life also operated on an individual human scale. Diversity in diet was considered essential to the maintenance of a healthy body. Food alternated from flesh to fish in order that the food stocks might not be depleted and nature might "never [be] incommoded nor weakened."[18] The number of meals per day varied with age in order that nature might be satisfied:

> Their food consists of flesh, butter, honey, cheese, garden herbs, and vegetables of various kinds. They were unwilling at first to slay animals, because it seemed cruel; but thinking afterward that it was also cruel to destroy herbs which have a share of sensitive feeling, they saw that they would perish from hunger unless they did an unjustifiable action for the sake of justifiable ones, and so now they all eat meat. Nevertheless they do not willingly kill useful animals, such as oxen and horses.[19]

Like the holistic health movements of today, which draw on ancient and Renaissance organismic principles, the natural harmonies of the body were maintained through nature's medicines. Herbs were used as mild natural cures for illness, harsh purgatives being advocated only rarely. The body itself was treated as a whole rather than a sum of separate parts. It was kept in tune through a program of regular gymnastics. Health, a delicate balance of the four humors, could be strengthened through the use of natural teas made from wild thyme, mint, basil, crushed garlic, or vinegar. Pleasant baths, removal to the country, and mild exercise were recommended for the curing of fevers.

People ate only according to the requirements of their own bodies, as determined by age and activity. Thus the general community took meals twice a day while the aged ate three times, consuming more easily digestible foods. Boys, while young and active, required four meals a day. Proper diet, sufficient exercise, and cleanliness formed the basis of a program of health care based on preventive medicine.

The stress on organic unity in agriculture, the community, and the human body was also manifested in the relation between cosmos and city. City planning and environmental design formed an important aspect of the holistic character of the City of the Sun. The city itself was a miniature replica of the larger macrocosm. The Temple of the Sun on a hill in the center of the city was surrounded by seven circular walls representing the seven circular planets and named for them. Four streets and four gates oriented the community to the four points of the compass. The circular temple in the center was a small model of the plan of the heavens.

Science, often viewed today as a separate and isolated discipline, was an integral part of the daily lives and education of the Solarians. Both sides of the city walls were painted with representations and diagrams from the natural sciences. Thus the mechanical arts, their uses, and inventors were depicted on the innermost wall, with mathematical propositions and definitions on the outermost. On the walls between were paintings of all the known precious metals and stones, trees and herbs, fresh- and salt-water fishes, birds, mammals, insects, worms, and serpents. Accompanying the picture of each specimen was a description of its properties, locations, habits, methods of production or breeding, medical uses, and value to the human race.

The harmonies and influences of the macrocosm could be absorbed by each person, not only during worship in the temple but also in daily life within the city itself. Renaissance Neoplatonic theory held that pictorial and symbolic representations of the heavenly bodies, infused by the influences of the stars through the *spiritus mundi* (or spirit of the world) could transmit their powers to the beholder.[20] In this way, the soul of each individual could be made more consonant with the cosmos and integrated with the larger universe. The plan of the city was thus consciously designed to bring people into greater harmony with their larger organic environment.

These same holistic principles integrated Valentin Andreä's ideal city, Christianopolis. Christianopolis was located on a triangular island that, like the City of the Sun, was a "world in miniature" and was inhabited by people with common ideals and principles. The island was ". . . rich in grain and pasture fields, watered with rivers and brooks, adorned with woods and vineyards, full of animals, just as if it were a whole world in miniature. One might think that here the heavens and the earth had been married and were living together in everlasting peace."[21]

In contrast to the circular plan of the City of the Sun, Christianopolis was set out in the form of a square oriented to the four corners of the world. Fresh air and ventilation, flowing water and springs contributed to its healthy atmosphere. Moats were stocked with fish, and open, unused spaces for wild animals were maintained. The city itself was a compact unit, each part serving a specific function necessary for the good of the whole. Fresh, clear water flowed through the town and supplied the houses. Underground canals removed daily wastes from the house for the sake of public health and pleasant surroundings.

Andreä believed that people were not beasts sent to "merely devour the pastures of the earth."[22] They were admonished to use it moderately, with gratitude, and with exact observation. The science laboratory therefore taught the details of natural history through a visual memory system. Painted on the walls of the hall of physics were detailed pictures of animals, plants, rocks and gems, the human races, and natural regions of the earth. Children learned the names, classes, and uses of herbs through play so that through exact observation they would be able to recognize them in nature.

Other pictures on the walls of the pictorial art shop showed the various regions of the earth, sketches of machines and statistics, perspective, and engineering fortifications. Astronomical instruments like those developed by Tycho Brahe were used for observations of the stars. Diagrams of the heavens illustrated their motions, harmonies, and the locations and shapes of individual planets and stars. In the chemical science laboratory, the properties of metals, minerals, and vegetables were studied that they might be used for the improvement of health and for the human race. Around the college were gardens containing the "living herbarium" of a thousand kinds of plants used for medicines, cooking, and decoration, as well as extensive gardens outside the walls for food crops.

In Andreä's community, nature was "aped" in order that her principles might be emulated and the earth and sky married together. Science served the human interest in Christianopolis because it was taught to all people in the community and because it was used in conjunction with natural harmonies rather than for exploitative purposes. Nature was not altered and tortured, as it would be in the laboratories of Bacon's *New Atlantis*, but observed and emulated. Natural science and the mechanical arts in the City of the Sun and in Christianopolis served the artisans and peasants because of the high status accorded them in the community and because of the great regard paid to the crafts and manual arts.

WORK.   One of the most serious human problems brought about by industrial capitalism has been the psychological alienation caused by a person's daily labor for wages in a business or industry owned by another individual who reaps both money and a higher standard of living as a result. A not-too-different alienation is found in modern socialist and communist societies in which the extensive bureaucracy required to manage industry also intrudes between worker and product. The unity of a meaningful life with productive "hands-on" work, lost to thousands of people at the bottom of today's society, was integral to Andreä's ideal organic city. In Christianopolis, manual labor was called "employment of the hands." All people spent equal amounts of time at work and at leisure. Work benefited rather than harmed the body. "For among us—i.e., the outsiders—one is worn out by the fatigue of an effort, with them the powers are reinforced by a perfect balance of work and leisure so that they never approach a piece of work without alacrity.... They have very few working hours, yet no less is accomplished than in other places as it is considered disgraceful by all that one should take more rest and leisure time than is allowed."[23]

The rise of modern science has been attributed by some historians to the attention given by Renaissance scholars to the empirical and mechanical skills of the artisan class. Andreä's *Christianopolis* reflects this interest but goes beyond the idea of merely making use of their techniques to honoring the people themselves. Craftspeople were the backbone of the community and greatly respected. They included:

> ... workers in brass, tin, iron; knife-makers, turners, makers of jewel cases, of statuary, workers in gypsum, fullers, weavers, furriers, cobblers; and, among the nobler crafts, sculptors, clockmakers, goldsmiths, organ-makers, engravers, goldleaf-beaters, ringmakers, and innumerable other like trades not to be despised. Tanners, harness-makers, blacksmiths, wagonmakers, trunk-makers, stonecutters, glassmakers, all these you will find here.[24]

The city plan itself was based on the total integration of all aspects of life with productive work. One-third of the city was designed for the production and storing of food, one-third for exercise, and one-third for scenic beauty. Of the portion designated for production, the farming and agricultural lands faced the east, the mills and bake shops south, the meat preparation and kitchen areas were in the northern districts, while the metal forges lay to the west. "For the whole city is, as it were, one single workshop, but of all different sorts of crafts."[25]

In Christianopolis, all citizens, including artisans and workers, received the same education that they might understand the theoretical foundation for their own work. Because they were trained in the scientific basis of their trade, they understood and found "delight in the inner parts of nature." Each individual engaged in both manual and mental activity. Mind and body were of equal importance, manual and intellectual work inseparable:

> This feature . . . is entirely peculiar to them, namely that, their artisans are almost entirely educated men. For that which other people think is the proper characteristic of a few . . . this the inhabitants argue should be attained by all individuals. They say neither the subtleness of letters is such nor yet the difficulty of work, that one man, if given enough, cannot master both. And yet there are some who incline more to this or to that occupation who, if they prefer to make a craft a specialty are masters over their fellows that they may in turn train up others and still others.[26]

In the context of sixteenth-century society, in which literary skills and education were available mainly to the upper classes and some of the bourgeoisie, advocating equal education for workers as well as intellectuals was a revolutionary and innovative ideal.

Within a framework in which the good of the whole took priority over that of the economic gain of any individual parts, Andreä recognized the importance of individual incentive, pride in one's work, and the rewards of competition. Thus some workers received recognition as master craftspeople. Inventive genius was valued and encouraged, and the human need for competition and reward among artisans was recognized and provided for.

In the City of the Sun, as in Christianopolis, the manual arts were held in high esteem. The greatest value was placed on the metal and building occupations, because these required the most labor. Slightly less esteem was accorded to anyone skilled in the manual arts, while the lowest, although still a noble grade, was assigned to those having knowledge of military and agricultural affairs.

All inhabitants participated in the production of food and goods. Everyone worked in the fields that they might understand the care and raising of animals. The European nobility who had no mastery of skills and who considered manual labor ignoble were to be ridiculed.

The equal sharing of workloads in the City of the Sun provided an opportunity for every person to engage in cultural and literary activities. People worked only four hours a day that the remainder might be spent in "learning joyously, in debating, in reading, in reciting, in writing, in walking, in exercising the mind and body, and with play."[27] As in Christianopolis, the democratization of work, education, and leisure for all people represented a marked contrast to sixteenth-century practices and norms.

COMMUNAL SHARING.

Both Christianopolis and the City of the Sun were communities in which goods, property, and knowledge were shared among the inhabitants. There was no differentiation between rich and poor, for wealth did not have meaning in real economic terms but rather a dialectical meaning in terms of values. "But with them [the inhabitants of the City of the Sun] all the rich and poor together make up the community. They are rich because they want nothing, poor because they possess nothing; and consequently they are not slaves to circumstances, but circumstances serve them."[28]

The City of the Sun and Christianopolis both reflected the needs of the society's poor for equal distribution of wealth and the philosophy that the whole is greater than the sum of its parts. In the City of the Sun, "no one wants either necessities or luxuries." Private property and ownership were considered to give rise to self-love and to foster private gain. By giving up private property, love was directed toward the whole community rather than toward the self. Likewise in Christianopolis, "No one has any money; yet the republic has its own treasury. And in this respect the inhabitants are especially blessed because no one can be superior to the other in the amount of riches owned." [29]

In both these communal societies, natural resources were utilized only for items necessary to the community as a whole. Nature was not raped for profits, the methodology on which the rising capitalist economy was to increasingly depend. The needs of the community took precedence over the special interests of the merchant class.

Thus in Christianopolis, goods were produced for use rather than for profit. The raw materials, "metals, stones, woods, and the things needed for weaving," were brought together for the production of communal goods.

> All things made are brought into a public booth. From here every workman receives out of the store on hand, whatever is necessary for the work of the coming week. For the whole city is, as it were, one single workshop but of different sorts of crafts. No one has any money nor is there any use for any private money; yet the republic has its own treasury.[30]

WOMEN.

In both Christianopolis and the City of the Sun, women were more liberated than in the real sixteenth-century society. However, their higher status was in each case due to different factors, and in each state there were negative aspects to the treatment of women.

In the City of the Sun, women were taught means of physical defense and military strategy in the manner of the Spartans and Amazons by female instructors and by magistrates. Like the men, they wore togas suitable for war but at a length below rather than above the knee. They received instruction in the manufacture and firing of cannon balls, the slinging of stones from precipices and the methods of ground attack. Any women who showed fear were severely punished. In times of greatest danger in battle, the armed women and boys in training were allowed to find a place of retreat, but afterward both were expected to help by treating the wounded men and praising them.

Yet the role of women in the City of the Sun was rather restricted. Both women and men worked at mechanical and theoretical occupations, but the more physically demanding tasks were assigned to men. Women were occupied by weaving, spinning, sewing, milking, cheese making, barbering, medical care, and music. They were excluded from "working in wood and the manufacture of arms."[31]

Because women in the City of the Sun would work and retain excellent bodily health and strength through exercise and military training, they would not "lose their color and have pale complexions and become feeble and small" as did the women of leisured sixteenth-century society. Norms for beauty were strength, agility, and tallness rather than the use of facial rouge, "high-heeled boots," and "garments with trains" in common usage in the real world. Although natural beauty was emphasized over artificial trappings, Campanella was too fanatic about its importance: women who used such artificial methods to attain beauty were "condemned to capital punishment."[32]

Utopian authors wished to make scientific learning available to all women in a society, not just those of the upper classes and court salons. In Campanella's City of the Sun the children of both sexes were to be taught natural science, followed by mechanical science, beginning at the age of six:

> After [two years] the weaned child is given into the charge of the mistresses if it is a female and to the masters if it is a male. And then with other children they are pleasantly instructed in the knowledge of the pictures, and in running, walking, and wrestling; also in the historical drawings, and in languages; and they are adorned with a suitable garment of different colors. After their sixth year, they are taught natural science, and then the mechanical sciences.[33]

The children of Andreä's Christianopolis were to be given over to the state at the age of six where they were divided into classes for education. Both men and women would instruct them; boys were taught in the morning and girls in the afternoon. "I know not why this sex [women], which is naturally no less teachable, is elsewhere excluded from literature. The rest of their time is devoted to manual training and domestic art and science as each one's occupation is assigned according to his natural inclination."[34] Young women were to be liberally educated, taught by the wives of the men who had been placed in charge of the boys and by widows who in Christianopolis were honored like mothers.

Married women would be expected to use the knowledge they acquired in college; however, they had no political voice:

> Whatever scholarship they have, being mentally gifted, they improve diligently, not only to know something themselves, but that they may sometime teach. In the church and in the council hall they have no voice, yet none the less do they shine with the gifts of heaven. God has denied this sex nothing, if it is pious of which fact the eternally blessed Mary is a most glorious example.[35]

If men married women who were too domineering for them, it was the man's fault for being too weak. Each of the two must be responsible for their own duties equally, without secret rule by the woman. Wife beating in Christianopolis would be considered a disgrace and a rarity.

Midwives were to be held in highest regard, and only those who were well versed in scientific knowledge, medical skills, and religious piety would be considered for this esteemed position. Women's accomplishments in childbirth and the strength with which they bore the pains of labor were given greater precedence than the achievements of all the earth's athletes.

Of the two authors, Campanella's views on the family were the more radical. People shared common dormitories and ate at common tables. After two years, children were weaned from their mothers and placed in the care of mistresses and masters. Andreä retained the family structure. Christianopolans lived in 264 separate simply furnished homes, each containing three rooms, a fragrant garden, and a small private cellar all owned by the community as a whole.

POLITICS AND COMMUNITY. Andreä's and Campanella's utopias can be seen as a political and religious response to the breakup of community taking place under sixteenth-century commercial expansion and continuing feudal disintegration. Both utopias were set in the religious framework of corporate Christian community. Campanella, a member of the Catholic Dominican order, was influenced by Platonist ideas. Although Andreä, a Protestant, was a Lutheran pastor, he advocated a social order dominated by Calvinist morality. Both rejected monarchy as a political form and drew up governments under collective rule.

Although he did not subscribe to Calvinist religious teachings, Andreä was very much influenced by the morals and customs of Calvinist Geneva, which he visited in 1610. Calvin's legacy to Andreä was his perception of the need to restore a lost sense of community combined with a strong church government. For the Protestant church to survive in a world of power, it needed an institutionalized cohesive organization. Calvin's contribution was to recognize the political possibilities of participation without a papal head. The need for order and structure made the Calvinist church a political society; the priesthood of all believers created a sense of equality in a mass movement against the established order.[36] In the Genevan system of church government, both elected lay people and ministers formed the consistory, in which the ministers had the greater power. The congregation approved or rejected the decisions of the consistory but did not formulate policy.

In Christianopolis, the two elements of participation by the citizenry and decision making by a ruling group were both present. Andreä's utopia was governed by a triumvirate—a Christian priest, a judge, and a director of learning. These three leaders each had separate duties, which were performed with the knowledge of the others and with their common consent. Each was responsible to his own senate and announced its decrees to the rest of the citizenry. The leaders qualified for office by working upward through the virtues to loyalty, prudence, and wisdom. This form of governance was judged by Andreä to be safer than monarchy.

But Andreä's zeal for Calvinist moral reform proved to be one of the weakest aspects of his utopia. The free atmosphere for artisans in Christianopolis was tainted by rigid moral controls that Andreä substituted for corporal punishment. He considered it more humane to prevent vice than to inflict harsh penalties. "As the Christian citizens are always chary of spilling blood, they do not willingly agree upon the death sentence as a form of punishment." The real world tended to make scapegoats out of undesirables for the sake of mollifying God—"thieves [out of] dissolute characters, adulterers of the intemperate, homicides of loafers [and] witches of courtesans." But in his attempt to prevent misdeeds, both on the part of citizens and enforcers of laws, Andreä turned to a rigid moralism. Crimes against property were punished most lightly, those against other people more severely, while those against God received harsh judgment. Infidelity was severely punished, and the state-owned houses were inspected so that nothing could be destroyed or damaged. Visitors were examined as to their morals, past lives, and personal culture, while attendance at sermons was compulsory. Throughout his treatise, Andreä moralized about the evils of Satan and the wickedness of human nature.[37]

The model for these moral strictures was Geneva. On his 1610 visit, Andreä was impressed by the fact that "all cursing, gambling, luxury, quarreling, hatred, conceit, deceit, extravagance, and the like, to say nothing of greater sin, are prevented." But this "glorious adornment" was achieved in Geneva through a "censorship of morals in accordance with which investigations are made each week into the morals and even into the slightest transgressions of the citizens."[38] The use of moral strictures to maintain political order was the major failing of Andreä's utopian model and stemmed in part from the inherent difficulties of reconciling the problems of power and order with community.

Whereas Andreä's utopia bore the theoretical stamp of Calvinism, Campanella's was influenced by Platonism. For Plato, ideal rule was expressed by the unity of the trinity of virtue, knowledge, and power. The City of the Sun was ruled by a philosopher king, Metaphysic (Hoh) assisted by the trinity—Power, Wisdom, and Love (Pon, Sin, and Mor). All four together made decisions and settled business matters but deferred to Hoh in event of disagreement. Nine magistrates responsible to each of the three ministers were elected by the people. Periodically, councils of all people over twenty years of age were called in which each person was asked to say "what [was] wanting in the state and which of the magistrates [had] discharged their duties rightly and which wrongly."[39]

Mor (Love) was in charge of the breeding of both animals and human beings. Eugenic controls were used to produce the "best offspring" for the "preservation of the species."[40] Those qualities were desirable that enhanced the natural strengths and beauty of the people and made them more consonant with their natural surroundings. Since war was emphasized as a protective measure, physical rather than intellectual qualities were the criteria for breeding. Therefore, if there were cases in which a union between two people might endanger the future of the race, friendship, conversation, flowers, and verses were substituted for sexual love.

The evils of eugenic programs for the breeding of peoples of superior intelligence have become apparent in the twentieth century. Although Campanella's controls were directed toward the enhancement of natural rather than intellectual qualities, the eugenics program was one of the weakest aspects of his utopian scheme.

The utopias of Campanella and Andreä had a number of shortcomings in the midst of plans for important social reforms. Yet their ideas are worth examining because they proposed concrete schemes for a better society in times of despair and oppression. They were not mere visionaries, but activists who attempted to translate their ideas into community change. Most importantly, they envisioned societies whose philosophies of science were consistent with the integrity of the natural environment and with human equality.

In these utopias, all parts of the natural and social community were interrelated in an organic unity in which both human and natural components were of equal value in the functioning of the whole. The whole organism was greater than the sum of the individual parts; the collective good was greater than the advancement of any one part over the others. Change came from within the community rather than as a directive from without and was determined by the needs of the entire human-nature system.

ECOLOGY AND UTOPIA.

In the tradition of the ancient connections assumed between nature and society, the ecology movement of the last several years has emphasized the small, ecologically-balanced community over the more conservative hierarchical model. By viewing people as integral components of the larger ecosystem, it has tried to break down the dualism between humans and nature and has challenged the idea of human superiority and independence from nature. The environmental crisis, it argues, is the result of human arrogance toward nature and of the continued use, without replenishment, of its resources.

The communal and utopian variants of the organic model are compatible with the ecology movement's emphasis on the integrative aspects of small communities in which people observe the limits of ecosystem resources. Using the concept of harmonic balance in the natural community, urban communal households and ecotopian farming communes have challenged the competitive ideals of capitalist society and are attempting to live within the laws prescribed by ecology. "Holistic houses" constructed of wood, operating on the energy of the sun and wind, demonstrate the possibility of urban living based on renewable resources and the recycling of wastes. The back-to-the-land movement creates rural communes by sharing human and natural resources and "appropriate" technologies, while employing organic farming techniques and biological pest controls.

Some environmentalists have ascribed an intrinsic ethic to the ecosystem itself. Abrogating the distinction between descriptive and prescriptive statements, that is, between "is" and "ought," ecological ethics boldly asserts that not only is it morally right to recycle resources for the sake of future generations, it is also (because human beings are integral to and dependent on the environmental fabric for survival) obligatory to live within the framework of ecological law. The environment prescribes human behavior and its limits, since exploitation will ultimately lead to extinction. Environmental ethics is thus not merely an ethic *about* the environment but an ethic determined by it as well.[41]

Until the appearance of Ernest Callenbach's *Ecotopia*, in 1976, modern utopian authors had not explored the possibility that a society could live in harmony with its environment while continuing to utilize many of the advances made through modern technology.[42] Most utopias, as reactions to the repressive possibilities of technology, had been technological dystopias, typified by Aldous Huxley's *Brave New World* and George Orwell's *1984*. Positive utopias such as Huxley's *Island* and A.T. Wright's *Islandia* (with the exception of B.F. Skinner's *Walden Two*, 1948) rejected technology in favor of a return to an era of pretechnological simplified living. Both types reflected the increasing despair of sensitive authors toward machines and their implications for the future of society.

In Callenbach's *Ecotopia*, based on the ideals of the ecology movement, northern California, Oregon, and Washington secede from the United States in the year 1980 and remain isolated until 2000, when a New York reporter is allowed to visit their steady-state society. In Ecotopia, social structure is based on an ecological philosophy of nature. Significantly, it is women who in this back-to-nature society organize the secession and then hold the positions of power in the major political party, the presidency, and the ministry. Farms, factories, and stores are owned and operated collectively, private property has been abolished, and the former downtown San Francisco offices of corporations have been turned into apartment houses. People live in mini cities separated by reforested wild areas and joined by electrically operated, high-speed mass transit systems, or in small communities in rural areas. City streets have been turned into malls planted with trees, flowers, ferns, and bamboos, through which creeks flow over little waterfalls (resurrected from water culverts). Intracity free transportation is by bicycle, electric taxi, and minibus; interurban trains connect Ecotopia to the rest of the country. Trains, freighters, electric trucks, and a few diesels are used to transport farm produce and manufactured items.

Buildings are constructed mainly of wood from carefully replanted forest plots in the Pacific Northwest, although new biodegradable plastics have been developed out of which modules are produced for imaginative do-it-yourself homes. All synthetic fibers were banished after independence in favor of recyclable cotton, wool, leather, and furs. Metals left over from presecession days were collected and reused in transport vehicles and electric motors. A small amount is still imported to replace exported electric machinery.

Electrical energy in Ecotopia is supplied by geothermal, hydroelectric, and solar sources, together with a few inherited, temporarily tolerated but carefully controlled fission plants. Technology in general is highly sophisticated, with picture phones, television, and computers used in communication, ecological monitoring devices for control of pollution levels, and radar-guided rockets, infrared homing missiles, and hidden mines in major American cities for defense.

In this ecological society, reverence for trees, water, and wildlife forms the basis of an ecological religion expressed in the prayers, poetry, and little shrines of the Ecotopians. Decentralized communities, extended families, spontaneity, freedom of emotional expression, and the practice of ritual war games to deal with competitive instincts characterize the cultural norms and values. As in Andreä's Christianopolis and Campanella's City of the Sun, Ecotopia's social structure is an expression of its holistic philosophy of nature. Like them, it reflects the need for social changes and the most idealistic aspirations of contemporary society.

As economic changes irrevocably altered the ecology of the farms, fens, and forests of Europe, as the organic hierarchical order was undermined by social mobility, and as the earth's womb and communal resources became the basis for a market in money and property, so too the larger cosmos as a living organism was being transformed. As a projection of the self and society onto the celestial spheres, the macrocosm, like the human microcosm, was for most Renaissance philosophers a tightly knit unity of body, soul, and spirit, within which all things were arranged in a hierarchical order. To see how these elements were reordered and assimilated or radically rejected by an emerging mechanical framework is the task ahead.

CHAPTER 4

The World an Organism

The Scientific Revolution of the sixteenth and seventeenth centuries has been treated by most historians as a period of intellectual enlightenment in which a new science of mechanics and a mechanical world view laid the foundation for modern scientific, technological, and social progress. But, in the face of the current crisis over the depletion of natural resources, Western society is once more beginning to appreciate the environmental values of the premechanical "world we have lost." Today the ecological consequences of exploitative attitudes toward the four elements—earth, air, water, and fire—the ancient sources of life and energy, are beginning to be fully recognized.

Along with current challenges to mechanistic technology, holistic presuppositions about nature are being revived in ecology's premise that everything is connected to everything else and in its emphasis on the primacy of interactive processes in nature. All parts are dependent on one another and mutually affect each other and the whole. Each portion of an ecological community, each niche, exists in a dynamic relationship with the surrounding ecosystem. The organism occupying any particular niche affects and is affected by the entire web of living and nonliving environmental components. Ecology, as a philosophy of nature, has roots in organicism—the idea that the cosmos is an organic entity, growing and developing from within, in an integrated unity of structure and function.

Organismic thought contributed the rudimentary philosophical framework out of which ecological science and the conservation of natural resources developed. The Romantics of the early nineteenth century, reacting against the mechanism of the Scientific Revolution and the Enlightenment, turned back to the organismic idea of a vital animating principle binding together the whole created world. American Romantics such as Emerson looked to wildness as a source of spiritual insight, while Thoreau found evidence of a vital life permeating the rocks, ponds, and mountains in pagan and American Indian animism. Such influences were an inspiration to the preservation movement led by John Muir in the late nineteenth century, and to such early ecologists as Frederick Clements, whose theory of plant succession held that a plant community grew, developed, and matured much like an individual organism.[1]

Variations of the organic framework of the Renaissance shared certain presuppositions about nature. The Renaissance cosmos was a living unit, of which all parts were interconnected in a tightly organized system. The orthodox view inherited from the medieval interpretation of Aristotle was an earth-centered hierarchical cosmos extending upward from the four inanimate elements, which were mixed together to form the minerals, vegetables, and animals found in the sublunar region of change, to the unchanging ether-filled spheres of the seven planets, with their associated hierarchies of angels, above the moon. Beyond the planets was the sphere of the *primum mobile*, source of the daily rotation of the heavens, then the sphere of the fixed stars and zodiacal constellations, and finally the Empyrean heaven of God. Together they comprised a living chain of being, each member a step in a stable, ordered, spherically-enclosed world, each member sharing some particular feature with the steps below and above, yet excelling in some unique characteristic. Man was linked to the animal world below, with which he shared sensation, and to the angels above, with whom he shared rationality. Each part of his body was governed by one of the zodiacal

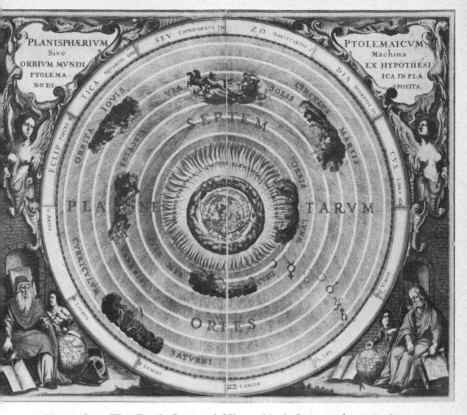

Figure 9. The Earth-Centered Hierarchical Cosmos, from Andreas Cellarius, *Harmonia Macrocosmica* (Amsterdam, 1661). In the hierarchical cosmos, the earth was at the center of a series of nesting spheres. The stars and the planets affected life on earth, as could magical manipulation of natural objects by human beings.

signs, so that as a microcosm, he was a miniature replica of the celestial spheres, or macrocosm. Human society, as discussed in the preceding chapter, was also stratified according to status, with peasants at the bottom, the king and pope at the apex, and women below the men of their particular status group. Nature as the involuntary agent of God was the immanent manifestation of God's law in the world (Fig. 9).[2]

Within this hierarchical system, Renaissance Neoplatonism revived a slightly less orthodox but widely held interpretation of nature as the soul of the world, a voluntary and immanent source of change. Neoplatonic magicians such as Marsilio Ficino (1433–1499), Giovanni Pico della Mirandola (1463–1494), and Henry Cornelius Agrippa (1486–1535), who attempted to draw down celestial influences to produce changes on earth, asserted a hierarchical system based on a distinction between matter, soul, and spirit. The activity of the world soul (*anima mundi*) was transmitted by the world spirit (*spiritus mundi*) to passive matter, infusing material objects with life and form.

As cultural developments challenged Aristotelian authority and economic changes undermined the established social order, more radical organic philosophies emerged, stressing change over structure and force over form. Naturalists including Bernardino Telesio (1509–1588), Tommaso Campanella (1568–1639), and Giordano Bruno (1548–1600) unified the world's soul and spirit into a single all-pervasive living entity, distinct from matter but coeternal with it, while asserting change as the dialectical opposition of contraries. The vitalist view put forward somewhat earlier by Paracelsus (1490–1541) further reduced the explanation of change to a monistic unity of vital spirit and phenomenal matter. Neoplatonism and naturalism thus made a distinction between matter and its activity, while vitalism unified matter and spirit into single, active evolving substances.

But the cluster of ecological, commercial, technological, and social changes evolving during the sixteenth and seventeenth centuries continued to differentiate among these philosophies, with the result that some of the above assumptions about the organic world were criticized and transformed by the emerging mechanical philosophy of the mid-seventeenth century, while others were rendered implausible and rejected. Mechanism, which superseded the organic framework, was based on the logic that knowledge of the world could be certain and consistent, and that the laws of nature were imposed on creation by God. The primacy of organic process gave way to the stability of mathematical laws and identities. Force was external to matter rather than immanent within it. Matter was corpuscular, passive, and inert; change was simply the rearrangement of particles as motion was transmitted from one part to another in a causal nexus. Because it viewed nature as dead and matter as passive, mechanism could function as a subtle sanction for the exploitation and manipulation of nature and its resources. (See Chapters 8 and 9.) By the late seventeenth century, however, the organicism of the Renaissance had begun to achieve a new synthesis with the mechanical philosophy of the mid-century, resulting in a managerial perspective concerning the future use of resources. (See Chapter 10.)

This mechanical philosophy and its managerial point of view have also contributed to the science of ecology. The concept of the ecosystem, which by the 1950s had replaced the idea of the biotic community (rejected as being too anthropomorphic), is based on the mathematical modeling of nature. Data are abstracted from the organic context in the form of information bits and then manipulated according to a set of differential equations, allowing the prediction of ecological change and the rational management of the ecosystem and its resources as a whole. The organic and mechanical philosophies of nature cannot, therefore, be viewed as strict dichotomies, nor can most philosophers be placed solidly in one camp or the other. The tensions between these two perspectives on nature have continued to be influential ever since the Scientific Revolution.

ORGANIC UNITY. Organic thought in the Renaissance had its roots in Greek concepts of the cosmos as an intelligent organism, which when revived and modified were assimilated into the consciousness of the fifteenth and sixteenth centuries. Three root traditions became the basis for later syncretic forms of organicism—Platonism, Aristotelianism, and Stoicism. Each of these organic traditions differed in important respects, so that when synthesized with other systems, such as Hermeticism, gnosticism, Neoplatonism, and Christianity, they produced a spectrum of Renaissance organismic philosophies.

Common to all was the premise that all parts of the cosmos were connected and interrelated in a living unity. From the "affinity of nature" resulted the bonding together of all things through mutual attraction or love. All parts of nature were mutually interdependent and each reflected changes in the rest of the cosmos. The common knitting of the world's parts implied not only mutual nourishment and growth but also mutual suffering. "When one part suffers, the rest also suffer with it," wrote Giambattista della Porta (1535–1615).[3] Or as Paracelsus expressed the idea, "If anything suffers from the error of the elements other things grow uncertain too . . . and the defects and errors of the firmament can be observed by us, no less than the firmament observes our defects."[4]

Astrologer John Dee (1527–1628) presupposed a harmonious universe in which celestial rays from the stars and zodiacal signs interacted with each other to produce different effects in each natural object. The coalescence and unification of natural forces as they flowed into each body produced a unique effect in that object, dependent on both source and receptor.

The organic unity of the cosmos derived from its conception as a living animal. A vast organism, everywhere quick and vital, its body, soul, and spirit were held tightly together. As Della Porta put it, "The whole world is knit and bound within itself: for the world is a living creature, everywhere both male and female, and the parts of it do couple together . . . by reason of their mutual love."[5]

All parts of this world, even the metals, contained life and were nourished by the earth and sun. Bernardino Telesio wrote that

> those things which are made in the depths of the earth, or those which derive or grow therefrom: the metals, the broken sulfuric, bituminous or nitrogenous rocks; and furthermore those sweet and gentle waters, as well as the plants and animals—if these were not made of earth by the sun, one cannot imagine of what else or by what other agent they could be made.[6]

His follower Campanella affirmed the vitality of the elements and the pervasive life and feeling of the entire cosmos:

> Now if animals have, as we all agree, what is called sense or feeling, and if it is true that sense and feeling do not come from nothing, then it seems to me that we must admit that sense and feeling belong to all elements which function as their cause, since it can be shown that what belongs to the effect belongs to the cause. Consider, then, the sky and earth and the whole world as containing animals in the way in which worms are sometimes contained in the human intestines—worms or men, if you please, who ignore the sense and feeling in other things because they consider it irrelevant with respect to their so called knowledge of entities.[7]

As the sixteenth century organic cosmos was transformed into the seventeenth century mechanistic universe, its life and vitality were sacrificed for a world filled with dead and passive matter. By examining variations in Renaissance philosophies of nature and their social contexts we can see the process by which some assumptions were transformed and retained while others were criticized and rejected.

NEOPLATONIC NATURAL MAGIC.

Neoplatonic natural magic presupposed a hierarchical cosmic structure and assumed that earthly changes were influenced by the celestial heavens and could be produced artificially by the human manipulation of natural objects in which these influences inhered. It originated as an elite aristocratic form of the magical world view in the Florentine Platonic Academy in the late fifteenth century.

A revival of Neoplatonic philosophy and an interest in the writings of Hermes Trismegistus took place under the sponsorship of the wealthy Medici family, who had obtained an aristocratic status through the commercial manufacture and trade of wool and silken goods, banking operations, and mine management. The Florentine Academy, which they funded, was a private community of scholars pursuing (outside the university structure) studies of a hierarchical cosmos in which changes could be effected by the manipulation of natural objects. The academy supported Ficino and visiting Neoplatonic scholars such as Pico della Mirandola.[8]

The study of Neoplatonic magic was consistent with a conservative interest in the reassertion of status in the hands of the prestigious and powerful class of wealthy Florentine bankers, merchants, and international traders in an economy that had peaked and ceased to expand after the mid-fifteenth century. As increasing competition in foreign markets began to reduce profits, the elite began to exhibit a preference for gentlemanly rather than entrepreneurial lifestyles. They broke away from traditional ascetic restraints and began to indulge in extravagances that asserted an aristocratic status within their society. The Medici academy reflected an increasing tendency in late fifteenth-century Florentine culture for learning to be supported by the private rather than public sector and for education to become exclusive and aristocratic.

Neoplatonic magic postulated a hierarchical universe that extended from the base matter of the earth upward to the divine intellect. It accepted the tripartite division of the macrocosmic world into body, soul, and spirit, the components of a living organism. The divine mind beyond the visible cosmos was the seat of the Platonic forms, the pure Ideas of which sensible corporeal objects were merely imperfect copies. The female soul of the world was everywhere present and, as in Plato's *Timaeus*, was the source of motion and activity in the macrocosm. It contained the celestial images of the divine Ideas. The world's body was its matter, the elements out of which corporeal objects were generated. Linking the celestial images in the world soul to the matter in the body was the world spirit. The *spiritus mundi* was the vehicle by which the influences of superior powers in the celestial realm could be brought down and joined to the inferior powers in the terrestrial region. As Agrippa put it,

> In the soul of the world there be as many seminal forms as ideas in the mind of God, by which forms she did in the heavens above the stars frame to herself shapes also, and stamped upon all these some properties. On these stars therefore, shapes and properties, all virtues of inferior species, as also their properties do depend; so that every species hath its celestial shape, or figure that is suitable to it, from which also proceeds a wonderful power of operating, which proper gift it receives from its own idea, through the seminal forms of the soul of the world.[9]

This Neoplatonic conceptual scheme was common to natural philosophers such as Ficino, Pico della Mirandola, Agrippa, Della Porta, and Thomas Vaughan. The hierarchical arrangement of the parts of the universe was a great chain linking inferiors to superiors: "For so inferiors are successively joined to their superiors, that there proceeds an influence from their head, the first cause, as a certain string stretched out to the lowermost things of all, of which string if one end be touched the whole doth presently shake."[10] Della Porta illustrated the role of the golden chain in the operations of the magus who "marries and couples together inferior things" by means of the powers they receive from their superiors:

> Seeing then the spirit cometh from God, and from the spirit cometh the soul, and the soul doth animate and quicken all other things in their order . . . so that the superior power cometh down even from the very first

cause to these inferiors, driving her force into them, like as it were a cord platted together and stretched along from heaven to earth, in such sort as if either end of this cord be touched, it will wag the whole; therefore we may rightly call this knitting together of things *a chain* ... wherein he feigneth, that all the gods and goddesses have made a golden chain, which they hanged above in heaven, and it reacheth down to the very earth.[11]

Thomas Vaughan, a seventeenth-century Neoplatonic alchemist, likewise held that the world's soul, spirit, and ethereal water were all connected together like the links of a chain. The attraction of the spirit for the soul moved the first link followed by the attraction of the water for the spirit. The soul thus became imprisoned in the liquid crystal of the waters.

In every frame, there are three leading principles. The first is this soul, whereof we have spoken ... already. The second is that which we have called the spirit of the world, and this spirit is "the medium whereby the soul is diffused through and moves its body." The third is a certain oleous, ethereal water. This is the menstruum and matrix of the world, for in it all things are framed and preserved.[12]

At the basis of Neoplatonic hierarchical magic, therefore, was a causal chain linking elemental and celestial objects and making it possible for bodies above the terrestrial sphere to affect and alter those on earth.

In the Neoplatonic scheme, the cosmic world soul was the source of life and activity in the natural world. The soul was immanent within nature, vivifying it like a cosmic animal. Matter was distinct from both the world's soul and its spirit. Agrippa held that the soul was the source of the world's power, while matter was inactive: "Now seeing the soul is the first thing that is moveable and as they say, is moved of itself; but the body or the matter, is of itself unable and unfit for motion and doth degenerate from the soul."[13]

Likewise, for Thomas Vaughan, the principle of motion was the soul of the world, trapped by matter and struggling for freedom. He considered the Aristotelian notion of a substantial form too limiting and absurd to be the source of motive power. Motion was caused by a principle internal to the macrocosmic world, the *anima mundi*. But like the other Neoplatonists, Vaughan considered matter to be "merely passive and furnished with no motive faculty at all."[14]

Although the ultimate source of activity in the Neoplatonic world picture was the *anima mundi*, which was connected to earthly objects by the *spiritus mundi*, changes in particular natural objects were induced through occult properties. The natural magician drew a distinction between *elementary* qualities, the properties of matter, and *occult* properties, those derived from the stars and infused into natural objects by the *spiritus*. These occult virtues were more powerful than elementary virtues because they contained more form and less materiality. The occult properties had the power to "generate their like," to make the objects "like and suitable to" themselves.[15] Since an excess of occult virtue in any object could generate a like quality in another, plants or animals containing strong virtues could be utilized to produce the desired property. For example,

> Any animal that is barren causeth another to be barren, and of the animal especially the generative parts . . . if at any time we would promote love, let us seek some animal which is most loving, of which kind are pigeons, turtles, sparrows, swallows, wagtails and in these take those members or parts in which the vital virtue is most vigorous such as the heart, breast, and also like parts. . . . In like manner, to increase boldness, let us look for a lion, or a cock, and of these let us take the heart, eyes or forehead.

An occult property had the power not only to generate its like in another object, but also to "shun its contrary and drive it away out of its presence." These enmities or antipathies between occult properties could be used by the magus to effect cures and produce changes in natural objects. According to Della Porta, "Amongst all the secrets of nature, there is nothing but hath some hidden and special property; and moreover that by this their consent and disagreement, we may conjecture, and in trial so it will prove, that one of them may be used as a fit remedy against the harms of the other."[16]

For the Neoplatonists, therefore, the opposites, or sympathies and antipathies, were the properties of natural objects. They were powers or forces within the material object, but distinct from it, deriving from the world soul in the celestial heavens and ultimately from the ideas in the divine mind. The tripartite distinction between matter, spirit, and soul was the foundation of the Neoplatonic hierarchical structure. Operating within this hierarchy, the magus could draw down the celestial powers to marry inferiors to superiors, and therefore to manipulate nature for individual benefit.

Condemned by the Catholic Church in the sixteenth century as heretical, natural magic was based on assumptions such as the manipulation of nature and the passivity of matter; these assumptions were ultimately assimilated into a mechanical framework founded on technological power over nature for the collective benefit of society. The Renaissance magus as an operator and arranger of natural objects became the basis of a new optimism that nature could be altered for human progress.

In the organic world view, the concept of nature as a living entity had limited the scale of power to individual needs and group benefits such as spiritual fulfillment, healing, the growing of crops, and the manufacture of tools. For the Neoplatonic magician, the upward gnostic ascent aimed at greater intellectual insight and spiritual regeneration. Knowledge and power could be obtained through a union with the understanding and intellect of God: "No one has such powers but he who has cohabited with the elements, vanquished nature, mounted higher than the heavens, elevating himself above the angels to the archetype itself, with whom he then becomes cooperator and can do all things."[17] But power obtained by such methods was restricted to each individual. It was an experience which could not be shared or transferred except through initiation.

Della Porta portrays the magician as nature's assistant in the cultivation of crops and breeding of animals, nature being the operator, the magician preparing the way:

> Wherefore as many of you as come to behold magick, must be persuaded that the works of magick are nothing else but the works of nature whose dutiful hand-maid magick is . . . as in husbandry it is nature that brings forth corn and herbs, but it is art that prepares and makes way for them. Hence it was that Antipho the poet said, *that we overcome those things by art wherein nature doth overcome us*; and Plotinus calls a magician such a one as works by the help of nature only, and not by the help of art.[18]

But although the magician is depicted here as nature's helpmate, the idea of altering and changing nature is also important to Della

Porta's natural magic (Fig. 10). Much of his book is devoted to the production of new plants, the generation of animals, and the changing of metals—how "an oak may be changed into a vine," how to generate "an apple compounded of a peach-apple and a nut-peach," and how to breed "new kinds of living creatures . . . of diverse beasts, by carnal copulation." He writes,

> Art, being as it were, nature's ape, even in her imitation of nature, effecteth greater matters than nature doth. Hence it is that a magician being furnished with art, as it were another nature, searching thoroughly into

Figure 10.   The hermetic philosopher following in the footsteps of nature, engraving by Johann Theodore de Bry, from *Atalanta Fugiens, Hoc Est, Emblemata Nova de Secretis Chymica* (1618). The magician as a model for human relations with nature changed from that of a follower, student, and helper to that of manipulator. Note the staff and light—both can help a traveler find a path, but the staff can also be used to prod, and the light to probe.

110

those works which nature doth accomplish by many secret means and close operations, doth work upon nature ... and either hastens or hinders her work, making things ripe before or after their natural season, and so indeed makes nature to be his instrument.

Although Della Porta considered himself to be the humble servant of nature working within its seasons and growing periods, aping and emulating its organic processes in order to perfect and hasten them, such manipulations, when assimilated into the utilitarian framework of Francis Bacon, would become instead techniques for control. Mechanism removed the organic substratum and substituted a mechanical framework for the same operations. And although the mechanists, too, were limited by the laws of nature and operated within them, "commanding nature by obeying her," they were free of the ethical strictures associated with the view of nature as a living being.

The process of mechanizing the world picture removed the controls over environmental exploitation that were an inherent part of the organic view that nature was alive, sensitive, and responsive to human action. Mechanism took over from the magical tradition the concept of the manipulation of matter but divested it of life and vital action. The passivity of matter, externality of motion, and elimination of the female world soul altered the character of cosmology and its associated normative constraints. In the mechanical philosophy, the manipulation of nature ceased to be a matter of individual efforts and became associated with general collaborative social interests that sanctioned the expansion of commercial capitalism. Increasingly it benefited those persons and social classes in control of its development, rather than promoting universal progress for all. It was intimately connected to an empirical philosophy of science and a concept of the human being as a designer of experiments who by wresting secrets from nature gained mastery over its operations.

NATURALISM.   Whereas natural magic tended to operate within a structure that conserved cosmic order in the form of hierarchy, the second organic variant, naturalism, laid greater stress on a concept of change that challenged the hierarchical structure of both nature and society. Renaissance naturalism, developing from within

111

the Aristotelian framework, exposed it to a radical critique.[19] The ultimate terms of philosophical explanation were reduced to two—the material substratum and the dialectical opposition of contraries. Naturalism differed from Neoplatonism in that the contraries were principles of change rather than properties of matter. The lack of a distinction between the world soul and spirit broke down the Neoplatonic hierarchies, utilizing only one category to account for natural changes.

Naturalism differed from traditional Aristotelianism in that activity was not accounted for through the actualization of the potential by means of the form; instead, the contraries were the agents of change. They were active principles; matter was a passive principle that received specification through the activity of the opposites.

Telesio, in his book *De Rerum Natura* (*On the Nature of Things According to Their Own Proper Principles*) (1565), reduced the explanatory entities to two substances or natures, a corporeal material substratum and an incorporeal dialectical activity that produced individuation in matter. Throughout changes in individual objects, the same body and matter remained. Matter was dead and passive, completely uniform throughout and lacking the capacity to act or operate. Its function was to receive and conserve the activity of the incorporeal substance.[20]

The distinctive feature of Telesio's natural philosophy was to define activity as a dialectic, the conflict between contraries. Active agents "perpetually oppose one another; forever disturbing or destroying each other. They do not desire to be together, nor can they remain together in any way." The primary opposites were hot and cold, and from these followed the operations of the other opposites: density and rarity, darkness and whiteness, lightness and obscurity, mobility and immobility, bringing the "active natures into perpetual conflict."

The two fundamental active principles, hot and cold, appeared in corporeal garb as the sun and the earth, sensible manifestations of the opposites. The sun was "supreme heat, whiteness, light, and motion," the earth "supreme cold, darkness, and immobility." These principles also appeared in all things generated out of the earth and sun (or sky) in a reduced or diminished form. In generated objects, the opposites interpenetrated and caused change, while in the sun and earth they were primary, supreme, self-constituted, and independent. Each natural organism developed in accordance with its own nature, while its motion benefited and maintained the harmony of the whole.

Telesio's naturalism was an important formative influence on the early ideas of Campanella. Campanella asserted that the earth, the plants, and the metals were living beings with sense and feelings. Plants and animals derived their matter from the earth and their activity and motion from the sun. The sun was an "active, diffusive, and incorporeal power." Sense and feelings were characteristic of active causes. "That the sun and the earth feel is undeniable," he asserted.[21]

Following Telesio, Campanella argued that change occurred through the opposition of active contraries. "Hot and cold, I say, understood as active and wholly free of atomic passivity, are not born without active power." The modes of being were produced by the opposing actions of these dynamic causes. All things were produced from the matter of the earth and the activity of the sun, arising from the opposition of the two contraries heat and cold.

Campanella criticized the atomic theory of the ancient philosophers Democritus and Lucretius on the basis that the mingling of inert, passive, insentient particles could not give rise to beings with feelings and sensations. For the atomists, he observed, heat and cold were not active principles within matter, but were produced instead by mechanical coupling: "Heat is born from those atoms which are sharper, and cold from those which are obtuse, while the soul is born of the round ones."

The basic dynamic of the opposition of hot and cold was extended to a general theory of dialectical process in the philosophy of Giordano Bruno, who synthesized Neoplatonic and Stoic ideas. From an early Neoplatonist phase, he moved to the view, in his *Expulsion of the Triumphant Beast* (1584), that two universal substances, one corporeal and material, the other incorporeal and spiritual, explained change.[22] The soul and spirit of the Neoplatonists were fused into a single active substance, a world soul or inner principle of motion, while prime matter was its passive corporeal opposite. Matter was not created *ex nihilo*, nor could it return to nothingness; it was "ingenerable and incorruptible," "arrangeable and fashionable," and a divine mother of all things. The active substance, or universal spirit, did not mix by composition with matter, but had the power to hold matter intact, keep its parts united, and maintain its composition: "It is exactly like the helmsman on the ship, the father of the family at home, and an artisan who is not eternal but fabricates from within, tempers and preserves the edifice. . . . It winds the beam, weaves the cloth, interweaves the thread, restrains, tempers, gives order to and arranges and distributes the spirits. . . ." On the highest level, matter and spirit achieved an absolute unity as a single universal substance.

Change was the unification and opposition of contraries. An efficient formative principle within the universal spiritual substance acted to unite the contraries and to arrange discordant qualities in harmonies. And then, "necessitated by the principles of dissolution, abandoning its architecture [the efficient and formative principle] causes the ruin of the edifice by dissolving the contrary elements, breaking the union, removing the hypostatic composition."

Bruno's character of Sophia, ancient priestess of gnostic wisdom, puts forth his ideas on the unification and dissolution of contraries in *The Expulsion of the Triumphant Beast*. The transit between states defines the reality in change. One condition has meaning only in terms of its opposite. Pleasure becomes meaningful in terms of past boredom, walking in terms of previous sitting, satiety with respect to hunger: " 'Association with one food, however pleasing,' " says Sophia, " 'is finally the cause of nausea. . . . Motion from one contrary to the other through its intermediate points come[s] to satisfy [us]; and, finally we see such familiarity between one contrary and the other that the one agrees more with the other than like with like.' " Responding to Sophia, Bruno's Saulino pointed out that it is no small thing to have discovered the principle of the coincidence of contraries and that it is the magician who knows how to look for them. Everything comes "from contraries, through contraries, into contraries, to contraries. And where there is contrariety there is action and reaction, there is motion, there is diversity." Reality was thus defined in terms of activity and process. Cosmic unity was maintained through the coming together and dissolution of opposites. The source of activity in nature was the universal spirit, the immanent activity of God within nature.

Bruno's dialectic stressed the unity rather than the struggle of opposites, anticipating idealist rather than materialist dialectics.[23]

He emphasized the harmony of the whole, pointing out that an organic whole is always more than the sum of its parts. His plurality of worlds within the infinite universe formed a living whole. "It is not reasonable," he wrote, "to believe that any part of the world is without soul life, sensation, and organic structure."[24] In his claim for the existence of innumerable other worlds, Bruno assigned no prime position to the human species and held that nature was everywhere uniform. "The ruler of our earth is not man, but the sun, with the life which breathes in common through the universe." In questioning the uniqueness of the earth-sun system, in emphasizing change, and in unifying the Neoplatonic soul and spirit into a single active principle, Bruno challenged the hierarchical conception of the cosmos.

In the final phase of his philosophy, Bruno focused on individual active substances or minimal units in nature. These soul-driven atoms or monads of different degrees applied not only to corpuscles of matter, but to planetary systems, the world soul, God, and the universe as a whole. The monads of one degree could include those of another degree within them, and all were parts of the same underlying substance.[25]

The distinctive feature of the naturalist philosophy was the dialectical process as the key to both the organic unity of nature and its immanent self-motion. Nature was a constantly growing, changing, and evolving organism. Naturalism thus postulated a more radical interpretation of change than Neoplatonism and more strongly reflected the breakup of the hierarchical social order and the movement to question the received authority of Aristotle.

Toward the end of the sixteenth century, the desire for a political overthrow of Spanish tyranny in Naples and emerging class conflicts contributed to the interest in change developed by Neapolitan intellectuals. The philosophy of the naturalists arose during troubled changing times in the province of Calabria in Southern Italy. Telesio, Campanella, and Bruno were all products of Naples and its surrounding towns during the era of Spanish control. A period of economic insecurity, popular political movements, and insurrections against Spanish oppression created an atmosphere conducive to the development and acceptance of the radical theories of nature espoused by the naturalists. According to historian Rosario Villari,

Hunger riots, appeals to the example of Flanders, religious anxiety, banditry, messianic visions, free thinking, and libertinism—all these are expressions of an extraordinary tension during these years. They are not individual aspects of a single historical process consciously directed toward a precise goal. But they do have something in common: their radical rejection of [the entire established order] and their reflection of a feeling of helplessness among the 'oppressed.'[26]

In the latter decades of the sixteenth century, the availability of jobs in Naples had decreased, wages had declined markedly with respect to prices, and from mid-century on all vagabonds had been classified as delinquents. Reforms introduced by the Spanish viceroy in 1548 had already reduced the accustomed participation of artisans and plebians in city affairs. Rentiers, grain merchants, and farmers of the privileged middle class were gradually being assimilated into the aristocracy that ruled the city.

During the earlier years of this period of oppression and unrest, Bruno was in Naples, living at first with his uncle and then, from 1563 to 1576, studying in the Cloister of St. Dominic and visiting libraries in neighboring monasteries. Telesio, who published three editions of his *De Rerum Natura* between 1565 and 1586, had begun his lecturing career at Naples and eventually founded his Academy in Cosenza, the city of his birth, south of the provincial capital. Here his system of naturalism was studied and expanded.

Under the rule of the Spanish viceroys, matters in Naples grew steadily worse. An increase in the price of bread in 1585 brought about by an alliance between the city's ruling nobility and the provincial grain producers touched off an insurrection. The organizer was Giovanni Leonardo Pisano, a pharmacist and the teacher of Della Porta. People who met in Pisano's "laboratory" "engaged in various conversations concerning the viceroy," and Pisano himself was associated with the political social circles around the "naturalists."

From 1585 until the 1599 revolt planned and executed by Campanella, the entire area was in a state of restless upheaval, tension, and agitation. Campanella prepared for his attempt to overthrow Spanish tyranny by lecturing to the people of the region on his vision of a new society based on the spiritual immanence of God within nature, communal ownership of property, and rule by brotherhood. The revolt failed when, as noted earlier, Campanella was arrested and thrown into prison in 1599. In 1600, Bruno was burned at the stake in Rome by the Inquisition for his challenges to the medieval earth-centered cosmology and for his assertion that other sun-planet systems existed.

Like the idealist dialectic of the gnostics, Shakespeare's Cordelia, the alchemists, and the millenarian utopists, naturalism looked forward toward a more egalitarian society based on the leveling of hierarchies. Mechanism would reject this radical dialectical vision of change as internal to the cosmos, the body, and the body politic by substituting external forces and a new set of cosmic and social hierarchies for the old. At the same time, the mechanists would assimilate and transform ideas such as Campanella's and Bruno's manipulative magic into a philosophy of power over nature and would ultimately accept the Copernican sun-centered cosmos, the infinite universe, other sun-planet systems, and the uniformity of space.

VITALISM. The most radical analysis of activity in nature was put forward by Paracelsus and later refined by Jean Baptiste Van Helmont (1577–1644), his son Francis Mercury Van Helmont, and Anne Conway. In this theory, matter and spirit are unified into a single, active vital substance. Paracelsus' cosmos was infused by Neoplatonic, gnostic, Stoic, and Christian ideas, yet his philosophy of matter and activity was a monistic idealism. Here the term *vitalism* designates the unity of matter and spirit as a self-active entity, in which the spiritual kernel is considered the real substance and the material "cover" a mere phenomenon.

Paracelsus' theory of the four elements as active entities rather than passive substrata was expounded in his *Archidoxis* (published in 1570). Although the four elements might all exist in a given object, only one of them attained perfection as the "ruling power" of that object, growing yet remaining invisible within it. The other three were so imperfect as not to warrant being called *elements*, in the true sense of "active substances." The observed individual object was merely a cover for the real immanent active soul.[27]

The theory of the elements was elaborated in a treatise attributed to Paracelsus and published posthumously in 1564, entitled *The Philosophy Addressed to the Athenians*. Although some have questioned the authenticity of this treatise on purely textual grounds, the doctrine itself is regarded as genuine and was accepted as such by the generation following Paracelsus. Here the four elements were essentially spiritual self-active forces with a self-determining principle *(archeus)* guiding their unfolding lives through time. The observable sensible elements and material objects were merely gross manifestations of the subtle soul that was the element itself. By a cosmic separation, the elements were generated from the uncreated *mysterium magnum*—the great mystery or "first mother of all creatures"—and folded back into it at the end of created time.[28]

Each element formed a world of its own, and each of the four worlds developed and evolved independently of the others through a consensus of actions. The elements did not mix in composition, but existed simultaneously and independently in each individual object. The predominant element in the object determined the world to which it belonged and became its guiding kernel or soul.

Each element was a matrix or mother of one of the four worlds emanating from it. Thus, from water, a unique world was created: fish of all forms and kinds; fleshy animals; marine plants such as corals, trina, and citrones; marine monsters; the elementals (nymphs, sirens, dramas, lorinds, and nesder); and stones such as beryl, crystal, amythyst, and amber. New growths were continually being produced as the separation was perfected. Waters existed as separate kinds rather than degrees—springs, streams, rivers, seas—none precisely like the others.

From the terrestrial separation sprang a second world—the metallic minerals, gems of manifold forms, stones, sands, and chalk; fruits, flowers, herbs, and seeds; sensible animals, and men—the partakers of eternity. The terrestrial separation included the earthy elementals: gnomes, sylvesters, lemurs, and giants.

Air, like the other elements, generated only things of its own kind—invisible and impalpable according to the principle that like produces like. Aerial creatures like witches had aerial speech, thoughts, and actions but ultimately returned to the element air.

The fourth element, fire, put forth stars, celestial objects, and the sun as its daughters, together with its own floral growths and mineral products. As an element, fire was responsible for an object's growth. In green wood, fire existed as an elemental soul producing growth, whereas in burning fire it appeared as a living eternal soul.

"Whatever grows is of the element fire, but in another shape. Whatever is fixed is from the element earth. Whatever nourishes is from the element air; and whatever consumes is from the element water."[29]

The harmonious unfolding of the four worlds formed from the four elements operated on the principle of mutual consensus. In this sense, all four elements derived strength and nourishment from harmony with the others or conversely were weakened by the "errors" of the others.

Since all the four Paracelsian elemental worlds contained both rational and irrational creatures, mankind, in the *Philosophia ad Athenienses,* was only one among the other beings of the natural world. This text presented a view that people existed within nature, in harmony with the whole rather than above it: "That philosophy then is foolish and vain which leads us to assign all happiness and eternity to our element alone, that is, the earth, and that is a fool's maxim which boasts that we are the noblest of creatures. There are many worlds and we are not the only beings in our world."[30]

Paracelsian epistemology, as a reflection of holistic cosmology, was based on the power of the imagination, as the link between body and spirit. As microcosms, human beings were miniature replicas of the greater macrocosm; both were composed of a physical body, a soul that was the life and breath obtained from God, and, uniting the two, a sidereal or astral spirit. The astral spirit came initially from the stars and was the source of divinity in all sublunar life, including animals, plants, minerals, and stones. In human beings, the astral spirit was located in the heart, circulated throughout the body, and formed the imaginative faculty joining the physical with the spiritual world.[31]

The imagination, as the sum of the astra within each individual, bestowed the power to create visible images of the astra through the application of art. It was the source of an inner knowledge of the divine plan, vital action in the mundane world, and was so powerful that it could reciprocally affect the heavens whence the astra initially derived. But because the astra acted out of necessity, reason must be used to control their power so that nature could be used purposively.

In the organic world, magicians, metallurgists, and healers viewed themselves as the servants of nature, assisting, mimicing, and perfecting natural processes through art *(technē)* for human benefit. Thus Paracelsus wrote in his *Credo* concerning healing: "There is nothing in me except the will to discover the best that medicine can do, the best there is in nature, the best that the nature of the earth truly intends for the sick. Thus I say, nothing comes from me; everything comes from nature of which I too am part." But elsewhere he wrote that nature exists for human use: "It is God's will that nothing remain unknown to man as he walks in the light of nature; for all things belonging to nature exist for the sake of man."[32] Viewed from within the organic context, there is nothing inconsistent in the two statements, yet when the organic bond between nature and human beings is severed, nature becomes a passive object rather than an active partner.

With his philosophical theories and medical practices, Paracelsus challenged the orthodoxy of the establishment. He advocated the freedom of ordinary people to study nature for themselves and believed in a self-active natural world and individual liberty. The bulk of his knowledge was obtained from lay people; his sources were women healers, barbers, bathkeepers, miners, and his own empirical observations, in addition to those of learned physicians; he visited universities only as a wandering journeyman scholar. He belonged to the guild of grain merchants rather than that of doctors and acted as an army surgeon, a position unacceptable to academic physicians. His life became a series of clashes with orthodox authorities and ruling officials. In city after city, he was able to cure princes and public officials pronounced incapable of recovery by the orthodox physicians. In each case, his success and fame procured him friends among the wealthy and influential town leaders. But soon his opinions and actions would alienate them and he would be forced to leave town secretly in order to escape arrest.[33]

Incensed by enslaving traditions and moved by the poverty and misery of the people, he identified with the peasants as he moved from spas to mines to towns attracting crowds and healing the sick. Often he refused to take money from the poor and sick, while at the same time he was again and again cheated by the rich. His violent reactions against these injustices would then make it necessary for him to leave the town.

Paracelsus' medical and chemical contributions became a stimulus to numerous sixteenth- and seventeenth-century followers working toward a new empirical methodology advocating the direct study of nature itself, instead of the books of the ancient Greeks. But his philosophy of the activity of nature became affiliated with revolutionary pantheistic and political ideas that surfaced in later neo-Paracelsist movements.

The animistic concept of nature as a divine, self-active organism came to be associated with atheistical and radical libertarian ideas. Social chaos, peasant uprisings, and rebellions could be fed by the assumption that individuals could understand the nature of the world for themselves and could manipulate its spirits by magic.[34] A widespread use of popular magic to control these spirits existed at all levels of society, but particularly among the lower classes. The raising of spirits, the construction of magical apparatus, the manufacture of talismans and charms, the preparation of love potions, the exorcism of demons, fortune telling, and hunts for treasures, lost relatives, and lovers were operations directed or performed by the village wizard. Every European village had its popular healers, sorcerers, and magicians, whose magical procedures had resulted from years of transmitting verbal recipes and cures handed down from medieval and ancient times. Articulation of the presuppositions of magical theory may be attributed to the natural magic tradition, but the practice of magic itself was an ancient folk tradition.

The natural philosophy of Jean Baptiste Van Helmont, the seventeenth-century follower of Paracelsus, likewise emphasized the activity of matter. He called the dynamic principle the *archeus*, the organizer of specificity within matter. It originated ultimately from the divine light and organized the living developing seeds of matter—the *semina*. Van Helmont transformed Paracelsus' four elements into a plurality of seeds each containing its own inner activity: "In the whole order of natural things nothing of new doth arise which may not take its beginning out of the seed, and nothing to be made which may not be made out of the necessity of the seed."[35]

He strongly took issue with Aristotle's doctrine of the four causes and held that two causes "joined or knit together" were sufficient to explain nature. The two unified principles were the efficient cause, or *archeus* (the chief workman), and the material cause, or plurality of seeds. "Wherefore, after a diligent searching, I have not found any dependence of a natural body but only on two causes, on the matter and the efficient, to wit, inward ones." The two were joined into a single unit, the generating seed. The efficient cause was the inward agent of the *semina*— the moving principle or immanent active principle in the material seed.

Thus Aristotle erred in stating that "the thing generating cannot be a part of the thing generated." For Van Helmont, the inward agent was the generative principle; nothing new arose in nature that did not begin from the efficient power in the material seed.

The inward agent was also called the *chief* or *master workman*. In vegetables, it was a juice; in metals, it was thicker and homogeneous with the material; in other things, it was an "air." It was the inward spiritual kernel of the seed through which the seed developed and grew to full fruitfulness. "The *archeus,* the workman and governor of generation, doth clothe himself . . . with a bodily clothing . . . and begins to transform matter according to the perfect act of his own image."

The formal cause postulated by Aristotle, said Van Helmont, was not really a cause but an effect, the end of generation. For Aristotle, the more perfectly the form was attained, the more actuality the object had achieved. For Van Helmont, the actuality or activity was the inner agent within the matter. Form and matter could not be separated from one another. Secondly, the efficient or moving cause could not be external to matter, but must be within it. Thirdly, the final or teleological cause, which contained the "instructions of things to be done," could not be external to nature nor should it be considered distinct from the efficient cause, the *archeus*. But, although the efficient seminal cause was postulated to be within nature, Van Helmont admitted that external occasional causes could function as outward awakeners. These outward agents operated according to art.

Matter and the efficient cause were therefore sufficient as explanatory entitles fused together into one unit. "Every natural definition is to be fetched . . . from the conjoining of both causes, because both together do finish the whole essence of the thing."

The activity within each particle of matter meant that activity and process were primary in the order of things. Within the matter of the world, there existed an inherent spontaneity, and within each individual being an inner-directedness determined its destiny.

During the upheaval of the English Civil War and its aftermath in the 1640s and 1650s, the ideas of Paracelsus and Van Helmont were revived in the radical doctrines of the dissenting sects. The breakdown of censorship permitted the translation and circulation of Paracelsian and Helmontian texts. Prior to the Civil War, the religious and political dimensions of mysticism, magic, vitalism, and pantheism had not been emphasized in England, but in the 1640s and 1650s Paracelsian challenges to authority, orthodoxy, and rationalism were held up as an alternative to both the Aristotelian philosophy taught in the universities and the new mechanical philosophy being discussed at Oxford.[36]

The pantheistic and Paracelsian ideas of the sects were part of a radical social philosophy that for some groups meant the seizing of common lands and the establishment of egalitarian communal societies like those attempted by the medieval millenarian utopists. Gerrard Winstanley, organizer of the Diggers, who in 1649 took possession of St. George Hill in Surrey and began to cultivate the commons and wastes, believed that by working together the poor could make the earth "a common treasury" for all. "Let everyone quietly have earth to manure," he wrote. The Diggers advocated cooperative farming of the commons and intensive cultivation of crops. When "the great creator, Reason, made the earth to be a common treasury, to preserve beasts, birds, fishes and man . . . not one word was spoken . . . that one branch of mankind should rule over another."[37]

Other sects advocated that every person should have "an equal share of earthly goods," as in accord with the law of nature and of God.[38] Until "the world returns to its first simplicity or . . . to a Christian utopia . . . covetousness will be the root of all evil." George Foster, member of a pantheistic religious sect known as the Ranters, believed that "the whole earth will be a treasury for all and not for some" (1650). God was a "mighty Leveller," who would reduce all political powers to the same level and return everything to communal ownership.

These ideas were set within a vitalistic philosophical framework which presupposed that nature was active, filled with God, and therefore good. Gerrard Winstanley wrote, "To know the secrets of nature is to know the works of God . . . how the spirit or power of wisdom and life, causing motion or growth, dwells within and governs both the several bodies of the stars and planets in the heavens above; and the several bodies of the earth below, as grass, plants, fishes, beasts, birds, and mankind." An early Ranter belief held that "every creature is God, every creature that hath life and breath being an efflux from God, and shall return into God again, be swallowed up in him as a drop is in the ocean." Jacob Bauthumley, a Ranter writing in 1650, agreed that "God is in everyone and every living thing, man and beast, fish and fowl, and every green thing, from the highest cedar to the ivy on the wall. . . . God is in this dog, this tobacco pipe, he is me and I am him." Richard Coppin, whose *Divine Teachings* (1649) influenced Ranter beliefs, held that "God is all in one and so is in everyone."

The sects pushed their radical ideas even further by undermining the basis for the patriarchal family. For the Familists, members of the Family of Love, a pantheistic sect emphasizing tenderness and quiet sympathy, founded in 1580, marriage was changed from a sacrament to a contract, and divorce became simply a matter of dissolution before the congregation. The Diggers advocated marriage based solely on love regardless of station or property. The Ranters challenged monogamy and believed that wives could be held in common. At Ranter and Quaker meetings, stripping to nakedness in church as a symbol of resurrection was not uncommon, and sexual freedom for both men and women was broached.[39]

Through membership in the new sects, women enjoyed significantly more religious freedom than in either the Church of England or the Catholic Church. The Puritans had elevated the female to helpmeet in the family, assigned her the task of instructing the family in religious matters, and argued for an end to wife beating, though women were still considered subordinate partners in the marriage relationship. In the sects, however, equality before God was stressed; women could be admitted to membership without their husbands and were free to preach, prophesy, and participate in governance. In many congregations, women outnumbered men, and in London women preaching on Sundays became common. Those women who joined the new religious sects were undoubtedly attracted by the freedoms and equalities found there.[40] (Such freedoms assumed by women were especially pronounced in Quakerism, a sect that claimed the allegiance of Francis Mercury Van Helmont and Anne Conway, who continued the vitalistic philosophy of the fusion of spirit and matter. See Chapter 11.) After the Restoration of 1660, with the reassertion of Anglican authority, Paracelsian and pantheistic ideas were denounced and refuted.

Similarly, in seventeenth-century France, Robert Fludd (1574–1637), a follower of Paracelsus and the naturalists, became identified with the "Rosicrucian scare" (a neo-Paracelsist movement), resulting in an examination and denunciation of his entire philosophy by the mechanists—Marin Mersenne, Pierre Gassendi, and René Descartes. France in the 1620s–40s and England in the 1650s–80s were at the center of the mechanical reconstruction of the cosmos. In both countries, ideas associated with the organic world view and with animistic and pantheistic philosophies were severely criticized.

Fludd's philosophy presented a synthesis of ideas from the preceding Neoplatonic, naturalist, and vitalist traditions. In his Neoplatonic heirarchical cosmos, founded on the microcosm-macrocosm analogy, God infused the world with his eternal spirit, which was housed in the sun and transmitted through the angels in the four corners of the world to the winds. The winds, in turn, representing the contrary principles of hot and cold, acted through a dialectical process of contraction and expansion, conveying activity to the clouds and air. The two contrary active principles, heat and cold, produced the opposites observed in nature. The hot winds caused dilation, mollification, rarefaction, volatility, and transparency; cold winds produced contraction, hardening, condensation, fixation, and opacity. Matter for Fludd was passive; not the source of activity, it was made active through the indirect operation of the winds. The angelical winds were "endued with most essential internal agents" and had "an essential and inward act, form, and principle." The clouds likewise moved through the internal agency of the spirit and represented its vehicle. God's activity in the world was the ultimate source of these dialectical tensions between contraries and the basis of cosmic unity and animate life.[41]

From the spectrum of Renaissance organicist philosophies outlined above, the mechanists would appropriate and transform presuppositions at the conservative or hierarchical end while denouncing those associated with the more radical religious and political perspectives. The rejection and removal of organic and animistic features and the substitution of mechanically describable components would become the most significant and far-reaching effect of the Scientific Revolution.

The breakup of the old order in western Europe was not only a period of challenge—a time when a broad spectrum of new ideas found articulation—but also a period of uncertainty and anxiety.

Fear that nature would interdict her own laws, that the cosmic frame would crumble, and that chaos and anarchy would rule lay just beneath the sheen of apparent order. Fostered by the competitive practices of the new commercialism and reinforced by the religious wars of the Reformation period and the growing stress on individualism and the senses over the authority of the ancients, the perception of disintegration increased. The ecological deterioration of the earth, changing images of the cosmic organism, and a sense of disorder within the soul of nature reflected an underlying realization that the old system was dying.

CHAPTER 5

Nature as Disorder

Women and Witches

The image of nature that became important in the early modern period was that of a disorderly and chaotic realm to be subdued and controlled. Like the Mother Earth image described in Chapter 1, wild uncontrollable nature was associated with the female. The images of both nature and woman were two-sided. The virgin nymph offered peace and serenity, the earth mother nurture and fertility, but nature also brought plagues, famines, and tempests. Similarly, woman was both virgin and witch: the Renaissance courtly lover placed her on a pedestal; the inquisitor burned her at the stake. The witch, symbol of the violence of nature, raised storms, caused illness, destroyed crops, obstructed generation, and killed infants. Disorderly woman, like chaotic nature, needed to be controlled.

Concurrently, the old organic order of nature in the cosmos, society, and self was symbolically giving way to disorder through the discoveries of the "new science," the social upheavals of the Reformation, and the release of people's animal and sexual passions. In each of these three realms, female symbolism and activities were significant.

DISORDER IN NATURE. That nature's order might break down was a persistent concern of Renaissance and Elizabethan writers. If order were taken away, chaos would reign, feared the English scholar Thomas Elyot (1531). In Shakespeare's *Troilus and Cressida* (produced 1601–1602), Ulysses worried that if hierarchical gradations were removed, the whole "enterprise [would be] sick." "Take but degree away, untune that string and hark what discord follows." Richard Hooker (1594) was concerned that if nature ceased to observe her own laws, the celestial frame might dissolve, the moon wander from her orbit, the wind die out, the clouds dry up, the earth's fruits wither, and then chaos would ensue.[1]

Contributing to the sense of sickness and decay in the organic order of nature were the discoveries of the "new science." The old hierarchical structure of the macrocosm had been disrupted by the cosmology of Nicolaus Copernicus, published in 1543, which advanced the heliocentric hypothesis and challenged the Ptolemaic geocentric model of the universe. As Bernard Fontenelle perceived later in his *Plurality of Worlds* (1686), Copernicus displaced the female earth from the center of the cosmos and replaced it with the masculine sun.

> He snatches up the earth from the center of the universe, sends her packing, and places the sun in the center, to which it did more justly belong. . . . All now goes round the sun, even the earth itself; and Copernicus to punish the earth for her former laziness, makes her contribute all she can to the motion of the planets and heavens; and now deprived of all the heavenly equipage with which she was so gloriously attended, she has nothing left her but the moon, which still turns round her.[2]

Whereas in the sixteenth century Elizabeth I had been compared to the Ptolemaic *primum mobile* encircling the female earth, in the seventeenth century Louis XIV became the Sun King.

The failing plausibility of the organic model of the macrocosm, and thus the need for a new metaphor, was perceived by Johannes Kepler (1571–1630), who wrote to a friend in 1605, "My aim is to show that the celestial machine is to be likened not to a divine organism but to a clockwork." The sightings of the new star of 1572 and the comet of 1577 by Tycho Brahe had provided evidence of the corruptibility of the unchanging heavens above the moon. The telescopic observations of Galileo Galilei in 1609–10 of craters on the moon, spots on the sun, moons around Jupiter, the phases of Venus, and the multitudes of stars in the Milky Way had further strengthened the belief that the frame of nature was blemished.

In the microcosm, or human body, the hierarchical structure of the three organ systems of liver, heart, and brain with their associated vital, natural, and animal spirits, as described by the ancient physician Galen (ca. A.D. 130–200), was being thrown into doubt by the discoveries of Andreas Vesalius (1514–1564), Michael Servetus (1511–1553), Andrea Cesalpino (1519–1603), Realdo Columbus (1510?–1559), and the early work of William Harvey (1578–1657), which would culminate in 1628 with his book *On the Motion of the Heart and Blood in Animals,* describing the circulation of the blood.[3] Finally, the geocosm, or living earth, was being defiled by pollution of "its purest streams" and by the mining of its entrails for metals.

Symptomatic of the decay and sickness perceived to pervade the whole organic world were the anniversary poems on the death of the world by John Donne in 1611–12 and *The Fall of Man* written in 1616 by Godfrey Goodman, bishop of Gloucester. Donne's "Anatomy of the World" (written to commemorate the first anniversary of the death of the young Elizabeth Drury) lamented the death not only of Elizabeth's soul but of the world soul itself:

> Sicke Worlde, yea, dead, yea putrified, since shee
> Thy intrinsique balme, and thy preservative,
> Can never be renew'd, thou never live.

The dirgelike refrain of the poem, "Shee, Shee is dead; shee's dead," reiterated the state of sickness and decay permeating the organic universe: not only was man a poor and trifling thing; the world was a cripple, an ugly monster, a wan ghost and a dry cinder.[4]

Sixteenth- and seventeenth-century writers exhibit a growing tendency to view both nature and society as wilderness and to advocate control over the forces and fates dealt by nature. For Niccolò Ma-

129

chiavelli (1469–1527) society appeared as a wilderness, and fortune, like nature, was unpredictable, violent, and must be subdued.

> Fortune is the ruler of half our actions. . . . I would compare her to an impetuous river that when turbulent, inundates the plains, casts down trees and buildings, removes earth from this side and places it on the other; everyone flees before it and everything yields to its fury without being able to oppose it; and yet though it is of such a kind, still when it is quiet, men can make provision against it by dikes and banks, so that when it rises it will either go down a canal or its rush will not be so wild and dangerous. So it is with fortune which shows her power where no measures have been taken to resist her, and directs her fury where she knows that no dikes or barriers have been made to hold her.[5]

But the potential violence of fortune could be mastered by aggression,

> for fortune is a woman and it is necessary if you wish to master her, to conquer her by force; and it can be seen that she lets herself be overcome by the bold rather than by those who proceed coldly, and therefore like a woman, she is always a friend to the young because they are less cautious, fiercer, and master her with greater audacity.

By adopting animal-like cunning and manipulative methods, a prince could meet the enemy on its own terms. "A prince being thus obliged to know well how to act as a beast must imitate the fox and the lion, for the lion cannot protect himself from traps, and the fox cannot defend himself from wolves. One must therefore be a fox to recognize traps, and a lion to frighten wolves."

In *The Tempest,* William Shakespeare recalls the primal forces of the original island desert before it is transformed, through the manipulative magic of Prospero, into a civilized land. The violence of nature contrasts with the ease of life made possible through human skill and ingenuity. If the forces of disorder could be harnessed, the tempest itself could be created and manipulated—a lesson taught by such Renaissance handbooks on natural magic as those of Agrippa and Della Porta. Neither physical nature nor human nature should be allowed to go untamed like the bestial nature of the cannibal Caliban. True civilization can be achieved only through the full exercise of power and control in both the physical and mental realms.

Nature and society appeared as a wilderness in picaresque literature, where the adventures of roguish heros who struggled for survival in a hostile environment were portrayed. The individual, weak and alone, was pitted against the forces of nature. The solution was not to escape to pastoral simplicity but to engage in an individualistic, competitive encounter with others for the sake of advancement and survival. Lazarillo, in *The Life of Lazarillo de Tormes* (1554), survives by taking advantage of others whenever possible and avoiding destructive circumstances in order to stay alive. Hans Jakob von Grimmelhausen's (1625–1676) picaro, Simplicius, adopts the role of a talking animal in order to deal with changing hostile circumstances in the social wilderness rampant during the Thirty Years' War, (1618–1648).[6]

Nature as a wilderness, an important element in the Judeo-Christian tradition, had been central to Old Testament interpretations of the desert. The inhospitable arid wilderness contrasted sharply with the bountiful, fruitful Garden of Eden and with the promised land of milk and honey. The expulsion from the Garden into the wilderness equated the wilderness with the evil introduced when Eve submitted to the temptation of the serpent. The desert represented a land to be subdued and made arable, a land whose fertility was tied directly to the amount of rainfall. In contrast to the Greek tradition, which tended to emphasize the benevolence of nature, the Judaic tradition viewed it as a condition to be overcome and subdued. While softer interpretations could be found in the books of the Old Testament, it was the perception of nature as a wilderness that became important in the early modern era. For Protestants such as John Locke, John Calvin, and the New England Puritans, God had authorized human dominion over the earth.[7]

Tales of wilderness in European and Anglo-Saxon folklore were dramatized by fifteenth- and sixteenth-century explorations of the New World. Voyagers brought back reports of wild, desolate, chaotic lands hostile to human settlement. The savages of the new lands became symbols of the wildness and animality that could gain the upper hand in human nature. Although many of the early accounts by explorers had reported Indians to be peaceful and loving, albeit at a hierarchical level somewhere between beasts and Europeans, by the early seventeenth century that image had hardened and calloused. At first, Indians were described variously as "courteous, gentle of disposition," "civil and merry . . . a people of exceeding good invention, quick understanding, and ready capacity." But other early descriptions reported that they "lived like wild beasts without religion, nor government, nor town, nor houses, without cultivating the land, nor clothing their bodies." People had been found "living yet as the first men, without letters, without laws, without kings, without commonwealths, without arts." They were "not civil by nature nor governed by discipline . . . living without houses, towns, cities."

After the Virginia massacre of 1622, the positive descriptions disappeared as the English image of Indians changed in a negative direction. Indians became wild, savage, slothful, and brutish outlaws. They had "little of humanity," were "ignorant of civility, of arts, of religion," and were "more brutish than the beasts they hunt."[8] These animal passions believed to govern the Indian and to be present in all human beings were also symbolized by the lust of women and the disorder wrought by the witch.

DISORDER, SEXUALITY, AND THE WITCH. Symbolically associated with unruly nature was the dark side of woman. Although the Renaissance Platonic lover had embodied her with true beauty and the good, and the Virgin Mary had been worshipped as mother of the Savior, women were also seen as closer to nature than men, subordinate in the social hierarchy to the men of their class, and imbued with a far greater sexual passion. The upheavals of the Reformation and the witch trials of the sixteenth century heightened these perceptions. Like wild chaotic nature, women needed to be subdued and kept in their place.

Robert Burton in his *Anatomy of Melancholy* (1621) played up the power of lustful love in both young and old women: "Worse it is in women than in men; when she is . . . an old widow, a mother long since . . . she doth very unseemly seek to marry; yet whilst she is so old a crone . . . a mere carcass, a witch, and scarce feel, she caterwauls and must have a stallion, a champion; she must and will marry again, and betroth herself to some young man."[9] Young women, he said, sought sex as soon as puberty was reached: "Generally women begin . . . at fourteen years old, then they do offer themselves and some plainly rage. . . . In Africa a man shall scarce find a maid at fourteen, and many among us after they come into the teens do not live without husbands." "Mother and daughter sometimes dote on the same man; father and son, master and servant on one woman." *Aristotle's Masterpiece* (first edition 1684), the popular handbook on sex used in the eighteenth century, still asserted the earlier common belief that women obtained more pleasure from sex than men. Females were seen to be so eager for sex after the age of fourteen that, "they care not how soon they are honestly rid of [their virginity]." During the "controversy over women" which raged in England in the late sixteenth and early seventeenth centuries, the greed and lust of tradesmen's wives were satirized. Joseph Swetnam's antifeminist attack of 1615 reflected the presumed lust of women in the title, "The Arraignment of Lewd, Idle, Forward, and Unconstant Women." In vindication, a play called "Swetnam, the Woman'Hater, Arraigned by Women" portrayed him muzzled and dragged through the streets after he had been convicted in court for female abuse.[10]

The blame for the bodily corruption of the male was attributed directly to lust and temptation by the female in the popular Renaissance belief that each completed male sexual act, or "little death," shortened the life by one day. Donne drew on this image in "The Anatomy of the World" when he described Eve's temptation of Adam and the consequences for ensuing generations of male lovers as they enter into the reproductive act:

> For that marriage was our funerall:
> One woman at one blow, then kill'd us all.
> And singly, one by one, they kill us now.
> We doe delightfully our selves allow
> To that consumption; and profusely blinde,
> Wee kill our selves to propagate our kinde.[11]

In works of art, women of the sixteenth and seventeenth centuries were depicted as disorderly and insolent, beating and tricking their husbands, drinking wine, and lustfully dragging them off to bed. The title page of *The Decyte of Women* (anonymously published in 1560) depicted a woman riding the back of her husband while beating him onward with a whip. A domestic quarrel showed an old woman about to beat over the head a young man fighting with his wife. The woodcut "Aristotle and Phyllis" (1513), by the German artist Hans Baldung Grien, featured Aristotle, who had assigned a passive role to the female, being ridden and driven like a horse as Phyllis exhibited the power of women (Fig. 11). The Swiss engraver Urs Graf depicted (in 1521) two women beating up a monk who had probably molested them (Fig. 12). Pieter Bruegel (1525?– 1569), Flemish painter, used in his "Dulle Griet" a woman, Mad Meg, to symbolize violence and chaos, while beside her St. Margaret tied up the devil. In village festivals, floats featured wives hitting their husbands, throwing stones and garbage at them and blasting them with insults.[12]

The supposed disorder wrought by lusty women was apparent in the hundreds of paintings and graphics on witchcraft. Based on a fully articulated doctrine emerging at the end of the fifteenth century in the antifeminist tract *Malleus Maleficarum* (1486), or *Hammer of the Witches,* by the German Dominicans Heinrich Institor and Jacob Sprenger, and in a series of art works by Hans Baldung Grien and Albrecht Dürer, witch trials for the next two hundred years threatened the lives of women all over Europe, especially in the lands of the Holy Roman Empire. Extant representations of the witches' Sabbath present an image of widespread chaos and uncontrolled nature dominated by women engaged in exuberant, frenzied activity. A 1610 print by the Polish engraver I. Ziarnko "palpitates with headlong movement and a restless turbulence, and the personages are swept along in a wild infernal dance by the feverish stir of all this frantic and disordered tumult." It depicted Satan in the form of a goat preaching to his congregation flanked by two witches. Wild, "indecent, and obscene" stamping and dancing of the witches with their demon consorts followed the meal, accompanied by musicians playing on recorders, harps, theorbos, and arched bows, while other witches prepared poisons in a cauldron from dismembered snakes and toads to be used in the killing of men and cattle. Hans Baldung Grien's "The Witches" (1510) illustrates three witches' revenge on society as they brew potions at the sabbath amid a chaos of clouds of smoke, animal familiars, and a witch arriving on a devil goat (Fig. 13).[13]

The supposed sexual lust of women provided one of the bases on which women were accused and tried for witchcraft. Twelve editions of Henri Boguet's *Discours des Sorciers* (ca. 1590), describing the trials of six women accused of sexual acts with the devil, appeared in a span of twenty years. In a chapter on "The Copulation

Figure 11.   Aristotle and Phyllis, woodcut by Hans Baldung Grien (1513). Aristotle, who held that the male was superior to the female, is shown subjected to the power of women in this popular sixteenth century illustration. Phyllis, according to the legend, had sought revenge on Aristotle after he accused her husband, Alexander the Great, of devoting too much attention to her. Phyllis, after promising Aristotle her favors, insisted he first give her a ride—a victory witnessed by Alexander.

Figure 12. Two women beating up a monk who has probably molested them, pen drawing by Urs Graf (ca. 1521). In the Renaissance, women were portrayed as capable of violence, revenge, and self-defense.

of the Devil with Male and Female Witches," Boguet, who was chief justice of St. Claude, France, explained, "The Devil uses them so because he knows that women love carnal pleasures, and he means to bind them to his allegiance by such agreeable provocations."[14] The devil was thought to appear to men in the form of a woman, or *succubus,* and to women as a man, or *incubus.* He worked by taking the semen from a man while disguised as a *succubus* and then delivered it to the woman in the form of an *incubus.* According to Boguet, the body of the devil assumed during the act

Figure 13. The Witches, woodcut by Hans Baldung Grien (1510). Three witches brew potions at a sabbath meeting, while a fourth arrives on a goat. The goat and cat were associated with women and witchcraft because of their presumed sexual lust and slyness.

was made from dense, palpable air, so that he would seem physically real.

Witches, reported Boguet, copulated with the devil at home in their beds and at the Sabbath and were often discovered in woods and fields engaged in such acts. He, like others, maintained that they were sucked by their familiars, a form in which Satan often appeared to them. Once imprisoned, they were stripped of their clothing, searched for the marks of familiars on their body, and examined for signs of intercourse with the devil. At the trials, women were made to confess that intercourse with the devil was painful because of the ugly form in which the devil appeared and because of the burning in the stomach felt from his "icy cold member." In England, several hundred women identified as having witch marks, often within the *labia majora,* were put to death in the years 1644–45 by Matthew Hopkins, an English lawyer whose campaign to exterminate witches earned him the title "the witch-finder general."

The control and the maintenance of the social order and women's place within it was one of the many complex and varied reasons for the witch trials. Women comprised the bulk of witchcraft prosecutions: Jean Bodin in 1580 set the ratio at fifty women to one man; James I in his *Daemonologie* (1597) placed it at twenty to one; Alexander Roberts (1616) believed it to be one hundred to one. A calculation of English Home Circuit Court executions showed that 102 out of 109 persons put to death were women. In 1585, two villages in the Bishopric of Trier in Germany were each left with only one female inhabitant. Combined modern statistics for several European countries indicate that of the total tried (some 100,000), women comprised approximately eighty-three percent. The most famous sexist denunciation of witches, the *Malleus Maleficarum,* reprinted fourteen times before 1521 and another fifteen times after 1576, used the feminine form *Maleficarum* rather than the masculine *Maleficorum* in its title. Not only were the majority of accused persons women, the victims were primarily people in the lowest social orders, even though witchcraft beliefs were popular on all levels of society (Fig. 14). Of 600 witches accused in England on the Home Circuit, 596 were tradespeople, husbanders, laborers, and their wives. Religious, social, and sexual attitudes toward women and their role in contemporary society played a significant part in delineating the victims.[15]

Figure 14.   The Four Witches, engraving by Albrecht Dürer (1497).
The four witches are from different social classes, as shown by their
head attire.

The reality and extent of witchcraft practices behind the perceptions of the inquisitors is unclear. Available sources were written primarily by the accusers and defenders of witches. The bulk of those accused were illiterate and did not leave a written record of their actual beliefs and activities; those who did write accounts defended their innocence. But witchcraft was a method of revenge and control that could be used by persons both physically and socially powerless in a world believed by nearly everyone to be animate and organismic.

The view of nature associated with witchcraft beliefs was personal animism. The world of the witches was antihierarchical and everywhere infused with spirits. Every natural object, every animal, every tree contained a spirit whom the witch could summon, utilize, or commune with at will. Witches were thought to make individual pacts with the devil or a demon, usually through an animal familiar. They did not depend on the complicated hierarchical mechanisms of *pneuma,* or *spiritus,* to draw down celestial influences, as did the Neoplatonic magician. Nor did they depend on hierarchies of demons, as did a magus such as Agrippa. The immediacy of individual relationships with a spirit or demon and the possibility of revenge and control may account for the popularity of witchcraft among oppressed women. No hierarchies stood between the witch and the object of her will.

Through the power of the spells that summoned spirits, witches could control the forces of nature; they could make hail or rain, destroy crops, and bring plagues. They could take revenge on those whom they disliked, attract those whom they liked, or turn themselves or anyone else into ugly creatures. Illiterate women at the bottom of the social order had little other means of control or defense against the repression and injustices of hierarchical society. The release of passion and violence at the devil's Sabbath symbolized the witch's alienation from a world that offered her little human comfort or hope of salvation.[16]

"SCIENCE" AND THE WITCH.   "Science," in the form of naturalistic explanations of witchcraft, brought forth in defense of the accused during the 1560s, has been judged to be the only sane, "objective" viewpoint in an age blinded by superstition. But it has not been generally recognized that those who defended women against accusation and trial did so by utilizing the prevailing antifeminist arguments.

An early defense of witches was made by the physician Johann Weyer in his *De Praestigiis Daemonum* (1563). Despite its attempts to exonerate the behavior of witches through natural explanations (based on an excess of melancholic humor), Weyer's book reflected attitudes toward women inherent in the prevailing natural philosophy. Witches are women "who because of their sex are inconstant and of dubious faith, and because of their age incapable of clear thought. They are especially vulnerable to the devil's wiles." Thus old women confess to crimes really done by the devil "while their minds are wounded, troubled, and disturbed by phantoms and apparitions in brains already addled by melancholy or by its vapors." [17]

Weyer believed that demons actually existed, and that witchcraft could be explained as the diabolic delusions they produced. But it was his view of women as the weaker and more credulous sex that caused him to argue against corporal punishment of witches. Women were passively victimized by the devil because they were female, silly, and senile. However, male magicians acted voluntarily in making pacts with the devil and could therefore be punished.

Weyer expanded his views in his *Histoires Disputes et Discours* (1579). Because of the fragility of their sex, women were heedless in their beliefs, impatient, melancholic, and unable to control their emotions. Old women were particularly stupid and mentally debilitated. It was thus easy and natural for the female to be deceived by the devil, as was Eve originally. Weyer cited a long list of biblical and ancient authorities who agreed with this point of view: St. Peter, St. Matthew, Quintillion, and so on. Aristotle and his medieval translator, Albertus Magnus (1206?–1280), both maintained that women were more easily led into deception and despair than men. As a final piece of evidence, Weyer produced a verse of a hymn to Pallas Athena, the goddess of wisdom, proclaiming that she was born solely of a father, without a mother, and therefore, unlike women, could possess wisdom and foresight. This was testimony to the Aristotelian sexual theory that intelligence sprang from the father, whereas the mother supplied only the matter. Women, fragile in mind and spirit, could thus easily be depraved in sense by the melancholic humors that burdened their brain. [18]

Refutations of Weyer's defense of witches came from the legal and medical faculties—Jean Bodin (1530–1596), jurist, and Thomas Erastus (1524–1583), professor of medicine at Heidelberg. Bodin, lawyer at the French court, and advocate of witchcraft persecutions, was particularly unsympathetic toward the female sex. His 1680 refutation of Weyer's naturalistic explanations of witchcraft attempted to refute the argument from melancholy. Woman's humor, he said, was contrary to the bile or melancholic juice from which true adult melancholy proceeded. Melancholy was the result of excess heat and dryness, whereas women were naturally cold and wet:

> It is . . . gross ignorance to attribute melancholic sickness to women, which suit them as little as do the praiseworthy effects of a temperate melancholic humor, which makes a man wise, serious or contemplative. . . . All these qualities are as little compatible with woman as fire is with water. So abandon the fanatical error of those who make women into melancholics.[19]

Erastus (1578) accepted the concept of the melancholic imagination but argued that not all witches were melancholic. They were not passive instruments used by the devil, but active instigators of evil magic.

Reginald Scot, a second defender of witches, maintained in his *Discoverie of Witchcraft* (1584) that the melancholy that affected witches attacked their brains and depraved their senses and judgment. If a witch's mind and will had not become corrupted and confused, she would never voluntarily make confessions of such deeds. A woman so weakened in body and brain could then imagine herself to be a witch who could do strange, impossible things. This was particularly likely to happen after menopause, "upon the stopping of their monthly melancholic flux or issue of blood."[20]

The fears, superstitions, and imagined illusions occurred when the melancholic humor in the dregs of the blood affected the brain. The dreams suffered by witches convincing them of fantastical powers and visions were inward actions of the vapors and humors. Dreams differed in type according to whether they were produced by choler, phlegm, melancholy, or blood.

The *incubus* who appeared to women in the night was really a disease of the body, Scot argued. A thick vapor generated from the rawness of the gut moved up into the head, burdening and oppressing the brain with such power that those stricken in the night were unable to call for help. Wakening the muttering person or turning the individual over might effect a remedy.

It was commonly believed, said Scot, that every month women were filled with superfluous humors from which the melancholic blood could boil up carrying vapors to the nose and mouth, bewitching anyone meeting them. Women were naturally more fickle and intemperate than men. The difficulty of moderation often resulted in unbridled fury infecting and bewitching those who had opposed them.

A seventeenth-century authority on melancholy, Robert Burton, stated that students and those who were "solitary by nature," leading a life of contemplation, tended to be the most susceptible to melancholy. Men were affected more often, but when women obtained the disease they were "far more violent and grievously troubled."[21]

In contrast, feminist defenders had turned such arguments upside down. Human temperaments were governed by all four humors, and therefore the greater influence of the hot and dry in men and the cold and wet in women could have both positive and negative effects. The hot, dry humors of the male could cause excessive passion, blocking rational thought. But in women the tempering of heat by humidity made them more deliberate in their decisions and less given to outbursts of violent anger. This also made them superior to men in their capacity for understanding. Women's humid nature was necessary for the important biological function of bearing and nurturing children.[22]

WOMEN'S PLACE IN THE ORDER OF NATURE. At the root of the identification of women and animality with a lower form of human life lies the distinction between nature and culture fundamental to humanistic disciplines such as history, literature, and anthropology, which accept that distinction as an unquestioned assumption. Nature-culture dualism is a key factor in Western civilization's advance at the expense of nature. As the unifying bonds of the older hierarchical cosmos were severed, European culture increasingly set itself above and apart from all that was symbolized by nature. Similarly, in America the nature-culture dichotomy was basic to the tension between civilization and the frontier in westward expansion and helped to justify the continuing exploitation of nature's resources. Much of American literature is founded on the underlying assumption of the superiority of culture to nature. If nature and women, Indians and blacks are to be liberated from the strictures of this ideology, a radical critique of the very categories *nature* and *culture,* as organizing concepts in all disciplines, must be undertaken.

Anthropologists have pointed out that nature and women are both perceived to be on a lower level than culture, which has been associated symbolically and historically with men. Because women's physiological functions of reproduction, nurture, and childrearing are viewed as closer to nature, their social role is lower on the cultural scale than that of the male. Women are devalued by their tasks and roles, by their exclusion from community functions whence power is derived, and through symbolism.[23]

In early modern Europe, the assumption of a nature-culture dichotomy was used as a justification for keeping women in their place in the established hierarchical order of nature, where they were placed below the men of their status group. The reaction against the disorder in nature symbolized by women was directed not only at lower-class witches, but at the queens and noblewomen who during the Protestant Reformation seemed to be overturning the order of nature.

The assumption of English governance by Mary Tudor (Bloody Mary, queen from 1553 to 1558), was followed by the restoration of Catholicism and the persecution of Protestants. In 1558, Mary of Lorraine, acting as Scottish queen regent while Mary Stuart (Queen of Scots from 1561–1567) was being raised at the French court, forbade the preaching of the reformed doctrine in Scotland. The Scottish Protestant reformer John Knox (1505–1572) responded to the persecution of Protestants under these queens with his misogynist polemic *The First Blast of the Trumpet Against the Monstrous Regiment of Women* (1558).[24] But unfortunately for Knox, in that same year Elizabeth I succeeded to the English throne, restoring Protestantism in England and aiding the Scottish Protestants against their queen regent. Now in a difficult political situation, Knox was forced to apologize to Elizabeth for his views on women rulers, but she never really forgave him. With the death of Mary of Lorraine in 1560, Protestantism was proclaimed in Scotland, and in 1561 Mary Stuart became "Queen of Scots," ruling until her fall from power in 1567. Mary retained Protestantism, but to Knox's horror celebrated the Catholic Mass in her own household, making her a "monstrous" Catholic, ruler, and woman.

Although Knox was not a hater of all women, his *First Blast of the Trumpet* was symptomatic of a wider controversy surrounding women who were overturning the order of nature. In advocating that women should be kept in their place in the natural order, Knox drew on the microcosm-macrocosm theory of correspondence between the hierarchically stratified natural world and the little world of human beings. In the macrocosm theory, spirit and pure activity (associated with the male) increased toward the upper reaches of the closed spherical world, while the female earth lay at its center. Women's place was symbolized by the passive, base matter, which nurtured the active spiritual principle but was always below it and inferior to it. Knox held that, since flesh is subordinate to spirit, a woman's place is beneath man's. The laws of nature ordain that men should command women: "The order of God's creation" and "the light of nature" dictate against women's rule, for it subverts "good order." Women's role, according to Knox, was that of obedient servant. Since a woman was physically weaker than a man, her place was below his. Women were given to "natural weakness" and "inordinate appetites." If a woman was presumptuous enough to rise above a man, she must be "repressed and bridled." Societies in which untamed women were known to wage war were "monsters" within nature. Their lowly place was reflected in the animal kingdom: "For nature hath in all beasts printed a certain mark of domination in the male and a certain subjection in the female which they keep inviolate. For no man ever saw the lion make obedience and stoop before the lioness, neither can it be proved, that the hind taketh the conducting of the herd amongst harts."

Knox asserted that the order of nature was manifested in both the human body and society: God had established an order in the "politic or civil body" that corresponded to the "natural body of man." Man's eyes acted as mirrors to behold nature's order. Each part of the body had its natural place, with head on top and the other members in their proper places. In society, the ruling head must be a man, for a woman's rule creates a monster. A woman should therefore never be allowed to rule as queen for the "empire of a woman" overturned nature: "To promote a woman to bear rule . . . is repugnant to nature. . . . God by the order of his creation hath spoiled women of authority and dominion." In allowing a female to rule, men themselves became inferior to beasts: "It is a thing repugnant to the order of nature, that any woman be exhalted to rule over men. For God hath denied to her the office of a head, . . . I have made the nobility both of England and Scotland inferior to brute beasts, for [what] they do to women, . . . no male amongst the common sort of beasts can be proved to do their females." Knox concluded that "to promote a woman head over men, is repugnant to nature, and a thing most contrarious to that order, which God hath approved in his commonwealth, which he did institute and rule by his word."

Knox's views were supported in works published by Christopher Goodman (in 1558), Anthony Gilby (1558, possibly a pseudonym of Knox), George Buchanan (1582), and Jean Bodin (1576). But female nobility had among its defenders highly learned and articulate members of the upper classes. Knox was vehemently opposed by Elizabeth I, who succeeded Mary on the throne in 1558, and by her official John Aylmer (1558), John Leslie, Bishop of Ross (1559), and David Chambers (1579). Neither the death of Knox nor the execution of Mary Stuart in 1587 ended the struggle over the female's right to rule and the attempt at repression by male supremacists. The arguments over female governance continued until the death of Elizabeth in 1603 and the resumption of male rule under James I.[25] Such discussions of woman's place in society were part of the broader, concurrent "controversy over women" that involved women of all social levels.

That woman's place in the order of nature was properly below that of man was reinforced by other Protestant leaders. John Calvin, prominent in the Reformation on the Continent, argued in his "Commentary on Paul's First Epistle to the Corinthians" that when a woman assumed authority and a higher place in the natural order than had been assigned to her, she would be discovered in her insubordination by the angels. The man should function as the ruling intellectual head, while the woman is the body that assists him. Calvin's "Commentary on the Book of Genesis" stated: "The order of nature implies that the woman should be the helper of the man," and "She should study to keep this divinely appointed order." Eve's punishment for her sin was to be cast into servitude and subjected to her husband's authority and will, whereas before the fall hers had been "a liberal and gentle subjection."[26]

Woman's place in the natural order dictated her societal and religious roles, according to literalist interpretations of the Bible. In the 1570s, Protestant clergy cited various writings of St. Paul dictating women's silence in the church to express disapproval of females reading the Bible in the vernacular. The writings of St. Paul were taken to mean that a woman should never usurp authority from a man, should not teach, and should be silent in her subjugation. She should never wear bright clothing or costly jewelry but should attire herself in sober garments. She should be her husband's helpmeet in generation and daily living.[27]

Reactions such as those of Knox, Calvin, and other Protestant theologians toward women who were overturning nature's order were symptomatic of responses to the social upheaval of the Reformation. Although the major Protestant leaders did not advocate changes in women's societal roles, they did initiate changes of significance for women's domestic life. Huldrich Zwingli (1484–1531), the major Swiss reformer who preached at Zürich, and Martin Bucer (1491–1551), German reformer at Strasbourg, supported the extension of the medieval divorce laws (based on adultery and impotency) to include insanity and cruelty. Calvin advocated the woman's right to divorce and her equal responsibility in family worship and in the spiritual instruction of the children.[28]

In the reform movements of northern Europe, women joined the struggle for religious equality. In France, noblewomen assumed leadership in evangelical reform and were in the vanguard of the Huguenot (French Calvinist) movement from 1557 until the decline of French reform in the 1570s. Even before the French Reformation, literate women had begun reading the Bible and other religious literature printed in the vernacular, and had engaged in theological speculation. Learned women resented not being taken seriously by male clergy in these endeavors.

The more radical religious movements offered women greater opportunities for speaking out on questions of church government. The membership of the early sixteenth-century English Lollards (forerunners of the Reformation, who read the Bible in English, stressed scriptural authority, and lay governance), was about one-third women. The radical anarchistic Anabaptists, centered in Münster until 1535, whose following came from the lower classes, allowed women to preach and prophesy, yet insisted on polygamy, early marriage for women, and subordination. In England more women than men joined the first group of separatists in 1568. Some of them went to Holland, where more freedom was allowed for women, and associated themselves with the separatist churches there. The religious sects of the English Civil War, discussed in Chapter 4, also offered greater opportunities for female participation.[29]

The disorder symbolized in the macrocosm by the dissolution of the frame of nature and the uncivilized wilderness of the new world, in society by the witch who controlled the forces of nature and the women who overturned its order, and in the self by the bestiality of the Indian and the cannibal, the sexual lust of the female, and the animal passions of all humans heralded the death of the old order of nature. From the stirred ashes a new order was emerging, which would reconstruct the self, society, and the cosmos. The passivity of the female in the sphere of reproduction would be reasserted, sexual passion would eventually be repressed and the spirits would be removed from nature in coincidence with the waning of witch trials; female roles would increasingly be defined in terms of domestic functions as middle- and upper-class women became economic subordinates in the marriage relationship. A new experimental method designed to constrain nature and probe into her secrets would improve and "civilize" society. The macrocosm would become a machine consistent with new managerial modes of order and power. Not until the late seventeenth century would women's reaction to the new order take the form of feminism, as they themselves began to assert the rights and privileges of their own sex.

CHAPTER 6

Production,
Reproduction,
and the Female

The new economic and scientific order emerging in sixteenth- and seventeenth-century Europe would be of lasting significance for both nature and women, for at its ideological core were the concepts of passivity and control in the spheres of production and reproduction. Disorderly female nature would soon submit to the controls of the experimental method and technological advance, and middle- and upper-class women would gradually lose their roles as active partners in economic life, becoming passive dependents in both production and reproduction.

In hierarchical society, women's economic and social roles were defined by the class to which they belonged through birth or marriage. In the lower orders of society, peasant and farm wives were integral parts of a productive family unit. Married women carried a heavy burden of labor—childbearing and childrearing, gardening, cooking, cheese and soap making, spinning and weaving, beer brewing, and healing. Unmarried women worked as servants in another household, as unskilled labor in mowing, reaping, and sheepshear-

149

ing, or at spinning and weaving. Urban women worked in the crafts or trades, owned shops, were members of craft guilds, and even occasionally worked at construction and ditch digging. At the upper levels of society, noblewomen were busy supervising the economic activities of the estate, owning and managing property, and keeping accounts.

Within preindustrial capitalism, women's economic roles became more restrictive and their domestic lives came to be more rigidly defined by their sex as women, rather than by their class. The ideal developing for the upper-class and well-to-do bourgeois wife was a life of leisure, symbolizing the success of her husband's economic ventures. In countries on the cutting edge of the capitalist advance (such as Italy in the fourteenth and fifteenth centuries and, as will be elaborated in this chapter, England in the sixteenth and seventeenth centuries), the contraction and redefinition of women's productive and domestic roles was consistent with changes in the ideology of sexuality.[1]

In Renaissance Italy (ca. 1350–1530), where a mercantile economy had developed much earlier than in northern Europe, men's public lives were expanding, while urban women's were contracting into domesticity. Many medieval Italian girls had been raised in the court of an educated noblewoman—which functioned as a cultural center in determining social values and mores—but in the Renaissance their education emphasized morality, chastity, and readings in the newly recovered classics of Greece and Rome. This shift presupposed male public and female domestic spheres. Medieval courtly love had been based on a knight's service to a lady who gave him her love in mutual exchange, combining genuine sexuality and passion with Christian love; his Renaissance counterpart loved a platonic ideal whom he seemed to serve, but who in reality was a chaste and passive subordinate.

In seventeenth-century England, significant changes were taking place in the productive work of women in domestic life, in the home, in family industries producing and selling foods for the local market, in early capitalist industries (such as agriculture, textile manufacture, and the retail trades), and in professional employment. The direction of the change was to limit and curtail the married woman's role as a partner, so that she became more dependent on her husband. In the sphere of reproduction, women midwives were losing their monopoly over assisting at childbirth to male doctors. Simultaneously, the female's passive role in biological generation was being reasserted by physicians and natural philosophers. The witch and her counterpart, the midwife, were at the symbolic center of a struggle for control over matter and nature essential to new social relations in the spheres of production and reproduction.

WOMEN AND PRODUCTION.

Aristocratic women of the Elizabethan era managed the business affairs of their estates during their husband's absence and after his death; yet Restoration "ladies of quality" often had "nothing better to do, but to glorify God and to benefit their neighbors," the expectation that women should be trained in business affairs having by then markedly decreased.

Under subsistence agriculture, the wives of yeomen and husbanders had participated in the family farming operation, the profits of which benefited and were shared by the whole family. But as yeomen became wealthy agricultural improvers and market farmers, hiring more servants and day laborers, their wives withdrew from active participation in daily farm work, devoting more time to pleasure. These changes reduced the married woman's ability to support herself and her children on her husband's death.

Male day laborers in the new capitalist agricultural operations were completely dependent on the wages earned for their labor to support their families, having no land on which to grow family food. A wife's earnings, if she worked outside the home, were considerably less than her husband's, because her capacity for outside employment was reduced by the number of small children to be cared for at home, and the health and nutrition of the family suffered as a result. If her husband deserted her, a mother encumbered by children was unable to engage in sufficient productive work to support her family.

The English export market for woolen goods was one of the earliest and most important capitalist industries. Women played no role as clothiers or wool merchants, and little mention is made of them as assistants in the business ventures of their husbands, their employment being confined to wage work, primarily in the spinning branch of the industry. Women bought wool and, when not occupied in agricultural production, spun it at home and sold the yarn on the market. If, however, a wife was forced to work outside the home, the wages she earned were insufficient to provide both food and clothing for her and her children, although an unmarried spinster could support herself. By 1511, women had been forbidden to weave, because strength was needed to operate the looms. A widow, however, was allowed to continue the weaving industry of her husband, directing the servants employed by him as long as she remained unmarried. In addition to spinning, women also participated in the bleaching and fulling operations. In periods of depression in the industry, fathers all too frequently deserted their families to seek work elsewhere or to become "masterless" vagabonds, leaving women with their children as the objects of charity.

In the retail trades, women fared well as long as the family operated as a productive unit, the wife helping in her husband's business and taking it over if widowed. But as the trades and crafts began to adopt the capitalistic mode of employing wage workers, the wives of master craftsmen had less opportunity for participation, while the wives of journeymen, who had hitherto received guild privileges through marriage, were now excluded from the new journeymen's organizations, which sought to protect the man's position vis à vis the master. These wives either became more dependent on their husbands, who were said to "keep them," or were forced to enter the marketplace on their own. They were able to become apprentices in women's trades such as hat making and cloak making, but as a group gradually lost ground in trades such as baking, butchering, fishmongering, and brewing, as rules and statutes began to limit the numbers of persons engaged. By the end of the seventeenth century, women had lost control of the brewing trade, an occupation that in earlier times they had monopolized.

Until the seventeenth century, midwifery was the exclusive province of women: it was improper for men to be present at such a private and mysterious occurrence as the delivery of a child. Midwives were professionals, usually well trained through apprenticeship and well paid for their services to both rural and urban, rich and poor women. Yet no organization of midwives existed that could set standards to prevent untrained or poverty-stricken women from taking up the practice. Moreover, women were excluded from attending universities and medical schools where anatomy and medicine were being taught.[2]

Seventeenth-century London midwives, rightly or wrongly, considered themselves a responsible, well-trained group of women. But by 1634, the midwife profession was being threatened by the licensing of male surgeons who wished to practice midwifery with forceps, a technology that would be available only to licensed physicians. The midwives had complained to the bishop of London that such a practice was often marked by violence and that men had insufficient experience with deliveries. The Chamberlen family, which had invented the forceps, was attempting to establish educational and legal restrictions on their use. Earlier, in 1616, the Drs. Peter Chamberlen, elder and younger, had tried to form a corporation of midwives. The midwives doubted the Chamberlen's ostensible motives to educate and organize them because they feared that the latter would attempt to assume sole licensing authority. They favored the older delivery methods of which they had knowledge and called the new forceps method a violent practice. Their 1634 petition directed against Peter Chamberlen III stated, "Dr. Chamberlane . . . hath no experience in [midwifery] but by reading. . . . And further Dr. Chamberlane's work and the work belonging to midwives are contrary one to the other, for he delivers none without the use of instruments by extraordinary violence in desperate occasions, which women never practiced nor desired, for they have neither parts nor hands for that art."[3]

In addition to the Chamberlens, other doctors of the period were sharply critical of the practices of midwives. William Harvey, noted for his discovery of the circulation of the blood, and one of the four censors of the Royal College of Physicians responsible for enforcing the College's monopoly over licensing laws, took issue with some of their methods in his essay "On Parturition" at the end of his *Exercitationes de Generatione Animalium* (On Generation), 1651:

Hence it is that midwives are so much to blame, especially the younger and more meddlesome ones, who make a marvellous pother when they hear the woman cry out with her pains and implore assistance, daubing their hands with oil, and distending the passages, so as not to appear ignorant in their art—giving besides medicines to excite the expulsive powers, and when they would hurry the labor, retarding it and making it unnatural, by leaving behind portions of the membranes, or even of the placenta itself, besides exposing the wretched woman to the air, wearing her out on the labor stool, and making her, in fact, run great risks of life.

In truth, it is far better with the poor, and those who become pregnant by mischance, and are secretly delivered without the aid of a midwife; for the longer birth is retarded the more safely and easily is the process completed.[4]

Harvey's *De Generatione* did not deal extensively with human reproduction, and his examples of difficult births by women were not a significant contribution to the period's inadequate gynecological science. Yet in spite of obvious lacunae in the state of obstetrical knowledge, Harvey's contributions have been eulogized by historians of medicine as the work of a "grand broad intellect which could at the same time teach the profoundest physicians and the most ignorant midwives" 'and "whose beneficial influence . . . can scarcely be overestimated."[5]

During the 1630s and 1640s, male physicians in addition to Harvey wrote treatises that helped to discredit midwives, contributing to the decline of female midwifery. Male physicians who wrote disparagingly about female practices included Peter Chamberlen the Elder and the Younger, who had petitioned Francis Bacon in 1616 to incorporate the "ignorant midwives"; Peter Chamberlen III, who wrote *A Voice in Rhama: Or the Crie of Women and Children* (1647), probably in retaliation for the midwives' opposition in 1634; Hugh Chamberlen the Elder, Paul Chamberlen, author of *Dr. Chamberlain's Midwifes Practice: Or a guide for women in that high concern of conception, breeding and nursing children* (1665); and Percivall Willughby (1596–1685) whose *Observations of Midwifery* praised Harvey's obstetrical directions. Since most historical accounts of midwifery in this period are based on the data supplied by male writers and male midwives—some of whom, like the Chamberlens, had political motives—an accurate assessment of the state of midwifery as a woman's art is difficult to make.[6]

After the middle of the century, English midwives such as Jane Sharp (fl. 1671) and Elizabeth Cellier (fl. 1680), along with medical practitioner Nicolas Culpeper (1616–1654), wrote midwifery handbooks in an attempt to make obstetrical and gynecological training available to women in the profession. In France, a school of midwifery was established where anatomy was taught through dissection, and surgeons examined women apprentices. Yet despite these attempts by a few persons to upgrade and include women in the advancing medical and scientific knowledge of the period, women began to lose control over midwifery and thus over their own reproductive functions. By the end of the century, childbirth was passing into the hands of male doctors and "man-midwives."[7]

While women's productive roles were decreasing under early capitalism, the beginning of a process that would ultimately transform them from an economic resource for their families' subsistence to a psychic resource for their husbands, the cultural role played by female symbols and principles was also changing. The female world soul, with its lower component, *Natura,* and the nurturing female earth had begun to lose plausibility in a world increasingly influenced by mining technology essential to commercial capitalism. The older organic order of nature and society was breaking up as the new mercantile activities threatened the ideology of natural stratification in society.

Symbolic of these changes were the midwife and the witch. From the perspective of the male, the witch was a symbol of disorder in nature and society, both of which must be brought under control. The midwife symbolized female incompetence in her own natural sphere, reproduction, correctable through a technology invented and controlled by men—the forceps. But from a female perspective, witchcraft represented a form of power by which oppressed lower-class women could retaliate against social injustices, and a source of healing through the use of spirits and the regenerative powers of nature. For women, the midwife symbolized female control over the female reproductive function. But until medical training became available to women and licensing regulations were equalized for both women and men, women had no opportunity to compare the effectiveness of the older, shared traditions of midwifery as an art with the new medical science.

REPRODUCTION AND THE FEMALE. William Harvey's discovery of the circulation of the blood, published in 1628, is considered one of the outstanding achievements of early modern science. Although Harvey's cosmos and physiology were still infused by vital animistic principles, his analogy of the heart as a pump would soon be incorporated into a mechanistic physiology by René Descartes (1596–1650), Dutch physician Hermann Boerhaave (1668–1738), and French physician and philosopher Julien La Mettrie (1709–

1751). The machine metaphor based on a dualism between body and soul, with the soul as an external operator, contrasted with the older vitalistic view that the body was permeated and enlivened by an animating spirit.[8] This "incredible machine" (as it was called in a recent television documentary) must therefore be repaired by medical intervention, rather than left to the healing powers of nature.

In addition to his work on blood circulation, Harvey devoted much energy to the subject of biological reproduction, an area influenced by cultural ideology about female roles. Harvey's *On Generation,* was filled with rich metaphor and value-laden phraseology.[9] When his biological ideas are set in the broader context of the cultural changes of his time, the passive role he assigned to both matter and the female in reproduction is consistent with (although not caused by) the trend toward female passivity in the sphere of industrial production and with the reassertion of the passivity and inertness of matter by the new seventeenth-century mechanical philosophy. Harvey's theory of biological reproduction is compatible with new scientific values based on the control of nature and women integral to the new capitalist modes of production.

Harvey's ideas were formulated during the same period in which the controversies over women's role in society and their declining role in productive work took place in England. He was "Physician Extraordinary" to James I and doctor to Francis Bacon, who, he observed derisively, wrote philosophy like a lord chancellor. After the death of James I, in 1625, Harvey became "Physician Ordinary" to Charles I, who placed the royal deer parks at his disposal to study the generative processes in the doe and the hind. As a royal employee, Harvey identified with the political values of his sovereigns, his sympathies being royalist, at least until 1649 when Charles was beheaded.[10]

Ideological assumptions about the female sex permeated English culture during the years in which Harvey was collecting data and writing the *De Generatione.* A middle-class "controversy over women" which raged in Elizabethan England continued to inspire new works on women's nature in the 1630s–1650s. Charles I, who was more cultured than James I, might have allowed women more freedom in education had it not been for the intervention of the Civil War.[11]

The view of biological generation that guided William Harvey's experimental work and theoretical interpretations had originated with Aristotle and, as discussed in Chapter 1, supposed that the female contributed the matter, or passive principle, on which the semen, the active male principle, worked in creating the embryo.

Aristotle's association of the male with the perfection of the sphere was emphasized during the medieval and early modern periods. Albertus Magnus agreed that male chicks developed from the more spherical eggs because the sphere was the most perfect figure in solid geometry. Long pointed eggs produced females; the rounder ones produced the stronger more perfect animal, the male. People were advised to eat the long, thin, sweet and more nourishing eggs rather than the hard round ones which contained the yolk of the male.

During the sixteenth century, the theory of female passivity in reproduction became well established. Although some writers followed Galen in assigning both formal and material causes equally to the male and female, even Galen viewed the female as less perfect than the male, because her genitalia did not emerge externally and her "semen" was imperfect. Most sixteenth-century writers sided with Aristotle's theory that the female provided only the matter, while the active principle was attributed to the male semen. Gossip and popular opinion held the male seed to be the chief agent in generation, although the female's cooperation was needed for development of the embryo. Most texts supported the ancient idea that the male embryo was twice as hot and developed twice as quickly as the female. Embryological illustrations showed the female blood along with the semen in an egg-shaped mass in the womb. Harvey's Paduan teacher, Hieronymous Fabricius, maintained in his *Embryological Treatises* (1621) that the sperm never reached the egg. It merely vivified it by an immaterial fecundative faculty. As sole agent and efficient cause, it imparted quality to the egg. The female supplied the nutriment and warmth for the embryo's development. Thus the theory of the man as parent and the woman as incubator formed the prevailing sexual ideology of the latter sixteenth and early seventeenth centuries.[12]

Although Harvey's *De Generatione Animalium* dealt primarily with reproduction in the chick and the deer, comparisons to human reproduction were made throughout, and human conception and parturition were discussed in the final chapters. Harvey's work followed Aristotle and Fabricius, in many respects. But whereas Fabricius' work mainly described biological structures and generative processes, Harvey's writings were filled with sexual metaphor.[13] Although some historians have interpreted Harvey's ideas on generation as a cooperation between sperm and *ovum,* with the egg assuming a newly elevated place of dignity, close attention to his language reveals the influence of cultural sexual biases in his scientific work.[14]

Harvey's work on generation led him to differ from Aristotle in assigning an efficient cause to the mother hen, but this did not imply the primacy of the female line in procreation. The hen's egg developed through its own internal principle, its vegetative soul. This soul was not derived from the mother hen, and the egg did not live by the vital principle of the mother . . . but was "independent even from its first appearance."[15] Free and unconnected to the uterus, the egg perfected itself by its internal formative force. Following an inner developmental plan, structures, including the soul of the chick, were formed in succession after fertilization. The plastic virtue or formative principle was present in the matter of unfertilized eggs and could act without the male contribution, albeit imperfectly.

Because a hen could lay unfertilized eggs, she served as an efficient cause, a role Aristotle had denied to the female. Thus, not all creative force was derived solely from the male agent, as Aristotle had argued. Nevertheless, the male for Harvey was more excellent as an efficient cause than the female. A hen's egg obtained perfection only when fertilized by the male "in virtue of an authority . . . or power required of the cock. . . . Among animals where the sexes are distinct, matters are so arranged that since the female alone is inadequate to engender an embryo and to nourish and protect the young, a male is associated with her by nature, as the superior and more worthy progenitor, as the consort of her labor and the means of supplying her deficiencies."[16]

The male also supplied "reason" and excellence to the egg, "as if the hen received the art and reason, the form and laws of the future embryo from his address. . . . Nor should we so much wonder what it is in the cock that preserves and governs so perfect and beautiful an animal, and is the first cause of that entity which we call the *soul;* but much more what it is in the egg, . . . of so great virtue as to produce such an animal and raise him to the very summit of excellence." Harvey recognized, as had Aristotle, that the offspring was not the conscious result of the male mind; hence they both held the position that it was nature working through the parents that produced the young. The "skill and foresight" contributed by the male must ultimately proceed from God, since the vegetative, not the rational, portion of the soul was used in reproduction.[17]

Harvey argued against Galenist physicians "conversant with anatomy" that women excreted a seminal fluid necessary to generation and that the semina of both male and female had material and efficient faculties. Although he was correct that the female did not produce semen, his reasons were value-laden and based on the masculine bias that energy and perfection could proceed only from the male:

> The other argument is drawn from the genital organs of the woman, the testes, to wit, and *vas spermatica, praeparantia, et deferentia,* which are held to serve for the preparation of the spermatic fluid. I, for my part, greatly wonder how anyone can believe that from parts so imperfect and obscure, a fluid like the semen, so elaborate, concoct, and vivifying, can ever be produced, endowed with force and spirit and generative influence adequate to overcome that of the male; for this is implied in the discussion concerning the predominance of the male or the female, as to which of them is to become agent and efficient cause, which the matter and pathic principle. How should such a fluid [the female's] get the better of another concocted under the influence of a heat so fostering, of vessels so elaborate, and endowed with such vital energy?—how should such a fluid as the male semen be made to play the part of mere matter?[18]

Cultural bias also guided his research on the theory of impregnation, in which he argued, contra Aristotle, that fertilization could take place without material contact of sperm and egg, a theory testable only after the perfection of the microscope. His conclusion that the male semen was so powerful that it could act at a distance by a seminal aura, or magnetic emanation, was supported by prevailing cultural ideas of male superiority.

This cultural preconception guided the interpretation of his experiments on the king's deer. During the 1630s, he dissected large numbers of King Charles' does just after coition with bucks and found "no seed in their uterus":

> But when after repeated inspections, I still found nothing more in the uterus, I began to doubt, and to ask myself whether the semen of the

male could by any possibility make its way—by attraction or injection—to the seat of conception? And repeated examination led me to the conclusion that none of the semen whatsoever reached this seat.[19]

He had also reached the same conclusion regarding the cock and the hen, an opinion first put forth by his teacher Fabricius:

The semen which is emitted by the male during intercourse does by no means enter the uterus of the female, in which the egg is perfected; nor can it indeed (as I first announced and Fabricius agrees with me) by any manner or way get into the inner recesses of that organ.[20]

Since there was no contact between egg and sperm, Harvey endowed the male's semen with the prodigious power of acting at a distance. The semen acted through a sympathy or magnetic emanation in a way similar to that by which diseases were transmitted. It was the same as the manner in which

Epidemic, contagious, and pestilential diseases scatter their seeds and are propagated to a distance through the air, or by some 'fomes' producing diseases like themselves in bodies of a different nature and in a hidden fashion silently multiplying themselves by a kind of generation.[21]

The female was impregnated by contagion! Generation was analogous to a disease caused by the introduction of a parasitic or foreign cause, which played the role of an active, generative, vital principle. A contagion was an active power as opposed to putrefaction, which was the degenerative principle found in decay. Later, Jean Baptiste Van Helmont characterized disease as the introduction of a foreign semen that became superimposed on the organism's own plan of vital internal growth and development.

For Harvey, the male semen had a "plastic power" that made it prolific, a spiritual power analogous to the essence by which the stars influenced earthly beings. The semen carried within it the virtue of divine agency affecting the female like lightning from above, a spark from a flintstone, or the magnetic power of the lodestone.

But since it is certain that the semen of the male does not so much reach the cavity of the uterus, much less continue long there, and that it carries with it a fecundating power by a kind of contagious property; the woman after contact with the spermatic fluid *in coitu,* seems to receive influence and to become fecundated without the cooperation of any sensible corporeal agent, in the same way as iron touched by the magnet is endowed with its powers and can attract other iron to it.[22]

The male's semen was so powerful that it even affected the woman's mind: "The virtue which proceeds from the male *in coitu* has such prodigious power of fecundation, that the whole woman both in mind and body undergoes a change."[23] The theory that the brain had the power to affect the developing fetus in the uterus was not an uncommon notion in Harvey's time. Paracelsus had written that "the imagination of a pregnant woman is so active that in conceiving seed into her body she can transmute the fetus in different ways ... her interior stars are so strongly directed to the fetus that they produce impression and influence."[24]

The male sperm endowed the female uterus with the "plastic power" to create an offspring. Thus the uterus was similar to a brain about to create, in the same way that the painter about to produce a work of art pictured to himself the painting. But this uterine brain was not a free agent. It produced only what was impressed on it by the more perfect male:

> And just as a "desire" arises from a conception of the brain, and this conception springs from some external object of desire, so also from the male, as being the more perfect animal, and as it were, the most natural object of desire, does the natural (organic) conception arise in the uterus, even as the animal conception does in the brain.[25]

Through the imposition of the form or plan of the father on the uterus during coitus, "it results that the female produces an offspring like the father." The " 'form' of the father existing in the uterus generates an offspring like himself with the help of the formative faculty."[26]

Harvey's work on generation thus reflected the cultural sex biases of his society, especially in areas where he most strongly disagreed with received authority. He differed from Aristotle by claiming that all animal life was generated from eggs and that the egg was an efficient cause, but he maintained that the sperm was superior to the egg as an efficient cause and was the agent of perfection. Secondly, his argument against Galenist physicians concerning the existence of a female semen was based on the cultural assumption that semen was too elaborate, vivifying, and energetic to come from obscure, imperfect, female parts. Thirdly, he disagreed with both Aristotle and Galen that generation occurred through the mingling of contributions from female and male, and concluded instead that impregnation took place without contact. This meant that the male semen must be endowed with an extraordinary magnetic power capable of affecting the female uterus and mind from a distance. While his conclusions were consistent with Aristotelian sexual ideas, they were reinforced by the conservative social values of his time. Far from championing the equality of male-female principles in reproduction, Harvey's theories fall within the tradition of male superiority.

Mid-seventeenth-century natural philosophers continued the tradition of male superiority in generation, maintaining that the soul itself was distributed through the male lineage. Descartes believed that the egg was impregnated by the power of the male semen, which endowed it with soul. In Emilio Parisano's opinion (1621), the semen was infused with soul for the propagation of the species. Even the atomists, traditionally associated with democratizing tendencies in cultural history, differentiated between those contributed by the male and those from the female. Thus physician Nathaniel Highmore, in his *History of Generation* (1651), postulated two sorts of seminal atoms in the seed—material and spiritual atoms. "These seminal atoms fall from all parts of both parents, the spiritual ones from the male, the material ones from the female. Thus the atoms of Democritus are transmuted into 'substantial forms' endowed with the efficient cause of Aristotle . . . and permitted to remain material."[27]

As late as 1661, Anthony Everard still held that the male semen contributed the spiritual element and the mother the matter. Everard argued that the fetus was formed from the union of the male and female seminal spirits. The "female semen," as such, did not contribute to generation, while the masculine semen, acting only as an efficient cause, did not contribute anything material. After Anton Van Leeuwenhoek's (1632–1723) introduction of the microscope, the problem of generation began to take on new dimensions in the controversy between spermists and ovists over preformation.

The use of science as an ideology to keep women in their place was not confined to ancient and early modern times. In the nineteenth century, Darwinian theory was found to hold social implications for women. Variability, the basis for evolutionary progress, was correlated with a greater spread of physical and mental variation in males. Scientists compared male and female cranial sizes and brain parts in the effort to demonstrate the existence of sexual differences that would explain female intellectual inferiority and emotional temperament. Women's reproductive function required that more energy be directed toward pregnancy and maternity, hence less was available for the higher functions associated with learning and reasoning. The "adventurous sperm" and the "passive ovum" continued to serve as reproductive metaphors.

In the twentieth century, hormonal differences between men and women have been used to imply abnormal levels of androgen in women who displayed high intelligence, competitive behavior, leadership, and executive ability. Thus "scientific" authority could be used to keep women in their place as intellectually inferior and economically dependent.[28]

While in each of these cases, male and female critics exposed the theoretical and cultural assumptions underlying the leap from differences to inequalities, new fields and new scientific studies continue to generate "evidence" to maintain outdated assumptions about the male-female hierarchy. Reproduction—hormones, menstruation, and pregnancy—is used to infer and justify the female economic dependence brought about in the seventeenth-century transition from subsistence to capitalist modes of production. For women, this aspect of the Scientific Revolution did not bring about the presumed intellectual enlightenment, objectivity, and liberation from ancient assumptions traditionally accorded it.

CHAPTER 7

Dominion over Nature

Disorderly, active nature was soon forced to submit to the questions and experimental techniques of the new science. Francis Bacon (1561–1626), a celebrated "father of modern science," transformed tendencies already extant in his own society into a total program advocating the control of nature for human benefit. Melding together a new philosophy based on natural magic as a technique for manipulating nature, the technologies of mining and metallurgy, the emerging concept of progress and a patriarchal structure of family and state, Bacon fashioned a new ethic sanctioning the exploitation of nature.

Bacon has been eulogized as the originator of the concept of the modern research institute, a philosopher of industrial science, the inspiration behind the Royal Society (1660), and as the founder of the inductive method by which all people can verify for themselves the truths of science by the reading of nature's book.[1] But from the perspective of nature, women, and the lower orders of society emerges a less favorable image of Bacon and a critique of his program as ultimately benefiting the middle-class male entrepreneur. Bacon, of course, was not responsible for subsequent uses of his philosophy. But, because he was in an extremely influential social position and in touch with the important developments of his time, his language, style, nuance, and metaphor become a mirror reflecting his class perspective.

Sensitive to the same social transformations that had already begun to reduce women to psychic and reproductive resources, Bacon developed the power of language as political instrument in reducing female nature to a resource for economic production. Female imagery became a tool in adapting scientific knowledge and method to a new form of human power over nature. The "controversy over women" and the inquisition of witches—both present in Bacon's social milieu—permeated his description of nature and his metaphorical style and were instrumental in his transformation of the earth as a nurturing mother and womb of life into a source of secrets to be extracted for economic advance.

Bacon's roots can be found in middle-class economic development and its progressive interests and values. His father was a middle-class employee of the queen, his mother a Calvinist whose Protestant values permeated his early home life. Bacon took steps to gain the favor of James I soon after the latter's ascent to the throne in 1603. He moved from "learned counsel" in 1603 to attorney general in 1613, privy councillor in 1616, lord keeper in 1617, and, finally, lord chancellor and Baron Verulam in 1618. His political objectives were to gain support for his program of the advancement of science and human learning and to upgrade his own status through an ambitious public career.[2]

Bacon's mentor, James I, supported antifeminist and antiwitchcraft legislation. During the "controversy over women," females had challenged traditional modes of dress considered as appropriate to their place in society. In Holland, for example, young women were criticized for wearing men's hats with high crowns. In England, the title page of a work called *Hic-Mulier or The Man-Woman* (1620) showed a woman in a barber's chair having her hair clipped short, while her companion outfitted herself in a man's plumed hat (Fig. 15).[3] In an attempt to keep women in their place in the world's order, King James in that same year enlisted the aid of the clergy in preventing females from looking and dressing in

Figure 15. Title page from *Hic-Mulier or the Man-Woman* (London, 1620). In the clothing controversy of seventeenth-century England, women challenged society by wearing "inappropriate" dress. This 1620 polemic of anonymous authorship shows a woman modeling a man's plumed hat, while a second woman has her hair clipped short by a barber.

masculine fashions: "The Bishop of London had express commandment from the king to will [the clergy] to inveigh vehemently against the insolence of our women, and their wearing of broad-brimmed hats, pointed doublets, their hair cut short or shorn, and some of them [with] stilettos or poinards . . . *the truth is the world is very much out of order.*"[4] (Italics added.)

In 1616, Mrs. Turner, accomplice in the murder of Sir Thomas Overbury, had been sent to the gallows by James wearing the yellow, starched ruffs she had brought into vogue and that he detested. As the king's attorney general, Bacon participated in the controversy, since it was his role to bring charges for the poisoning of Overbury against the Countess of Somerset. Overbury had publicly (through a poem, "The Wife") opposed the romance between his close friend, subsequently Earl of Somerset, and the countess. The perfect wife, he said, was one who combined goodness, virtue, intelligence, and common sense but not too much "learning and pregnant wit," for "Books are a part of man's prerogative." Angered by his insults, and fearful of his influence, the countess contrived to poison Overbury through the help of a physician's widow, Mrs. Turner, and an apothecary named Franklin.

Bacon prepared two versions of his charge against the countess, one should she confess, the other should she plead not guilty. At the packed trial, at which some places sold for £10–50, the countess confessed, but was spared. Mrs. Turner, however, was convicted and sent to the gallows, and "as she was the person who had brought yellow starched ruffs into vogue, [it was decreed that] she should be hanged in that dress, that the same might end in shame and detestation."[5]

The Overbury case increased interest in the popular controversy over women and resulted in the publication of several editions of Overbury's poem and a number of reactions to the murder; for example, "A Select Second Husband for Sir Thomas Overburies'

HIC MVLIER:

OR,

The Man-Woman:

Being a Medicine to cure the Coltish Diſeaſe of
the Staggers in the *Maſculine-Feminines*
of our Times.

Expreſt in a briefe Declamation.

Non omnes poſſumus omnes.

Miſtris, will you be trim'd or truſſ'd?

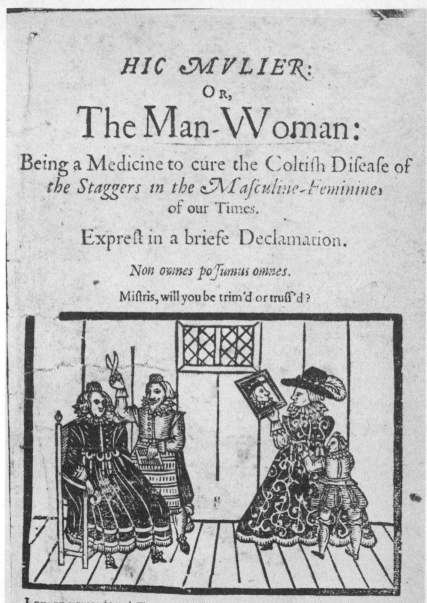

Loncon printed for J.T. and are to be ſold at Chriſt Church gate. 1620.

Wife, Now a Matchless Widow" (1616) and Thomas Tuke's "A Treatise Against Painting and Tincturing of Men and Women: Against Murder and Poysoning: Pride and Ambition" (1616).

Bacon was also well aware of the witch trials taking place all over Europe and in particular in England during the early seventeenth century. His sovereign, while still James VI of Scotland, had written a book entitled *Daemonologie* (1597). In 1603, the first year of his English reign, James I replaced the milder witch laws of Elizabeth I, which evoked the death penalty only for killing by witchcraft, with a law that condemned to death all practitioners.[6]

It was in the 1612 trials of the Lancashire witches of the Pendle Forest that the sexual aspects of witch trials first appeared in England. The source of the women's confessions of fornication with the devil was a Roman Catholic priest who had emigrated from the Continent and planted the story in the mouths of accused women who had recently rejected Catholicism.

These social events influenced Bacon's philosophy and literary style. Much of the imagery he used in delineating his new scientific objectives and methods derives from the courtroom, and, because it treats nature as a female to be tortured through mechanical inventions, strongly suggests the interrogations of the witch trials and the mechanical devices used to torture witches. In a relevant passage, Bacon stated that the method by which nature's secrets might be discovered consisted in investigating the secrets of witchcraft by inquisition, referring to the example of James I:

> *For you have but to follow and as it were hound nature in her wanderings, and you will be able when you like to lead and drive her afterward to the same place again.* Neither am I of opinion in this history of marvels that superstitious narratives of *sorceries, witchcrafts, charms,* dreams, divinations, and the like, where there is an assurance and clear evidence of the fact, should be altogether excluded. . . . howsoever the use and practice of such arts is to be condemned, yet from the speculation and consideration of them . . . a useful light may be gained, not only for a true judgment of the offenses of persons charged with such practices, *but likewise for the further disclosing of the secrets of nature. Neither ought a man to make scruple of entering and penetrating into these holes and corners, when the inquisition of truth is his whole object—as your majesty has shown in your own example.*[7] (Italics added.)

The strong sexual implications of the last sentence can be interpreted in the light of the investigation of the supposed sexual crimes and practices of witches. In another example, he compared the interrogation of courtroom witnesses to the inquisition of nature: "I mean (according to the practice in civil causes) in this great plea or suit granted by the divine favor and providence (whereby the human race seeks to recover its right over nature) *to examine nature herself* and the arts upon interrogatories."[8] Bacon pressed the idea further with an analogy to the torture chamber: "For like as a man's disposition is never well known or proved till he be crossed, nor Proteus ever changed shapes till he was *straitened* and *held fast*, so nature exhibits herself more clearly under the *trials* and *vexations* of art [mechanical devices] than when left to herself."[9]

The new man of science must not think that the "inquisition of nature is in any part interdicted or forbidden." Nature must be "bound into service" and made a "slave," put "in constraint" and "molded" by the mechanical arts. The "searchers and spies of nature" are to discover her plots and secrets.[10]

This method, so readily applicable when nature is denoted by the female gender, degraded and made possible the exploitation of the natural environment. As woman's womb had symbolically yielded to the forceps, so nature's womb harbored secrets that through technology could be wrested from her grasp for use in the improvement of the human condition:

> There is therefore much ground for hoping that there are still laid up in the womb of nature many secrets of excellent use having no affinity or parallelism with anything that is now known ... only by the method which we are now treating can they be speedily and suddenly and simultaneously presented and anticipated.[11]

Bacon transformed the magical tradition by calling on the need to dominate nature not for the sole benefit of the individual magician but for the good of the entire human race. Through vivid metaphor, he transformed the magus from nature's servant to its exploiter, and nature from a teacher to a slave. Bacon argued that it was the magician's error to consider art (technology) a mere "assistant to nature having the power to finish what nature has begun" and therefore to despair of ever "changing, transmuting, or fundamentally altering nature."[12]

The natural magician saw himself as operating within the organic order of nature—he was a manipulator of parts within that system, bringing down the heavenly powers to the earthly shrine.

Agrippa, however, had begun to explore the possibility of ascending the hierarchy to the point of cohabiting with God. Bacon extended this idea to include the recovery of the power over nature lost when Adam and Eve were expelled from paradise.

Due to the Fall from the Garden of Eden (caused by the temptation of a woman), the human race lost its "dominion over creation." Before the Fall, there was no need for power or dominion, because Adam and Eve had been made sovereign over all other creatures. In this state of dominion, mankind was "like unto God." While some, accepting God's punishment, had obeyed the medieval strictures against searching too deeply into God's secrets, Bacon turned the constraints into sanctions. Only by "digging further and further into the mine of natural knowledge" could mankind recover that lost dominion. In this way, "the narrow limits of man's dominion over the universe" could be stretched "to their promised bounds." [13]

Although a female's inquisitiveness may have caused man's fall from his God-given dominion, the relentless interrogation of another female, nature, could be used to regain it. As he argued in *The Masculine Birth of Time*, "I am come in very truth leading to you nature with all her children to bind her to your service and make her your slave." "We have no right," he asserted, "to expect nature to come to us." Instead, "Nature must be taken by the forelock, being bald behind." Delay and subtle argument "permit one only to clutch at nature, never to lay hold of her and capture her." [14]

Nature existed in three states—at liberty, in error, or in bondage:

> She is either free and follows her ordinary course of development as in the heavens, in the animal and vegetable creation, and in the general array of the universe; or she is driven out of her ordinary course by the perverseness, insolence, and forwardness of matter and violence of impediments, as in the case of monsters; or lastly, she is put in constraint, molded, and made as it were new by art and the hand of man; as in things artificial. [15]

The first instance was the view of nature as immanent self-development, the nature naturing herself of the Aristotelians. This was the organic view of nature as a living, growing, self-actualizing being. The second state was necessary to explain the malfunctions and monstrosities that frequently appeared and that could not have been caused by God or another higher power acting on his instruction. Since monstrosities could not be explained by the action of form or spirit, they had to be the result of matter acting perversely. Matter in Plato's *Timaeus* was recalcitrant and had to be forcefully shaped by the demiurge. Bacon frequently described matter in female imagery, as a "common harlot." "Matter is not devoid of an appetite and inclination to dissolve the world and fall back into the old Chaos." It therefore must be "restrained and kept in order by the prevailing concord of things." "The vexations of art are certainly as the bonds and handcuffs of Proteus, which betray the ultimate struggles and efforts of matter."[16]

The third instance was the case of art (techné)—man operating on nature to create something new and artificial. Here "nature takes orders from man and works under his authority." Miners and smiths should become the model for the new class of natural philosophers who would interrogate and alter nature. They had developed the two most important methods of wresting nature's secrets from her, "the one searching into the bowels of nature, the other shaping nature as on an anvil." "Why should we not divide natural philosophy into two parts, the mine and the furnace?" For "the truth of nature lies hid in certain deep mines and caves," within the earth's bosom. Bacon, like some of the practically minded alchemists, would "advise the studious to sell their books and build furnaces" and, "forsaking Minerva and the Muses as barren virgins, to rely upon Vulcan."[17]

The new method of interrogation was not through abstract notions, but through the instruction of the understanding "that it may in very truth dissect nature." The instruments of the mind supply suggestions, those of the hand give motion and aid the work. "By art and the hand of man," nature can then be "forced out of her natural state and squeezed and molded." In this way, "human knowledge and human power meet as one."[18]

Here, in bold sexual imagery, is the key feature of the modern experimental method—constraint of nature in the laboratory, dissection by hand and mind, and the penetration of hidden secrets—language still used today in praising a scientist's "hard facts," "penetrating mind," or the "thrust of his argument." The constraints against penetration in Natura's lament over her torn garments of modesty have been turned into sanctions in language that legitimates the exploitation and "rape" of nature for human good. The seventeenth-century experimenters of the Accademia del Cimento of Florence (i.e., The Academy of Experiment, 1657–1667) and the Royal Society of London who placed mice and plants in the artificial vacuum of the barometer or bell jar were vexing nature and forcing her out of her natural state in true Baconian fashion.[19]

Scientific method, combined with mechanical technology, would create a "new organon," a new system of investigation, that unified knowledge with material power. The technological discoveries of printing, gunpowder, and the magnet in the fields of learning, warfare, and navigation "help us to think about the secrets still locked in nature's bosom." "They do not, like the old, merely exert a gentle guidance over nature's course; they have the power to conquer and subdue her, to shake her to her foundations." Under the mechanical arts, "nature betrays her secrets more fully . . . than when in enjoyment of her natural liberty."[20]

Mechanics, which gave man power over nature, consisted in motion; that is, in "the uniting or disuniting of natural bodies." Most useful were the arts that altered the materials of things—"agriculture, cookery, chemistry, dying, the manufacture of glass, enamel, sugar, gunpowder, artificial fires, paper, and the like." But in performing these operations, one was constrained to operate within the chain of causal connections; nature could "not be commanded except by being obeyed." Only by the study, interpretation, and observation of nature could these possibilities be uncovered; only by acting as the interpreter of nature could knowledge be turned into power. Of the three grades of human ambition, the most wholesome and noble was "to endeavor to establish and extend the power and dominion of the human race itself over the universe." In this way "the human race [could] recover that right over nature which belongs to it by divine bequest."[21]

The interrogation of witches as symbol for the interrogation of nature, the courtroom as model for its inquisition, and torture through mechanical devices as a tool for the subjugation of disorder were fundamental to the scientific method as power. For Bacon, as for Harvey, sexual politics helped to structure the nature of the empirical method that would produce a new form of knowledge and a new ideology of objectivity seemingly devoid of cultural and political assumptions.

THE NEW ATLANTIS: A MECHANISTIC UTOPIA. Bacon's utopia, the *New Atlantis*, was written in 1624, shortly before his death.

In contrast to the organic egalitarian societies of Campanella and Andreä, in which women and men were to receive much the same education and honor, the social structure of Bacon's Bensalem was hierarchical and patriarchal, modeled on the early modern patriarchal family.

In seventeenth-century England, the largest community of individuals, with the exception of the church and the court, was the patriarchal family, consisting of the immediate family and its servants, usually not larger than a dozen people. Emotional sustenance, discontent, and reconciliation of differences were centered on the family. Love relationships, hatred, and tension existed within the family group but could be dealt with at the level of the individual. The large majority of people had little expectation of changing these social relationships and organizational patterns. The patriarchal family consisted of a baker, butcher, husbander, or master craftsperson, his wife, several children, paid journeypeople, apprentices, and maidservants. The man was master and father to the others, the woman a subordinate partner, manager of the women and mother to all.[22]

Recent research in family history indicates that the English family structure in the late sixteenth and early seventeenth centuries, was becoming more patriarchal and authoritarian. The status and rights of wives declined, the cult of the Virgin Mary paled with the weakening of Catholicism, and an emphasis on loyalty to sovereign authority in central government reinforced the power of the male head of the household in the nuclear family. Kinship ties, which had been economically and religiously important in the Middle Ages, were weakened by the new geographical mobility of artisans, the poor, and the young. As a result the nuclear family grew stronger, the new stress on the marital union of the couple weakening the ties to parents and cousins. The husband gained power over the wife, the father over the children. Protestantism subordinated the wife and children to the father, while the state emphasized the subordination of the subjects. James I asserted that "Kings are compared to fathers in families: for a king is truly *parens patriae*, the politic father of his people."[23]

Bacon's Bensalem in the *New Atlantis* illustrated a patriarchal family structure in which the "Father" exercised authority over the kin and the role of the woman had been reduced to near invisibility. The shipwrecked visitors to Bensalem were invited to a "Feast of the Family" headed by the "Father of the Family, whom they call the Tirsan."[24] The Father was described entering the room where the feast was to be held followed by "all his generation or lineage, the males before him and the females following him." The mother, however, although present, was kept hidden behind a glass partition. "If there be a mother from whose body the whole lineage is descended, there is a traverse [partition] placed in a loft above on the right hand of the chair, with a privy door and a carved window of glass, leaded with gold and blue, where she sitteth but is not seen." After the opening ceremony, the Father began the dinner alone, "served only by his own children, such as are male, who perform unto him all service of the table upon the knee, and the women only stand about him, leaning against the wall." After the guests had been served in the room below, hymns were sung to the fathers: Adam, Noah, Abraham, the Father of the Faithful and the Savior. Then in order of age the Tirsan blessed his children in the name of the "everlasting Father, Prince of Peace." The son chosen to be in the house with him was named Son of the Vine and given the family's golden cluster of grapes to bear before his father in public places.

The "Father" had the assistance of the governor of the state who "put in execution by his public authority the decrees and orders of the Tirsan." However, because of the respect and obedience paid by all to the "order of nature" this was only infrequently required.

The "Father" also settled family quarrels, and appeased "discords and suits." He decreed relief for any family members in financial distress or "bodily decay." He chastised those who had fallen into vice and approved impending marriages. Through these vivid descriptions, Bacon provided an illustration of the social role assigned to the patriarchal father by seventeenth-century society.

Another manifestation of social hierarchy in the *New Atlantis* was the attention Bacon gave to the clothing of state officials and scientists, men of high status in the Bensalem community. The dress of the "revered person of place" with whom the visitors first conversed, the governor of the House of Strangers, and the scientists from Salomon's House was described meticulously. This had interesting social implications.

In sixteenth- and early seventeenth-century England, clothing was an indicator of an individual's status group in a hierarchically

Figure 16. *Five Women,* from the Art Collection of the Folger Shakespeare Library, Washington, D.C. In the Middle Ages and the Renaissance, clothing was used to show social status: a merchant's wife, three noblewomen, and a countrywoman are shown here.

stratified society. A series of clothing laws extending back to the fourteenth century regulated the types of cloth and accessories permitted as wearing apparel for the individuals of each social stratum and forbade people in each group to wear clothing designated for those above them. The English Parliament continued passing such laws until the early seventeenth century, an attempt by the state to maintain existing social stratification and to prevent upward mobility (Fig. 16.)[25]

In his *Anatomy of Abuses* (1583), Phillip Stubbes (fl. 1581–1593) complained that

The greatest abuse . . . is the abominable sin of pride and excess in clothing . . . by wearing clothing more gorgeous, sumptuous and precious than our state, calling or condition of life requires. . . . The nobility . . . and the gentry may use a rich and precious kind of clothing (in fear of God) to ennoble, garnish, and exhibit their births, dignities, functions, and callings. . . . The magistrates and public officials of every title may also wear (according to their abilities) . . . costly ornaments and rich clothing, to dignify their callings and to demonstrate and display the excellence and worthiness of their offices and functions, so as to strike terror and fear into the hearts of the people who might offend against the majesty of their callings. . . . As for private subjects, it is not in any way proper for them to wear silks, velvets, satins, damasks, gold, silver, or whatever they like (however much they are able to afford it).[26]

In the *New Atlantis*, Bacon's clothing descriptions illustrate Stubbes' observation that the clothing of the nobility, the magistrates and public officials was a mark of the majesty of their conditions of life. Thus the "revered man" who welcomed the shipwrecked visitors to Bensalem was dressed in a wide-sleeved, glossy, azure-colored gown made of camlet—an expensive camel's hair cloth with a satin weave—and a daintily made turban.[27] The governor of the House of Strangers was "clothed in blue . . . save that his turban was white with a small red cross on the top" and wore "a tippet of white linen." The "scientist" from Salomon's House, who wore a robe of fine black cloth, excellent white linen undergarments, and velvet shoes, was an indicator of the high status Bacon accorded to science in the utopian community.

Except for the garments of the fifty attendants of the scientist and the heralds of the Tirsan, the clothing of the ordinary citizens and servants of Bensalem was not described, as it was presumably not proper for them to dress in fine materials. For example, the servant who came aboard the visitor's boat refused to accept "a piece of crimson velvet to be presented to the officer," and, as for the officer himself, only his tipstaff was mentioned.

Bacon's utopian society therefore accepted the period's existing social stratification, but displayed upward mobility in the case of the "scientist," who was placed in the highest stratum. This structure contrasted sharply with the utopias of Campanella and Andreä. The white togas worn by all the citizens of the City of the Sun can be viewed, not as a manifestation of uniformity or monastic severity, but as a protest against the stratification of status and wealth reflected in fine, rich clothing. Likewise, in Christianopolis all people wore gray or white clothing made of linen or wool, depending on the season. "They have only two suits of clothes, one for their work, one for the holidays; and for all classes they are made alike. . . . None have fancy, tailored goods."[28]

Hierarchy and patriarchy, which supported social inequality in actual seventeenth-century society, were not even questioned by Bacon. That they formed the very foundation of his ideal state was an indication of his nonegalitarian philosophy.

CAPITALISM AND SCIENTIFIC PROGRESS.

Bacon's *New Atlantis* postulated a program of scientific study that would be a foundation for the progress and advancement of "the whole of mankind." But, whereas Andreä and Campanella had in mind the improvement of the lot of the peasantry, beggars, cottagers, and artisans, Bacon can be identified with the interests of the clothier capitalists, merchants, mine owners, and the state.[29]

By the time Bacon wrote his *New Atlantis*, a significant cleavage existed in English society between wage laborers and merchants. The rift between middle-class society and the poorer sectors was developing in the textile industry, mining industry, and the crafts.[30]

In seventeenth-century England, the rural poor became servants for the families of gentleman landlords, husbanders, and yeoman farmers. The cottager's son or daughter who became a servant in husbandry left home around the age of ten and was cared for, fed, clothed, and housed by the surrogate family for the next ten to twenty years. After marriage, probably to another servant, the cottage-laborer might well face the rest of his or her short life in poverty, earning small sums for daily labor contracts.

Cottagers supplied much of the labor for the rural putting-out systems, which combined large numbers of households in the production of textiles under the direction of a clothier capitalist. When not employed in planting, plowing, and harvesting operations, farmers and their families engaged in the sorting, carding, and spinning of wool. The subsequent weaving, dying, and dressing of cloth was performed by craftspeople, journeypeople, and artisans who only secondarily helped with harvesting in the late summer and fall. The clothier supplied the raw materials and marketed the textiles in domestic and international trade.

During the sixteenth century, the relationship of the artisan to the clothier capitalist had been changing. When the rural weaver could not afford a loom, the clothier offered one for rent. Most looms were operated in individual homes, but in some rural communities the clothier's enterprise became concentrated in a group of houses or in one building. A larger clothier might own dye houses where the work was supervised, and fulling mills and workshops where the cloth was stretched and pressed. In 1618, an English writer estimated that a clothier who made twenty broadcloths a week provided work for 500 persons, counting wool sorters, carders, spinners, weavers, burlers, fullers, cloth finishers, dyers, and loom and spinning wheel makers. In the West Country, the capital investments were larger and the number of operations directed by the clothier more extensive than in the Yorkshire country of the north.

The transition from craft production to preindustrial capitalism taking place throughout the century was more pronounced in the rural rather than the urban putting-out systems. A clothier in the rural putting-out system was freer than his urban counterpart from municipal taxes and regulations and from restrictions on the quality of his product, the number of his employees, and his methods of production. "The expansive years between 1460 and 1560 are particularly important because the balance of tradition and innovation shifted gradually but decisively. . . . But by 1560 the cleavage between capital and labor . . . was firmly and widely established in many parts of industrial Europe."[31] Rising prices widened the separation between wages and profits with a larger share of community wealth going to the capitalist.

A second industry that employed the poor as wage workers was mining.[32] Large-scale operations were rare, with only about 100 workers being employed in each of the larger mines. In England, the coal industries at Newcastle-Upon-Tyne and Wear developed rapidly in the late sixteenth century, impelled by the increasing scarcity of timber.

In the British copper and brass industry that developed in the 1560s, large capital investments were necessary for opening and developing the mining shafts, smelting the ore, producing brass wire, and flattening ingots. Since neither the workers nor any single capitalist had the necessary funds, capital was supplied by English and German shareholders—members of the nobility, clergy, state officials, and merchants. Separation of worker and capitalist was thus a prerequisite for the start of this industry.[33] In the iron industry, foundry and forge were owned and products marketed by entrepreneurs. Free and independent miners and metal workers were a decreasing group.

A similar separation was taking place within the crafts, created by decreasing upward mobility for journeypeople. By hard and diligent work or by marrying the master's daughter, a journeyman might succeed.[34] But more and more masters tended to pass their craft to their sons, making the group hereditary. Masters became "small-scale industrial capitalists," and journeypeople became their paid workers, with less chance for independence. The journeyman weaver, for example, owned neither the material, as did the clothier, nor the looms, as did the master weaver.

Within the craft guilds, some masters accumulated money and extended the markets beyond their own towns. Lower craftspersons became more dependent on them. The same phenomenon of market extension and dependence also took place between one craft and another.

Francis Bacon's early interest in writing a "History of Trades" was a manifestation of his desire to discover those secrets of the craft workshops that could be applied to the practical needs and interests of middle-class society. Growth and progress could be achieved from the study of the mechanical arts, "for these ... are continually thriving and growing."[35]

The concept of scientific progress that Bacon developed as a program sanctioned the gap between journeyman and mastercraftsman. Much has been made of the concept of progress in Western society, through which standards of living for "all mankind" are presumably improved. But did the "public good" really include the cottager, journeyperson, and peasant, or did it function so as to benefit the master craftsman, clothier, and merchant?

The idea of scientific progress has been associated with the rise of technology and "the requirements of early capitalistic economy" by scholars who have argued that the idea of cooperation and the sharing of knowledge for both the construction of theory and the public good stemmed from the intellectual attitudes of sixteenth-century master craftsmen, mechanical engineers, and a few academic scholars and humanists. "The absence of slavery, the existence of machinery, the capitalistic spirit of enterprise and economic rationality seem to be prerequisites without which the ideal of scientific progress cannot unfold."[36]

The sixteenth-century groups that evolved the concept of progress are the same groups that right up until the present have pressed for increased growth and development: entrepreneurs, military engineers, humanist academics, and scientists and technicians. Sixteenth-century master craftsmen and technicians who embraced the idea of progress in written works included: Kaspar Brunner (1547), master of ordnance, locksmith, and clockmaker; Robert Norman (1581), instrument maker; Peter Apian (1532), mechanic, globe and instrument maker, and mathematics professor; Ambroise Paré (1575) master of the guild of barber surgeons and surgeon to the king and military; and Gerard Mercator (1595), instrument maker, map maker, and surveyor. Military engineers included William Bourne (1578), gunner and military engineer, who addressed his treatise to "all generals and captains"; Niccolò Tartaglia (1537), mathematical advisor to gunners and merchants; Simon Stevin (1608), military engineer and bookkeeper; and Bonaiuto Lorini (1597), military technician to the Medicis. Humanists and academics included Abraham Ortelius (1570), classical scholar and mapmaker; François Rabelais (1535), writer; Jean Bodin (1566), political philosopher; and Loys Leroy (1568). Humanist concerns are not only fully compatible with the improvement of the human condition through technological advance, but imply an environment filled with humans at the expense of nature.

What had been merely prefaces and statements advocating a utilitarian concept of progress in these sixteenth-century treatises became a whole program and ideology in the utopian thought of Francis Bacon. In the *New Atlantis*, progress was placed in the hands of a group of scientists and technicians who studied nature altered by "the mechanical arts" and "the hand of man" that her secrets might be utilized to benefit society.

MECHANISM AND THE NEW ATLANTIS.

The scientific research institute designed to bring progress to Bensalem, the community of the *New Atlantis*, was called Salomon's House. The patriarchal character of this utopian society was reinforced by designating the scientists as the "Fathers of Salomon's House." In the *New Atlantis*, politics was replaced by scientific administration. No real political process existed in Bensalem. Decisions were made for the good of the whole by the scientists, whose judgment was to be trusted implicitly, for they alone possessed the secrets of nature.

Scientists decided which secrets were to be revealed to the state as a whole and which were to remain the private property of the institute rather than becoming public knowledge: "And this we do also, we have consultations, which of the inventions and experiences which we have discovered shall be published, and which not: and all take an oath of secrecy for the concealing of those which we think fit to keep secret, though some of those we do reveal sometimes to the state, and some not."[37]

The cause of the visit to the governor by a scientist from the distant Salomon's House, which resulted in a conference with the visitors to Bensalem, was shrouded in secrecy. No father of the institute had been seen in "this dozen years. His coming [was] in state, but the cause of his coming [was] secret."

The scientist father was portrayed much like the high priest of the occult arts, the Neoplatonic magus whose interest in control and power over nature had strongly influenced Bacon. He was clothed in all the majesty of a priest, complete with a "robe of fine black cloth with wide sleeves and a cape," an "undergarment . . . of excellent white linen," and a girdle and a clerical scarf, also of linen. His gloves were set with stone, his shoes were of peach-colored velvet, and he wore a Spanish helmet.

The worship to be accorded to the scientist was further enhanced by his vehicle, a "rich chariot" of cedar and gilt carried like a litter between four richly velveted horses and two blue-velveted footmen. The chariot was decorated with gold, sapphires, a golden sun, and a "small cherub of gold with wings outspread" and was followed by fifty richly dressed footmen. In front walked two bareheaded men carrying a pastoral staff and a bishop's crosier.

Bacon's scientist not only looked but behaved like a priest who had the power of absolving all human misery through science. He "had an aspect as if he pitied men"; "he held up his bare hand as he went, as blessing the people, but in silence." The street was lined with people who, it would seem, were happy, orderly, and completely passive: "The street was wonderfully well kept, so that there was never any army [which] had their men stand in better battle array than the people stood. The windows were not crowded, but everyone stood in them as if they had been placed."

Bacon's "man of science" would seem to be a harbinger of many modern research scientists. Critics of science today argue that scientists have become guardians of a body of scientific knowledge, shrouded in the mysteries of highly technical language that can be fully understood only by those who have had a dozen years of training. It is now possible for such scientists to reveal to the public only information they deem relevant. Depending on the scientist's ethics and political viewpoint, such information may or may not serve the public interest.

Salomon's House, long held to be the prototype of a modern research institute, was a forerunner of the mechanistic mode of scientific investigation. The mechanical method that evolved during the seventeenth century operated by breaking down a problem into its component parts, isolating it from its environment, and solving each portion independently. Bacon's research center maintained separate "laboratories" for the study of mining and metals, weather, fresh- and salt-water life, cultivated plants, insects, and so on.

The tasks of research were divided hierarchically among the various scientists, novices, and apprentices. Some abstracted patterns from other experiments, some did preliminary book research, some collected experiments from other arts and sciences; others tried out new experiments, or compiled results or looked for applications. The interpreters of nature raised the discoveries into greater observations, axioms, and aphorisms. This differentiation of labor followed the outlines of Bacon's inductive methodology.

In the laboratories of Salomon's House, one of the goals was to recreate the natural environment artificially through applied technology. Large, deep caves called the Lower Region were used for "the imitation of natural mines and the producing of new artificial metals by compositions and materials."[38] In another region were "a number of artificial wells and fountains, made in imitation of the natural sources and baths." Salt water could be made fresh, for "we have also pools, of which some do strain fresh water out of salt, and others by art do turn fresh water into salt."

Not only was the manipulation of the environment part of Bacon's program for the improvement of mankind, but the manipulation of organic life to create artificial species of plants and animals was specifically outlined. Bacon transformed the natural magician as "servant of nature" into a manipulator of nature and changed art from the aping of nature into techniques for forcing nature into new forms and controlling reproduction for the sake of production: "We make a number of kinds of serpents, worms, flies, fishes of putrefaction, where of some are advanced (in effect) to be perfect creatures like beasts or birds, and have sexes, and do propagate. Neither do we this by chance, but we know beforehand of what matter and commixture what kind of those creatures will arise."

These examples were taken directly from Della Porta's *Natural Magic* (1558), the second book of which dealt specifically with putrefaction and the generation of the living organisms mentioned by Bacon—worms, serpents, and fishes. The chapter dealing with putrefaction had discussed the generation of canker worms from mud, so that "we may also learn how to procreate new creatures."[39] "Serpents," wrote Della Porta, "may be generated of man's marrow, of the hairs of a menstrous woman, and of a horsetail, or mane," while "certain fishes," such as groundlings, carp, and shellfish, "are generated out of putrefaction." New beasts and birds could be generated through knowledge and carefully controlled coupling.

Della Porta also set down instructions as to how to produce a new organism in a series of trials. Such creatures "must be of equal pitch; they must have the same reproductive cycle, and one must be equally "as lustful as the other." Furthermore "if any creatures want appetite . . . we may make them eager in lust."

The *New Atlantis* had parks and enclosures for beasts and birds where just such experiments were performed: "By art likewise we make them greater or taller than their kind is, and contrariwise dwarf them, and stay their growth; we make them more fruitful and bearing than their kind is, and contrariwise barren and not generative. Also we make them differ in color, shape, activity, many· ways."[40]

The scientists of Salomon's House, not only produced new forms of birds and beasts, but they also altered and created new species of herbs and plants: "We have also means to make divers plants rise by mixtures of earths without seeds, and likewise to make divers new plants differing from the vulgar, and to make one tree or plant turn into another."

Rather than respecting the beauty of existing organisms, Bacon's *New Atlantis* advocated the creation of new ones:

We have also large and various orchards and gardens, wherein we do not so much respect beauty as variety of ground and soil, proper for diverse trees and herbs. . . . And we make (by art) in the same orchards and gardens, trees and flowers to come earlier or later than their seasons, and to come up and bear more speedily than by their natural course they do. We make them by art greater much than their nature, and their fruit greater and sweeter and of differing taste, smell, color, and figure, from their nature.[41]

Della Porta had, again, given numerous examples of changing the colors and tastes of plants: a white vine could be turned into a black one, purple roses and violets could become white, and sweet almonds and pomegranates sour.

That such experimentation on animals and the creation of new species was ultimately directed toward human beings was intimated by Bacon: "We have also parks and enclosures of all sorts of beasts and birds, which we use not only for view or rareness but likewise for dissections and trials, that thereby we may take light [i.e., enlightenment] what may be wrought upon the body of man. . . . We also try all poisons and other medicines upon them as well of chirurgery as physic."[42]

Much of Bacon's strategy in the *New Atlantis* was directed at removing ethical strictures against manipulative magic, of the sort found in Agrippa's *Vanity of Arts and Science* (1530), a polemic probably written for Agrippa's own self-protection, containing important arguments against transforming and altering nature. Just as Agricola had been obliged to refute Agrippa's views on mining in order to liberate that activity from the ethical constrants imposed by ancient writers, so Bacon was obliged to refute the constraints against the manipulation of nature. Agrippa had argued against tampering with nature and maiming living organisms:

Those exercises appurtenant to agriculture . . . might in some measure deserve commendation, could it have retained itself within moderate bounds and not shown us so many devices to make strange plants, so many portentous graftings and metamorphoses of trees; how to make horses copulate with asses, wolves with dogs, and so to engender many wondrous monsters contrary to nature: and those creatures to whom nature has given leave to range the air, the seas and earth so freely, to captivate and confine in aviaries, cages, warrens, parks, and fish ponds, and to fat them in coops, having first put out their eyes, and maimed their limbs.[43]

Agrippa had further inveighed against the manipulators of nature who had tried to discover "how to prevent storms, make . . . seed fruitful, kill weeds, scare wild beasts, stop the flight of beasts and birds, the swimming of fishes, to charm away all manner of diseases; of all which those wise men before named have written very seriously and very cruelly."

Much of Bacon's program in the *New Atlantis* was meant to sanction just such manipulations, his whole objective being to recover man's right over nature, lost in the Fall. Agrippa had observed that after the Fall nature, once kind and beneficent, had become wild and uncontrollable: "For now the earth produces nothing without our labor and our sweat, but deadly and venomous, . . . nor are the other elements less kind to us: many the seas destroy with raging tempests, and the horrid monsters devour: the air making war against us with thunder, lightning and storms; and with a crowd of pestilential diseases, the heavens conspire our ruin."

In order to control the ravages of wild tempestuous nature, Bacon set as one of the objectives of Salomon's House the artificial control of the weather and its concomitant monsters and pestilences: "We have also great and spacious houses, where we imitate and demonstrate meteors, as snow, hail, rain, some artificial rains of bodies and not of water, thunder, lightnings, also generation of bodies in air, as frogs, flies, and diverse others." Tempests (like that produced by Shakespeare's magician, Prospero), could also be created for study by using "engines for multiplying and enforcing of winds."[44]

The Baconian program, so important to the rise of Western science, contained within it a set of attitudes about nature and the scientist that reinforced the tendencies toward growth and progress inherent in early capitalism. While Bacon himself had no intimation as to where his goals might ultimately lead, nor was he responsible for modern attitudes, he was very sensitive to the trends and directions of his own time and voiced them eloquently. The expansive tendencies of his period have continued, and the possibility of their reversal is highly problematical.

Bacon's mechanistic utopia was fully compatible with the mechanical philosophy of nature that developed during the seventeenth century. Mechanism divided nature into atomic particles, which, like the civil citizens of Bensalem, were passive and inert.

Motion and change were externally caused: in nature, the ultimate source was God, the seventeenth century's divine father, clockmaker, and engineer; in Bensalem, it was the patriarchal scientific administration of Salomon's House. The atomic parts of the mechanistic universe were ordered in a causal nexus such that by contact the motion of one part caused the motion of the next. The linear hierarchy of apprentices, novices, and scientists who passed along the observations, experimental results, and generalizations made the scientific method as mechanical as the operation of the universe itself. Although machine technology was relatively unadvanced in Bensalem, the model of nature and society in this utopia was consistent with the possibilities for increased technological and administrative growth.

In the *New Atlantis* lay the intellectual origins of the modern planned environments initiated by the technocratic movement of the late 1920s and 1930s, which envisioned totally artificial environments created by and for humans. Too often these have been created by the mechanistic style of problem solving, which pays little regard to the whole ecosystem of which people are only one part. The antithesis of holistic thinking, mechanism neglects the environmental consequences of synthetic products and the human consequences of artificial environments. It would seem that the creation of artificial products was one result of the Baconian drive toward control and power over nature in which "The end of our foundation is the knowledge of causes and secret motions of things and the enlarging of the bounds of human empire, to the effecting of all things possible."[45] To this research program, modern genetic engineers have added new goals—the manipulation of genetic material to create human life in artificial wombs, the duplication of living organisms through cloning, and the breeding of new human beings adapted to highly technological environments.

THE BACONIAN PROGRAM. The development of science as a methodology for manipulating nature, and the interest of scientists in the mechanical arts, became a significant program during the latter half of the seventeenth century. Bacon's followers realized even more clearly than Bacon himself the connections between mechanics, the trades, middle-class commercial interests, and the domination of nature.

Lewis Roberts lamented the unexploited state of Mother Earth in his *Treasure of Traffike, or a Discourse of Foreign Trade* (1641):

> The earth, though notwithstanding it yieldeth thus naturally the richest and most precious commodities of all others, and is properly the fountain and mother of all the riches and abundance of the world, partly . . . bred within its bowels, and partly nourished upon the surface thereof, yet is it observable, and found true by daily experience in many countries, that the true search and inquisition thereof, in these our days, is by many too much neglected and omitted.[46]

John Dury and Samuel Hartlib, followers of Bacon and organizers of the Invisible College (ca. 1645), forerunner of the Royal Society, connected the study of the crafts and trades to increasing wealth. One of Dury's objectives was to make observations of the inventions and sciences "as may be profitable to the health of the body, to the preservation and increase of wealth by trades and mechanical industries, either by sea or land; either in peace or war."[47]

The avowedly Baconian utopia "The Kingdom of Macaria, (1641), attributed to Hartlib but probably written by Gabriel Plattes, an English writer on husbandry and mining, was dedicated not merely to the "knowledge of causes and secret motions of things," as was the *New Atlantis*, but to the total agricultural, commercial, and medical improvement of society.[48] In Macaria, the king has improved his forests, parks, and lands "to the utmost"—bringing in huge revenues. Owing to the efforts of the council of husbandry, "the whole kingdom is become like to a fruitful garden, the highways are paved, and are as fair as the streets of the city." Any man who held more land than he could develop and improve was admonished and penalized for each year during which he continued to leave it unimproved, until at last "his lands be forfeited and he banished out of the kingdom, as an enemy to the commonwealth." A council of fishing was to establish laws "whereby immense riches are yearly drawn out of the ocean," while the councils of trade by land and sea were to regulate the number of tradespeople and encourage all navigation that "may enrich the kingdom."

The health of the inhabitants was maintained by a "college of experience, where they deliver out yearly such medicines as they find out by experience." As members of the Society of Experimenters, all were required to defend any new ideas before a Great Council, which judged the truth or falsity of the discovery. "If any divine shall publish a new opinion to the common people, he shall be accounted a disturber of the public peace and shall suffer death for it."

Dissent, not only in science but also in religion, would be avoided "by invincible arguments as will abide the grand test of extreme dispute." Rational scientific judgment would thus overcome the passions and individualism of religious sects and promote health, welfare, and commercial growth in Macaria.

The virtuosi of the Royal Society were interested in carrying out Bacon's proposal to survey the history of trades and augment their usefulness. The English divine Thomas Sprat, whose *History of the Royal Society* (1667) defended it against its critics, desired to extract from the "operations of all trades," their "physical receipts or secrets," their "instruments, tools, engines, [and] manual operations." He extolled "our chief and most wealthy merchants and citizens" who had added their "industrious, punctual, and active genius" to the "quiet, sedentary, and reserved temper of men of learning."[49]

Human dominion over nature, an integral element of the Baconian program, was to be achieved through the experimental "disclosure of nature's secrets." Seventeenth-century scientists, reinforcing aggressive attitudes toward nature, spoke out in favor of "mastering" and "managing" the earth. Descartes wrote in his *Discourse on Method* (1636) that through knowing the crafts of the artisans and the forces of bodies we could "render ourselves the masters and possessors of nature."[50] Joseph Glanvill, the English philosopher who defended the Baconian program in his *Plus Ultra* of 1668, asserted that the objective of natural philosophy was to "enlarge knowledge by observation and experiment . . . so that nature being known, it may be mastered, managed, and used in the services of humane life." To achieve this objective, arts and instruments should be developed for "searching out the beginnings and depths of things and discovering the intrigues of remoter nature."[51] The most useful of the arts were chemistry, anatomy, and mathematics; the best instruments included the microscope, telescope, thermometer, barometer, and air pump.

The harshness of Bacon's language was captured in Glanvill's descriptions of the methods of studying nature. Bacon had advocated the dissection of nature in order to force it to reveal its secrets. For Glanvill, anatomy, "most useful in human life, . . . tend[ed] mightily to the eviscerating of nature, and disclosure of the springs of its motion." In searching out the secrets of nature, nothing was more helpful than the microscope for "the secrets of nature are not in the greater masses, but in those little threads and springs which are too subtle for the grossness of our unhelped senses."

According to Glanvill, Robert Boyle's experimental philosophy had advanced "the empire of man over inferior creatures" by taking seriously "those things which have been found out by illiterate tradesmen" and by developing the "dexterity of hand proper to artificers." Glanvill advocated chemistry as one of the most useful arts, for "by the violence of [its] artful fires it is made [to] confess those latent parts, which upon less provocation it would not disclose." By chemical techniques, "nature is unwound and resolved into the minute rudiments of its composition."

In his "Experimental Essays" (1661), Boyle distinguished between merely knowing as opposed to dominating nature in thinly veiled sexual metaphor: "I shall here briefly represent to you . . . that there are two very distinct ends that men may propound to themselves in studying natural philosophy. For some men care only to know nature, others desire to command her" and "to bring nature to be serviceable to their particular ends, whether of health, or riches, or sensual delight."[52]

The new image of nature as a female to be controlled and dissected through experiment legitimated the exploitation of natural resources. Although the image of the nurturing earth popular in the Renaissance did not vanish, it was superseded by new controlling imagery. The constraints against penetration associated with the earth-mother image were transformed into sanctions for denudation. After the Scientific Revolution, *Natura* no longer complains that her garments of modesty are being torn by the wrongful thrusts of man. She is portrayed in statues by the French sculptor Louis-Ernest Barrias (1841–1905) coyly removing her own veil and

Figure 17.   Nature Reveals Herself, sculpture by Louis-Ernest Barrias (French, 1841-1905). This sculpture suggests the sexuality of nature in revealing her secrets to science. A similar statue by the same sculptor in the Ecole de Medecine, Paris, bears the inscription, "La Nature se devoilant devant la Science" ("Nature Revealing Herself to Science").

exposing herself to science (Fig. 17). From an active teacher and parent, she has become a mindless, submissive body. Not only did this new image function as a sanction, but the new conceptual framework of the Scientific Revolution—mechanism—carried with it norms quite different from the norms of organicism. The new mechanical order (Chapter 8) and its associated values of power and control (Chapter 9) would mandate the death of nature.

CHAPTER 8

The Mechanical Order

The fundamental social and intellectual problem for the seventeenth century was the problem of order. The perception of disorder, so important to the Baconian doctrine of dominion over nature, was also crucial to the rise of mechanism as a rational antidote to the disintegration of the organic cosmos. The new mechanical philosophy of the mid-seventeenth century achieved a reunification of the cosmos, society, and the self in terms of a new metaphor—the machine. Developed by the French thinkers Mersenne, Gassendi, and Descartes in the 1620s and 1630s and elaborated by a group of English emigrés to Paris in the 1640s and 1650s, the new mechanical theories emphasized and reinforced elements in human experience developing slowly since the late Middle Ages, but accelerating in the sixteenth century.

New forms of order and power provided a remedy for the disorder perceived to be spreading throughout culture. In the organic world, order meant the function of each part within the larger

192

whole, as determined by its nature, while power was diffused from the top downward through the social or cosmic hierarchies. In the mechanical world, order was redefined to mean the predictable behavior of each part within a rationally determined system of laws, while power derived from active and immediate intervention in a secularized world. Order and power together constituted control. Rational control over nature, society, and the self was achieved by redefining reality itself through the new machine metaphor.

As the unifying model for science and society, the machine has permeated and reconstructed human consciousness so totally that today we scarcely question its validity. Nature, society, and the human body are composed of interchangeable atomized parts that can be repaired or replaced from outside. The "technological fix" mends an ecological malfunction, new human beings replace the old to maintain the smooth functioning of industry and bureaucracy, and interventionist medicine exchanges a fresh heart for worn-out, diseased one.

The mechanical view of nature now taught in most Western schools is accepted without question as our everyday, common sense reality—matter is made up of atoms, colors occur by the reflection of light waves of differing lengths, bodies obey the law of inertia, and the sun is in the center of our solar system. None of this was common sense to our seventeenth-century counterparts. The replacement of the older, "natural" ways of thinking by a new and "unnatural" form of life—seeing, thinking, and behaving—did not occur without struggle. The submergence of the organism by the machine engaged the best minds of the times during a period fraught with anxiety, confusion, and instability in both the intellectual and social spheres.

The removal of animistic, organic assumptions about the cosmos constituted the death of nature—the most far-reaching effect of the Scientific Revolution. Because nature was now viewed as a system of dead, inert particles moved by external, rather than inherent forces, the mechanical framework itself could legitimate the manipulation of nature. Moreover, as a conceptual framework, the mechanical order had associated with it a framework of values based on power, fully compatible with the directions taken by commercial capitalism.

THE RISE OF MECHANISM IN FRANCE. Against the background of the disorder in society, religion, and cosmology of the late sixteenth and early seventeenth centuries, mechanism arose as an antidote to intellectual uncertainty and as a new rational basis for social stability. The emphasis on individual interpretation brought about during the Reformation had raised the specter of the anarchy of ideas and created the serious problem of the source of certainty and authority in the sphere of knowledge. Martin Luther (1483–1546) had challenged the authority of the pope and his councils to determine religious truth by asserting that all Christians had the power to discern truth or falsity for themselves through reading the scriptures. Each person's individual conscience should be the criterion for religious truth, not the old standards for faith set by the church hierarchy. Luther's search for criteria in religious truth was applied to all natural knowledge by the skeptics of the late sixteenth century.[1]

The skeptical movement, based on a renewal of interest in ancient writings on skepticism, was itself a response to the social uncertainty created by the Reformation. After 1560, skeptical ideas became widely known, making the search for certainty in knowledge a primary intellectual problem in the early seventeenth century. The dispute over standards for truth could not be resolved without criteria by which to judge the arguments, but until the dispute itself had been resolved, the criteria could not be decided. The dilemma created by this circular reasoning was symptomatic of the deep rifts in the older, accepted system of values.

In the 1620s the French mechanists Marin Mersenne (1588–1648), Pierre Gassendi (1592–1655), and René Descartes (1596–1650), initially a mutually supportive social group, began to construct a mechanical philosophy that ultimately presented a solution to the problems of certainty, social stability, and individual responsibility. In reinstating moral and intellectual order, they revived the corpuscular philosophies of the ancient atomists, but placed them in a Christian context, and attempted to devise criteria for certainty in knowledge. For Descartes, a metaphysical system based on the principle of identity, on the immutable forms and mathematical axioms of Plato, and the primacy of God's intellect, logic, and rationality could reflect the stability hoped for in the intellectual world. For Gassendi, because God's will, rather than intellect, was primary, the laws of nature were only contingent, but could nevertheless be reasonably certain.[2]

In addition to their reaction against an absolute skepticism, such as that espoused by the Greek skeptic Pyrrho (ca. 365–275 B.C.), the mechanists, as will be seen, sharply criticized the presuppositions of the organic world view. From the spectrum of Renaissance organic philosophies, mechanism took over ideas compatible with order, control, and manipulation, while rejecting those associated with change, uncertainty, and unpredictability. Conservative ideas, such as the passivity and manipulability of matter conducive to the domination and control of nature, were appropriated into the new philosophical framework, while the more radical vitalistic and animistic ideas were subjected to severe criticism and rejected. The mechanical philosophies of Mersenne, Gassendi, and Descartes all exhibited a strong common reaction against naturalism, vitalism, and animistic magic, which in a world of social upheaval and religious conflict, could be seen as aiding and abetting disorder and chaos.

The mechanists transformed the body of the world and its female soul, source of activity in the organic cosmos, into a mechanism of inert matter in motion, translated the world spirit into a corpuscular ether, purged individual spirits from nature, and transformed sympathies and antipathies into efficient causes. The resultant corpse was a mechanical system of dead corpuscles, set into motion by the Creator, so that each obeyed the law of inertia and moved only by external contact with another moving body.

Thirdly, they criticized ideas associated with social disorder and anarchy, such as uncontrolled passion and spontaneity, individual criteria for religious truths, control over the spirits of nature by ordinary people, and "subversive" secret sects, such as the Rosicrucians, who claimed to have occult knowledge and power and were believed to hold views dangerous to Christian religion and society. In place of such rampant individualism, they substituted a more sober form of self-control, temperance, reasonable judgment, and sovereign law.

While social and historical conditions did not determine the specific content of the mechanical philosophy or form an impetus to construct a philosophy in direct response to social circumstances, they helped to make plausible some presuppositions about nature and to invalidate others. The social context helps to explain why intellectuals in the early seventeenth century became interested in particular ideas of their predecessors, transforming and elaborating them to be consistent with their own historical experience. And within the groups of French and English natural philosophers who devised the new mechanical philosophy, social relationships strengthened and supported the new directions taken.

Mersenne's search for certain knowledge was combined with an attack on skeptics, naturalists, pantheists, cabalists, atheists, and deists. In his writings of the 1620s, he denounced magical and atheistical ideas, stating that he did not know which were the more detestable. In *La Verité des Sciences* (1624), he attempted to refute the claims of the skeptics that no knowledge was certain by enumerating in 800 pages what was known about arithmetic, geometry, mathematical physics, their source of truth, their relation to nature and God, and what conclusions could be obtained by the hypothetical deductive method. The study of mathematics, he argued, was a study of hypothetical relationships in which truths are based on the condition that, if (for example) such things as triangles exist, then conclusions can be drawn concerning their properties. For practical purposes, sufficient information is known about mathematical relationships to be useful in everyday life.[3]

During the 1620s, Mersenne sought a new philosophy of nature to resolve the skeptical problem, and by 1634 had found what he was looking for in mechanism. His answer was to use mechanism not as an absolute truth but as a useful way of ordering knowledge about the practical everyday world. Ultimate knowledge was not possible, but a pragmatic knowledge based on everyday experience and the appearances of the senses could be attained. Viewing the world as a machine was the most practical and useful way to organize information derived from the appearances.

In his effort to establish the validity of the mathematical method and mechanism, Mersenne challenged the presuppositions of the organic world view. His attack was directed at the Neoplatonic concept of the soul of the world, source of its vital activity. Deriving from Plato's *Timaeus,* the world soul was central not only to the Neoplatonic views of Ficino, Pico della Mirandola, and Agrippa, but also to certain church fathers who held God to be immanent in his creation. Mersenne avoided heresy by transforming the threatening idea of the soul of the world into a mechanism of the world.[4]

He argued that, since the earth did not have organs of sensation, it was not alive. In opposition to Campanella, he asserted that the earth was not a living animal. Against Bruno and Telesio, he wrote that the earth did not have animal intelligence nor animal instinct: the world soul was neither the formal nor efficient cause of all earthly beings.

Mersenne also criticized potentially dangerous social ideas associated with the magical tradition by attacking his contemporary, Robert Fludd, defender of Rosicrucianism, a neo-Paracelsist religious, occult secret society. The appearance of Fludd's defenses (1617) convinced Mersenne that this dangerous group was real and not chimerical.[5] The immediacy of the Rosicrucian threat symbolized the larger problem of social control over anarchistic groups of ordinary people and served as a stimulus for a critique of the radical form of the Paracelsist tradition described in Chapter 4.

The Rosicrucian manifesto, the *Fama Fraternalis* (1614), had portrayed a secret society of brothers who (like Paracelsus) were committed to a life of wandering and of healing the sick without pay, assuming the clothing and customs of the common people of any particular country. They were to separate and remain secret, but were to communicate to each other any problems and errors they perceived. And they were to search for and train successors to perpetuate the brotherhood.

The Rosicrucian movement bore similarities to the millenarian mystical anarchists. The reputed founder, Christian Rosenkreutz, was supposed to have been born in 1378 and his tomb to have been discovered in 1604, the year of general millenarian reformation signaled by Kepler's 1604 supernova. Like the original "people's messiah," crusader Emperor Frederick I, Christian Rosenkreutz had traveled to the East where he learned his wisdom from Eastern sages. Rosenkreutz, a Paracelsist physician, was dedicated to the life of an itinerant doctor who wore the garb of the people of each locale.

The second Rosicrucian manifesto, the *Confessio*, which appeared in 1615, linked the brotherhood to millenarian and religious movements that had called the pope the Antichrist: "As we do now altogether freely and securely and without hurt call the Pope of

Rome Antichrist, that which heretofore was held for a deadly sin and men in all countries were put to death for it."[6]

Like the millenarian movements of the preceding centuries, the *Confessio* heralded an end to the world and predicted that the course of nature would be overturned and the virtuous relieved of their toil. Simon Studion's 1604 prophecy that the pope would fall from power in 1620, marking the end of the reign of the Antichrist and the beginning of the millennium, seemed to have been fulfilled when in 1620 (the sole year of his reign as the "Winter King" of Bohemia) Frederick V emerged as the leader of an activist political and hermetic movement that hoped to resolve Germany's social and religious problems. Andreas Libavius (d. 1616) an Aristotelian alchemist, viewed the Rosicrucians as a secret political, as well as magical, society that intended to undermine state authority. The Jesuit father, François Garasse, had condemned them as evil magicians operating as an extremely dangerous magical society, in a treatise (1623) on groups pernicious to religion and the state.[7]

Mersenne's fears seemed to be confirmed when in that same year a proclamation appeared announcing the arrival of the Rosicrucian brothers in Paris and stating that they would make their stay visible yet invisible in the capital city by teaching without books, marks, or signs and by speaking the language of the country. Their objective was to lead the people away from error and death. Although no Rosicrucian groups were uncovered, they existed as a subterranean chimera whose invisibility and secrecy were all the more threatening to an unstable society.

In reaction, Mersenne attacked Fludd as a heretic and visionary in the beginning chapters of his *Quaestionnes in Genesim* (1623). Fludd responded to these insults in two works written in 1629 against Mersenne: "The Battle of Wisdom with Folly" and "On the Supreme Good Which is the True Subject of the Cabala, Alchemy, and Brothers of the Rosy Cross," the second written under the pseudonym Joachim Frisius. Two members of Mersenne's order, François de la Noue and Jean Durée, rose to his defense. But, considering these to be insufficient, he wrote to Pierre Gassendi for help in refuting Fludd's ideas.

Gassendi was closely allied with Mersenne in the search for a respectable rational religion and a new science. After receiving his doctorate in theology, Gassendi served for a period as lecturer in philosophy at Aix and then as canon of Grenoble. On moving to Paris in 1624, he became a part of an intellectual circle known for its skepticism concerning Aristotelianism and orthodox intellectual authority.[8] Members of this group had received their offices from chief ministers Richelieu (1585–1642) and Mazarin (1602–1661): Gabriel Naudé, librarian to Richelieu and Mazarin; Guy Patin, rector of the medical school of the Sorbonne; Leonard Marande, secretary to Richelieu; and François de la Mothe le Vayer, teacher of the king's brother. Later, in 1645, through the efforts of Richelieu's brother, Gassendi was appointed to the chair of mathematics at the College Royale in Paris.

By 1622, Gassendi had prepared a set of theses for and against Aristotle. His attitude reflected a new critical stance against skepticism not unlike that of Descartes in the 1620s. Both were antagonistic toward Aristotle, and both were looking for a way of establishing truth on a new basis.

Soon after his arrival in Paris, he had made the acquaintance of Marin Mersenne, who became his close friend and with whom he came to share similar attitudes toward Paracelsus, Fludd, and the magical sects. There he was exposed to the attacks of Mersenne and Garasse on the libertines and occultists. Like them Gassendi began to criticize the philosophy of naturalism permeated by alchemical and magical ideas.[9]

In 1631 Gassendi responded to Mersenne's request for help by writing a treatise, entitled *Examen Philosophiae Roberti Fluddi*, against Fludd's ideas, taking, however, a more moderate approach than had his friend. In contrast to Mersenne, who had branded Fludd a heretic, defiler of Christianity, and a propagator of vile magical ideas, Gassendi praised Fludd's erudition and admitted that Fludd at least professed Christianity. He declined to discuss the issues of Fludd's alchemy, of which he knew little, or of his atheism, but set out to study Fludd's philosophical principles. Only then could he examine the polemics Fludd had written against Mersenne.[10]

Gassendi's *Examen* summarized and criticized Fludd's ideas on the structure of the macrocosm and microcosm and the activity and passivity of light and darkness. He questioned the dialectical assumptions on which Fludd's universe was constructed. Fludd's analysis of the universe as composed of the "radical union" of the two first principles—active light and passive darkness—was an incorrect hypothesis. It separated out two principles that were really one and the same, since no fundamental distinction could be made between light and darkness.

Gassendi questioned the idea basic to emanation theories of the existence of time prior to created time and Fludd's theory that before the creation of the universe God was hidden light who became visible light in the alchemical distillation by which the world was created. He also opposed the idea that human souls were portions of God emanating from his light, an opinion he considered more dangerous than atheism. The theory of an active soul of the world housed in the sun and vivifying the whole universe was untenable. Gassendi defended Mersenne, who, unlike those Rosicrucians and alchemists looking for truth through the philosopher's stone, was searching for a more solid truth. Although Fludd responded to Gassendi's *Examen* and to Mersenne with his four-part *Clavis Philosophiae et Alchymiae Fluddanae* (1633), Gassendi made no further contributions, having turned to the development of his own philosophy.

Gassendi's critiques of Aristotle and Fludd in the 1620s were largely polemical. He had not yet committed himself to the ideas of the Greek philosopher Epicurus (341–270 B.C.), a task that took shape during the 1630s. The system of Epicurus was based on the ideas of the earlier Greek philosophers Democritus (born ca. 460 B.C.) and Leucippus (fl. ca. 460 B.C.) and later more fully elaborated by the Roman poet Lucretius (98–55 B.C.) in his *De Rerum Natura (On Nature)*. Epicureanism postulated the existence of an infinite number of unchanging atoms of different shapes and sizes moving ceaselessly through infinite void space, falling, swerving, combining, and separating to form the objects of the changing sensible world. Although Democritus and Epicurus were known during the Middle Ages, and the rediscovery of Lucretius's poem in 1417 had sparked interest and discussion, atomism was then too limited in generality to compete with the system of Aristotle. It was discussed by the Englishman Nicholas Hill in 1601 as a philosophical system, by the Frenchman Sebastian Basso in 1621 as an argument against Aristotle, by the German iatrochemist Daniel Sennert as a medical theory (1619 and 1637) and in a treatise of 1643 by the French philosopher Claude Berigard, who taught at Pisa and Padua. Gassendi, however, developed Epicureanism in great detail as a complete physical and philosophical system, setting it in a Christian framework.[11]

Although Gassendi had cited Epicurus in 1621, he had at that time shown no marked preference for the system. In Grenoble he had probably learned of the work of Basso which employed an atomic philosophy as an alternative to the philosophy of the medieval scholastics, whose substantial forms Basso believed to be too much like animistic deities. In 1626, he professed to be anxious to read a work that he believed misrepresented Epicurus' ideas and then, in March 1628, confessed that he was working on the life, ideas, and writings of Epicurus, which he intended to publish at the end of his critique of Aristotelianism. At this point, Gassendi began what was to be a lifetime of work on the mechanical atomism of Epicurus, culminating in his definitive *Syntagma Philosophicum,* published posthumously in 1658. His first treatise, *Philosophiae Epicuri Syntagma,* an exposition of Epicurean atomism, begun in 1631, finally appeared in 1649.[12]

Gassendi's mechanical philosophy supposed the unchanging constituents of reality to be solid, impenetrable, corporeal atoms that retained their identity through change. They had the properties of extension, shape, weight, and an internal force of motion. Although the atoms and corpuscles of later mechanists were inert and passive, Gassendi's atoms, like those of Epicurus, were endowed with an "internal energy" and were therefore active and in motion. For Gassendi, however, this principle of motion was material, not incorporeal and vital, as were the elements and seeds of Paracelsus and Van Helmont. Leucippus, Democritus, and Epicurus, he wrote, "did not consider atoms the matter of all things to be inert or motionless, but rather as most active and mobile."[13] Quoting Lucretius on "matter's teeming bodies . . . driven with what movement endowed to scoot across the great void," Gassendi concluded that "the internal principle of action that works in second causes is not some incorporeal substance [such as the forms and species assumed by the Aristotelians and scholastics] but a corporeal one." The atoms were self-moving, least parts of matter activated by an inherent corporeal principle. But they were created by an incorporeal Christian God who "pervade[s] and support[s] the universal machine of the world." The atoms were material, not the incorporeal parts of

God's substance "pulled apart as it were and cut into little pieces which become the individual souls, or forms," of men, beasts, plants, metals, stones, and "every single thing."

Although the atomic particles of matter were inherently active, the cumulative effect of their individual motions was such that gross visible bodies obeyed the principle of inertia. In a larger corporeal body, "atoms may indeed be restrained until they do not move, but not to the point that they do not strain and endeavor to disentangle themselves and renew their motion." The motion of gross bodies occurred through a nexus of efficient physical causes rather than through the tensions between sympathies and antipathies, attractions and repulsions assumed by the naturalists. Occult qualities could be explained by that universal law governing the whole natural world—the law of causality. Every effect had a cause, no cause acted without motion, and no cause acted at a distance.

Using political metaphor, Gassendi argued that God was not the soul of the world inherent within it, but its governor or director. The earth could not be said to be animated by a soul, as could plants, animals, or men. There was no place for spontaneity in nature itself, nor could it be sanctioned in the human soul. Spontaneous action was not to be confused with freedom of action. Spontaneity was an impulse of nature, of passion. Free action depended on man's reason, examination, judgment, and choice.[14]

Gassendi's answer to the problem of certainty, however, differed significantly from that of Descartes, and his criticisms of Descartes' 1641 *Meditations on the First Philosophy,* led to a rift between the two mechanists on the issue of certainty in science. Gassendi had accepted the skeptical argument of the true essences of things and the necessary connections between them. Because God's free will could abrogate the present order of nature, it was possible to attain only probable knowledge of the connections among phenomena. Yet these could be the basis for a practical knowledge useful in everyday life. Because observations of individual things made through the senses tell us nothing about their true essence, we can have knowledge only of appearances and the probable connections between them. Gassendi's answer to the search for certainty led to an empirical science based on observations and experimental laws. His influence was significant in England to which country his ideas were transmitted by Walter Charleton (1619–1707), later to be taken up by Robert Boyle and Isaac Newton (1642–1727).[15]

For Descartes, as for Mersenne and Gassendi, the search for certainty formed the central problem of the 1620s. Finding himself "saddled with so many doubts and errors" that he seemed to have gained nothing from his education, he was disillusioned by the failure of philosophy to produce "anything which is not in dispute and consequently doubtful and uncertain." As a consequence of his 1619 vision leading him to the idea that mathematics was the key to understanding, Descartes, too, began a search for the source of certainty in knowledge. He spent four years in southern Germany in quiet contemplation, during which time he reputedly tried to locate the Rosicrucians, believing that they had an answer to the problem. When he returned to Paris in 1623, he learned of the Rosicrucian threat and the fact that six Rosicrucians were reported to have arrived at the same time. Rumors were being spread that he was associated with them. Descartes at once reestablished contact with Mersenne in order to demonstrate that he was not a member of that invisible underground sect.[16]

Descartes' answer to the problem of certainty was to fuse the will and intellect of God into the concept of a sovereign creator freely and unconditionally creating *ex nihilo* a body of eternal truths, intelligible and accessible to the human intellect. By his absolute and transcendent will, God created nature and its laws and subsequently sustained them by recreating the world anew from moment to moment. But, owing to his eternal and immutable intellect, the laws of nature were unchanging and intelligible to the human understanding. "In God," he wrote, "willing and knowing are a single thing in such a way that, by the very fact of willing something, he knows it."[17]

The criteria for intelligibility were those of mathematics. Clear and distinct ideas, as in geometrical demonstrations, could be said to be true. Hence mathematical descriptions of the material world would yield true physical laws, the logic of the mathematical method becoming the key to valid knowledge.

For Descartes, unlike Mersenne and Gassendi, clear and distinct ideas were the basis for ascertaining the essence or true nature of things behind the appearances and their necessary connections. Because clear and distinct ideas could be formed in the mind, essences could be inferred to exist and to have quantifiable properties. The relations between quantities could be described by mathematics, acquiring the status of true physical laws. Descartes' rationalist answer to the problem of certainty led to a science of mechanics based on quantitative properties (or primary qualities), such as the amount of matter in a body and its motion. A body's quantity of motion could be transferred wholly or in part to another body, the amount being determined by a set of derived rules. Mechanism, or the science of matter in motion, could be used to describe the entire universe—the human body, the physical surroundings, and the larger cosmos.[18]

By 1629, he had resolved the problem of certainty sufficiently to begin the writing of his mechanical treatise, *Le Monde* (*The World*, published in 1664), the same period in which Gassendi had begun his work on mechanism. Then, from 1632 to 1633, after his retreat from Paris to Holland, he produced *L'Homme,* (*Treatise of Man*, first published in Latin in 1662), which analyzed the human body explicitly as a machine. Although Descartes' model was not based on new experimental evidence, it restructured what was known about physiology and the circulation of the blood in terms of the new machine metaphor, which competed with Galen's hierarchical three-organ system.[19]

Similarly, his *Principles of Philosophy* (published in 1644) restructured the cosmos as a mechanism, based on the motion of inert material corpuscles that transmitted motion consecutively from part to part through efficient causation. The force that produced the motion was not something vital, animate, or inherent in bodies, but a measure of the quantity of matter and the speed with which they moved. Motion was external to matter and was put into the universe at the moment of creation. It could be transferred among bodies, but its total amount was conserved from instant to instant by God. Change occurred through the rearrangement of inert corpuscles. The *spiritus mundi* of the Neoplatonists was translated into a subtle mechanical ether whose whirlpool circulations pushed the planets around, a sleight-of-hand not lost on subsequent critics such as Henry More and Henry Power.[20]

All spirits were effectively removed from nature. External objects consisted only of quantities: extension, figure, magnitude, and motion. Occult qualities and properties existed only in the mind, not in the objects themselves. Such qualities could not initiate motion because they could not possess force sufficient for putting objects into motion. Nor could the soul, whether the world soul of the occultists or the human soul, initiate motion, it was merely the seat of the will and of perception. In the *Principles of Philosophy,* Descartes argued against the existence of occult properties and sympathetic attractions: "No qualities are known which are so occult and no effects of sympathy and antipathy so marvellous and strange, and finally nothing else in nature so rare (provided it proceeds entirely from purely material causes lacking in thought or free will) for which the reason cannot be given by means of the same principles."[21]

Descartes reduced the imagination, source of universal knowledge in the holistic world view, to an individual operation of individual souls. Chimeras, sirens, hippogryphs, and spirits existed only in the soul's imagination, as a result of the action of the will.[22] Witches, monsters, nymphs, and satyrs became figments of the individual imagination.

In France, the rise of the mechanical world view was coincident with a general tendency toward central governmental controls and the concentration of power in the hands of the royal ministers. The rationalization of administration and of the natural order was occurring simultaneously. Rational management in the social and economic spheres helps to explain the appeal of mechanism as a rational order created by a powerful sovereign deity. As Descartes wrote to Mersenne in 1630, "God sets up mathematical laws in nature as a king sets up laws in his kingdom."[23]

In France by the seventeenth century, the centralized control and management of industry and natural resources was well under way. The king controlled large-scale industry (principally gunpowder, saltpeter, and salt) and claimed ownership of all metals and ores. Concessions were granted for mining and milling in return for revenue. The regulation of industry was conducted by salaried administrators who fixed the prices of products and prescribed manufacturing methods. Royal authority thus tended to discourage private industrial enterprises while strengthening state control. Under the rationalizing tendencies emerging in the governments of strong nation-states such as France and England, nature came to be viewed as a resource to be subjected to control with human beings as her earthly managers.[24]

HOBBES: MECHANISM AND THE SOCIAL ORDER. In the 1640s, the work begun by Mersenne, Gassendi, and Descartes that would ultimately restore order and certainty to the cosmos through a mechanical philosophy of nature was continued by Thomas Hobbes and a circle of English Royalists escaping the parliamentary faction of the English Civil War. Hobbes, a thoroughgoing materialist, further mechanized the cosmos by denying any inherent force to matter, by reducing the human soul, will, brain, and appetites to matter in mechanical motion, and by transforming the organic model of society into a mechanistic structure. Hobbes was well acquainted with many of the founders of the new science. He had been secretary and friend to Francis Bacon for a short period sometime during the years 1621–1626. On his third journey to the Continent, in 1634, at a time when the study of geometry and motion were both central to his interests, he had met Galileo. On the same visit, he was introduced to Mersenne and his circle of scientifically oriented friends. On his return to England in 1637, he began work on a system of philosophical thought based on mechanical principles. Then in 1640, with the dissolution of the English Parliament and the fear of reprisals by parliamentary leaders for his support of the monarchy, Hobbes fled to Paris, where he was again welcomed into the scientific group around Mersenne.[25]

During the 1640s, Hobbes was part of an active group of English Royalist emigrés, in close association with French savants and mathematicians—Mersenne, Gassendi, Claude Mydorg, and Giles Roberval—who discussed the atomic theories of Democritus, Epicurus, and Lucretius and developed the rudiments of a mechanical philosophy of nature. The group included Charles Cavendish, his brother William with his wife Margaret (later Marquis and Marchioness of Newcastle), William Petty, and through correspondence, chemist Sir Kenelm Digby in England and mathematician John Pell in Holland. Hobbes was "joined in a great friendship with Gassendi," during the period when the latter was developing his Epicurean atomism. In 1644, he saw a manuscript of Gassendi's *Animadversiones in Decimum Librum Diogenis Laertii,* which contained a long section entitled *"Pars Physica,"* almost identical to sections of the more complete *Syntagma Philosophicum* of 1658. The central problem addressed by this group of mechanical philosophers was the development of an explanatory framework based on the nature of matter, and the origin and transmission of motion among its parts.[26]

Already in his "Little Treatise" of 1630, Hobbes had developed a mechanical explanation of sensation based on the motion of contiguous particles of matter. Sensations, he held, were not in the objects but purely subjective. Thus the sound was not in the bell, for only its parts were in motion; or in the air, for only its parts were in motion. Nor was it in the ear, for the parts of the ear only moved the animal spirits in the nerves and brain. Sensation was therefore only the subjective experience of local motion in the brain. No initiating principle of motion was contained in the soul, the brain, the will, the appetite, or the animal spirits. But in the "Little Treatise," Hobbes had not yet resolved the important question of whether matter had some inherent form of motion (or species) that produces the motion of the medium on which sensation is based. The issue was whether matter itself was active and therefore sent out the "species" whose motion produced sensation. "Whatsoever moveth another moveth it either by active power inherent in itself, or by motion received from another. . . . Agents send out their species continually. For seeing the agent hath power in itself to produce such species, and is always applied to the patient, . . . it shall produce and send out species continually."[27]

During the early 1640s, Hobbes—probably influenced by Gassendi, with whose work on Epicurus he was acquainted—assumed that matter was composed of small, invisible parts or atoms in a vacuum. In a 1645 letter, Charles Cavendish reported that Hobbes believed that all change occurred as the result of the motions of invisible parts of bodies. The following year, Hobbes admitted the existence of vacuity, in a draft of his "optical treatise." A 1648 letter from Hobbes to Mersenne stated, "there exist very small spaces now here, now there, in which there is no body." Bodies are set in motion by collision with neighboring bodies, "an action which necessarily results in certain void spaces."[28] It is not clear, however, whether Hobbes, like Gassendi, assumed the atoms to be self-active through a material cause.

But between 1648 and 1655, when his *De Corpore,* on the physics of bodies was published, Hobbes had become a plenist like Descartes, denying the existence of a void or vacuum in nature and repudiating Epicurus and Lucretius. All motion was passed from one body to the next only through contact and was not the result of an inherent activity or power. The cosmos was made up of "small atoms disseminated throughout space between earth and stars," with a fluid ether filling in the vacancies.[29]

In 1648, Hobbes and Descartes had achieved something of a rapprochement after a long period of antagonism that had begun in 1641 as a result of Hobbes' adverse comments on Descartes' *Meditations*. In a letter to John Pell, Charles Cavendish reported that Hobbes and Descartes "have met and had some discourse, and as they agree in some opinions so they extremely differ in others, as in the nature of hardness, Mr. Hobbes conceiving the cause of it to be an extreme quick motion of the atoms or minute parts of a body which hinders another body from entering and Mr. Descartes conceives it a close joining of the parts at rest."[30] Their agreement on the mechanical basis for the corporeal world, however, did not extend to metaphysics, where Hobbes espoused a monistic materialism and Descartes a dualism.

In *De Corpore*, Hobbes asserted that all motion arises from contact with an external contiguous moving body. Moreover, by the law of inertia, a body once in motion will continue that motion with the same velocity unless hindered by another body. Whereas in the "Little Treatise," Hobbes had retained the possibility of an inherent active power, in *De Corpore* he said all action came from the prior motion of a contiguous body. Here, Hobbes referred to the active power of the mover, but this was not an inherent activity. When he wrote that "active power consists in motion," he meant that the moving body acted as agent or efficient cause, while the resistance of the patient was a passive power or material cause. Thus "the agent has power, if it be applied to a patient; and the patient has power if it be applied to an agent." All change occurred from the action of material and efficient causes, the Aristotelian formal and final causes being excluded. Motion was the result of necessary cause-and-effect relations; no effect could occur without a prior necessary cause.[31]

In *De Corpore*, Hobbes introduced the concept of endeavor or *conatus*, which in the hands of philosopher Gottfried Wilhelm von Leibniz (1646–1716) would become the basis for dynamics. Endeavor, however, was not an inherent, vital, spontaneous function of matter, but an impulse toward motion, a *conatus,* "made in less space and time than can be given," propagated through the surrounding medium by contact with continuous bodies, *ad infinitum.* Impetus was a measure of the velocity of its propagation.

In animals, endeavor was the beginning of animal motion but was nevertheless mechanical motion, an appetite. Sensation or the transmission of mechanical motion by the animal spirits either excited or slowed the action of the heart's motion, resulting in pleasure or pain. Found even in the embryo, endeavor was the first response of the animal to pleasure or pain, the first impulse toward motion "for the avoiding of what troubleth it, or the pursuing of what pleaseth it," the two being called *aversion* and *appetite.* In both man and beast, appetite was the result of a prior necessary cause—that is, motion—and was not free. Similarly, liberty was not freedom and spontaneity, but simply the power to carry out what previously had been willed.[32]

Mechanism thus reordered the world in terms of a new metaphor, the machine. The cosmos was operated from the outside by God, the bodily machine by the human soul, or, as for Hobbes, by the mechanical transfer of motion among material particles. To this restructuring of reality, Hobbes added, as a third essential ingredient, a mechanistic analysis of society. The body politic was composed of equal atomistic beings united by contract out of fear and governed from above by a powerful sovereign (Fig. 18).

In his *Leviathan* (1651), Hobbes developed a mechanical model of society as a solution to social disorder. The state of nature was a state of chaos, anarchy, and fear brought about by the material appetites of each individual for competition, domination, and glory. Like disorderly nature, disorderly humans were basically hostile, unfriendly, and violent. Hobbes astutely discerned that neither the communal model of organic society in which the commons was shared by all, nor the hierarchical model that every person was a link in a vast social chain extending from peasant to king was applicable to England's seventeenth-century market-oriented economy.

He attacked the foundations of both the hierarchical model based on natural inequalities and the communal model based on the sharing of natural resources. Unlike the social insects in the organic analogies of Aristotle and John of Salisbury, humans were in a continual state of competition for glory and found great joy in compar-

Figure 18. Frontispiece, Thomas Hobbes, *Leviathan* (first edition, 1651). Society as composed of atomic individuals governed by social contract with a sovereign. Note the imagery of force and dominion: these were Hobbes' solution to social disorder—by *nature,* people are lawless, violent, and unfriendly.

ing themselves with each other. The hierarchial theory of organic society was incorrectly based on natural inequalities:

> I know that Aristotle in the first book of his *Politics,* for a foundation of his doctrine, maketh men by nature, some more worthy of command, meaning the wiser sort, such as he thought himself to be for his philosophy; others to serve, meaning those that had strong bodies but were not philosophers as he; as if master and servant were introduced not by consent of men, but by difference of wit; which is not only against reason; but also against experience.[33]

Every man must therefore "acknowledge another for his equal by nature."[34]

Secondly, the communal theory was founded on competition, not sharing. In the state of nature, everyone has an equal right to everything: "Nature hath given all to all." But to Hobbes this did not mean that all people would share nature's gifts communally. Instead, everyone would be competing for the same natural resources, "for although any man might say of every thing, this is mine, yet he could not enjoy it, by reason of his neighbor, who having equal right and equal power, would pretend the same thing to be his."[35] Because of competitive self-interest, the commons cannot be shared, but must be fought over. As in the "tragedy of the commons," those who did not compete for resources would be secretly ridiculed by their peers: "For he that should be modest and tractable and perform all he promises, . . . should but make himself a prey to others, and procure his own certain ruin."[36] The commons, like a marketplace, was based on competition and self-interest, hence analogous to a battleground in need of law and order, not to a Garden of Eden. Hobbes' antidote was order, peace, and control through a set of accepted rules for the conduct of each citizen analogous to the rules governing the operation of a machine. His sovereign, who was the external embodiment of contracted unity and dispenser of these rules, operated from outside the machine like a technician. The

state, created *ex nihilo,* was an artificial ordering of individual parts, not bound together by cohesion, as in an organic community, but united by fear. The resultant Leviathan was a "mortal God, to which we owe under the immortal God our peace and defense."[37]

In his introduction to *Leviathan,* Hobbes transformed the organic metaphor of the state as a person-writ-large into a machine:

> For, seeing life is but a motion of limbs, the beginning whereof is in some principal part within; why may we not say that all automata (engines that move themselves by springs and wheels as doth a watch) have an artificial life: For what is the heart, but a spring; and the nerves, but so many strings; and the joints, but so many wheels, giving motion to the whole body, such as was intended by the artificer?[38]

Whereas in Salisbury's *Policraticus* the body was endowed with life and ruled by the prince who together with the clergy constituted its soul, the great Leviathan is "but an artificial man . . . in which the sovereignty is an artificial soul" giving not vital life but artificial motion.[39] In Salisbury's organism, the judges and governors who communicated with the people represented its sense organs; in Hobbes' machine state the magistrates and judges were, instead, its artificial joints. For Salisbury, the senate held the position of the heart and represented the good of the commonwealth; Hobbes assigned the heart (a spring) no human function whatever and dispensed rewards and punishment through the nerves. For Salisbury, the prince embodied wisdom and used it to restrain the arms of the organism; for Hobbes, equity and laws, formed an artificial reason. In *Policraticus,* the financiers in the intestinal tract dispensed the state's finances being wary of anal hoarding; in *Leviathan,* the accumulation of wealth and riches of all the members constituted the strength of the artificial man while business assured its safety. The feet, which were the multitudes of common people for Salisbury, having no riches to contribute were not even mentioned by Hobbes. The Hobbesian metaphor thus not only became completely mechanical but was also fully consistent with a description of a market economy, the strength and operation of which depended on money exchanges and quantitative calculations.

In England, by the time Hobbes wrote his *Leviathan,* competition for natural resources and enclosures of the commons were the basis for a growing domestic and international market economy. The most powerful socioeconomic groups were the large landholders and the merchants. The landholders comprised the gentry, larger freeholders, and those merchants who had acquired sufficient wealth to buy country estates; the merchants were traders who had acquired wealth on the international market, monopolized local markets, or had invested in industrial enterprises. The sons of these two groups frequently interchanged status, the younger sons of landholders moving to towns to enter business ventures, while those of the merchants took over management of newly acquired country estates and became local officials and justices of the peace.[40]

Because common-law decisions limited the English Crown's ownership of metals and ores to a few gold and silver mines, the development of the mining and metallurgical industries proceeded as a competitive rather than as a state-controlled enterprise. Attempts by the Crown to control new industries and manufacturing ventures (such as alum and glass making), through the granting of patents for exclusive development of new industry, were increasingly resented by the gentry and merchants because they limited competition. Before these attempts at royal control, English landholders, in contrast to the French, were already involved in pasturing sheep for the wool industry and in leasing land to the new mining and metallurgical industries for furnaces, water supplies, and timber sources, while merchants were already investing in mines and trading the manufactured products. As the source of status and wealth shifted from land to open-ended profit accumulation in the market, status was purchased at the expense of nature as a resource base.

These merchant adventurers and landholding agricultural improvers had begun to break away from the organic philosophy expressed by Aristotle and Hooker, that matter and the material had a subordinate place within the good of the whole. Their material wealth and self-interest gave them both the power and the incentive to oppose the king, eventually helping to propel the country toward civil war.

Hobbes' answer to the disorder and anarchy posed by the Civil War was to reassert sovereign power, a French solution to an English problem, and an answer not well received in England in the 1650s. The operation of the state, like the operation of a calculator, was to be based on a rational system of rules derived by the application of logic and deductive syllogistic reasoning to a set of defini-

213

tions, names, and meanings derived from experience. The first of these rules, which Hobbes called "laws of nature," was that men ought to seek peace. From this followed others that promoted the peaceful, rational, and civil behavior necessary to the conduct of business, industry, trade, navigation, arts, letters, and culture.[41]

What implications did these mechanical politics hold for women? Hobbes' assumption of equality in the state of nature in theory challenged the patriarchal family in as much as it was the mother and not the father who had dominion over the child in the state of nature, for "It cannot be known who is the father, unless it be declared by the mother." Yet, in his discussion of the family in *Leviathan,* the mother has vanished, the family being constituted of "a man and his children; or . . . a man and his servants; or . . . a man, and his children, and servants together." The father was given dominion over the household, since "for the most part commonwealths have been erected by fathers, not the mothers of families." In *De Cive* (1642, published 1647), a family is defined as "a father with his sons and servants, grown into a civil person by virtue of his paternal jurisdiction." Thus, while the state of nature would logically imply full equality for women, in democratic consent theories arising from Hobbes and Locke, women remained under the dominion and authority of men. Seventeenth-century "democratic" theories, based on atomistic equality for every individual, in reality meant equality for middle- and upper-class property-holding males.[42]

The rise of mechanism laid the foundation for a new synthesis of the cosmos, society, and the human being, construed as ordered systems of mechanical parts subject to governance by law and to predictability through deductive reasoning. A new concept of the self as a rational master of the passions housed in a machinelike body began to replace the concept of the self as an integral part of a close-knit harmony of organic parts united to the cosmos and society. Mechanism rendered nature effectively dead, inert, and manipulable from without. As a system of thought, it rapidly gained in plausibility during the second half of the seventeenth century. Its ascendancy to status as a new world view was achieved through combating some presuppositions of the older organic view of nature while absorbing and transforming others.

Both ideas and social conditions help to account for the rapid acceptance of the mechanical philosophy as a conceptual system in the mid-seventeenth century. Yet the mechanism did not gain ascendancy without struggle. Debates between organicists and mechanists continue even into the present. But mechanism as a metaphor ordered and restructured reality in a new way, eliminating some kinds of ideas and problems from its scope of explanation, and opening up new ones for investigation. Among its great strengths were that it served not only as an answer to the problem of social and cosmic order, but it also functioned as a justification for power and dominion over nature.

Mechanism as Power

The brilliant achievement of mechanism as a world view was its reordering of reality around two fundamental constituents of human experience—order and power. Order was attained through an emphasis on the motion of indivisible parts subject to mathematical laws and the rejection of unpredictable animistic sources of change. Power was achieved through immediate active intervention in a secularized world. The Baconian method advocated power over nature through manual manipulation, technology, and experiment. But mechanism as a world view was also a conceptual power structure.

The mechanical philosophy took over the concept of power from both experiential and intellectual sources. The natural magic tradition which relied on the power of the magus to manipulate the occult powers within natural objects through external operations and physical arrangements, depended on a new image of the human being as operator. A second experiential source of power derived from the increasing use of two types of machines in the everyday lives of sixteenth- and seventeenth-century Europeans. The nonautonomous machines (windmills, cranes, pumps, and so on) multiplied power through external operation by human or animal muscle or by natural forces such as wind or water. Autonomous machines such as clocks were internalized models of the ordered motions of the celestial spheres. The first were symbols of power, the second of order. Both were fundamental to the new value system of the modern world.

Intellectually, order and power were at the core of the two most important medieval concepts of the nature of God—intellectualism and voluntarism. Basic to the intellectualist tradition was the primacy of God's intellect and logic over his will. Voluntarism assigned primacy instead to God's will and immediate active power. In seventeenth-century mechanics, emphasis either on logic, order, and predictibility or on power and activity led to different styles of science and to different modes for dominating and controlling nature. Also of intellectual significance to the mechanization of the world picture, although not discussed in detail in this book, were internal developments within the sciences of mechanics and mathematics themselves, during the medieval and early modern eras.

MACHINE TECHNOLOGY. The transformations taking place in the farm, fen, and forest ecosystems not only affected the lives of those persons displaced, but brought with them new everyday experiences in the lives of those working in the growing numbers of new industries. The watermills and windmills that supplied these industries with energy from the natural forces of sun, wind, and water had a dual identity. On the one hand, they were a source of renewable energy, as opposed to the stored nonrenewable energy of coal, which, in accord with modern demands, can be stockpiled and held in reserve for immediate use. But, on the other hand, they were large, geared machines that, as necessary parts of the new industries, were foci around which new forms of daily life became organized and institutionalized. With the spread of capitalist economic forms, these mills, together with furnaces, forges, bellows, cranes, and pumps, became an integral part of the everyday experience of many Europeans.

The mechanization of the world picture as a conceptual scheme had foundations first of all in the institutionalization of machine technology as an integral ingredient in the evolution of early capitalist economic patterns. Secondly, both autonomous and nonautonomous machines played symbolic roles that emphasized to the human imagination their capacities for power and order. This symbolism mediated between experience and conceptual structures.[1]

Machine technology was not new to the economic expansion of the sixteenth century. Its basic elements—gears, chains, pulleys, linkages, and pumps—were known to the Greeks, Romans, and the Latin West. But geared machines operated by horse, wind, or water power, such as grainmills, sawmills, windmills, and fulling mills, became more widespread during the late Middle Ages and Renaissance. A geared mill was illustrated in the notebook of a Hussite engineer in 1430 and in the drawing of a sixteenth-century horse-operated mill. Sawmills driven by water power were referred to in German and French writings of the late fourteenth century.[2]

Windmills were of two types, due to the necessity of pivoting the sails so they could be presented to the wind. Small mills, as illustrated in a 1430 drawing, were attached to a central post around which they circulated; the sails of large mills were attached to a moveable turret mounted on a stable tower and, as described in 1556, by Agricola, could be controlled by a brake. Agostino Ramelli's *Various and Ingenious Machines* (1588) featured an even more sophisticated form (Fig. 19). The windmill's development was further advanced by sixteenth-century Dutch engineers.

Many fulling mills, driven by water power and used for shrinking and thickening cloth, had been built along streams by the late twelfth century. These operated by the rotary action of cams that lifted large hammers and then released them to deliver a blow to the cloth in the vat below. The fulling mill was also pressed into service as the village washing machine.

By the end of the sixteenth century, watermills were used extensively in the lead and tin industries for smelting and stamping and for hammering iron bars and drawing iron into rods. Large watermills of the type employed in paper and gunpowder manufacture required capital outlays in the neighborhood of £1,500.

One need only look carefully through Renaissance technological and engineering treatises or at the paintings of Dutch and Flemish artists to get a sense of the impressive size and variety of machines

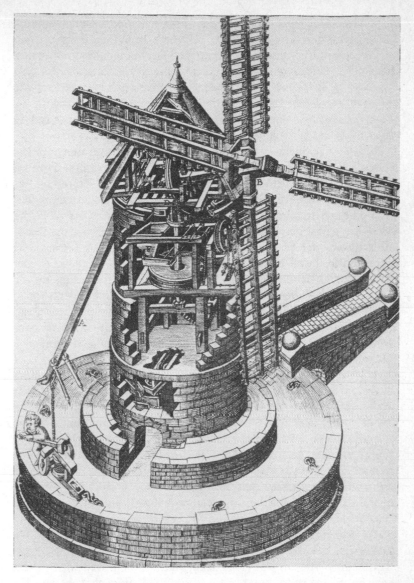

Figure 19. Windmill used for grinding, drawing by Agostino Ramelli, from his *Various and Ingenious Machines* (Paris, 1588). In the Renaissance, nature was increasingly harnessed to do men's bidding.

as extensions of human power. Georg Agricola (1556) and Vannuccio Biringuccio (1540) provide accurate representations of machines used in the mining, metallurgical and gunmaking industries. Ramelli's *Various and Ingenious Machines* and Jacques Besson's *Theater of Machines* (1569) were important encyclopedias of the types of machines used in pumping water, milling, lifting heavy weights, sawing wood, crossing streams, digging ditches, and in measuring and surveying (Fig. 20).[3]

But the machine also functioned symbolically as an image of the power of technology to order human life. For example, two of Pieter Bruegel the Elder's paintings, both entitled "Tower of Babel," depict huge twenty-foot waterfront cranes operated by human-powered treadmills (Fig. 21). The half-dozen types of lifting machines with their wooden wheels, hoists, and squirrel-cage treadmills tower over the human workers as they raise massive stone slabs and load and unload commercial products onto ships. A covered treadmill crane dwarfing the merchants in a painting by the Flemish artist Pieter Pourbus the Elder (1523–1584) was meant to illustrate the power of the machine to produce profits in trade and shipping (Fig. 22). Such paintings not only dramatize the human-overpowering qualities of machine technology but also symbolize the role of the machine in organizing human life.

In contrast to the power of the large force-multiplying machines of the early industries, the clock in the preindustrial era had only limited use in workplace applications and served instead as a symbol of cosmic order. Although rural communities continued to live in accord with day length, seasonal cycles, and work rhythms, by 1500 tower and cathedral clocks had begun to govern the civil life of most European towns, especially where commerce was flourishing, and most parishes had constructed church clocks by the late sixteenth century. But the early clocks had only an hour hand, were inaccurate, and needed sundials for resetting (Fig. 23). By 1700, clocks had become much more precise, due to the addition of the pendulum and minute hand, and appeared in middle-and upper-class homes. The accuracy of pocket watches improved with the refinement of the spiral balance spring in 1674. English workmanship in clock and watch making took the lead around 1680.[4]

Perhaps because of their expense as well as their limited industrial application before the eighteenth century, clocks and their asso-

Figure 20. Watermill and pump for raising water, drawing by Agostino Ramelli, from his *Various and Ingenious Machines* (Paris, 1588). As in the gears, wheels, and pulleys of machines, the mechanistic cosmos of Descartes postulated that motion was passed from part to part.

221

222

Figure 21. Human-powered treadmill crane, from *Tower of Babel,* by Pieter Bruegel the Elder. The legendary disorderliness caused by the Tower of Babel contrasts with the mechanical orderliness of Bruegel's representation.

ciated mechanical automata primarily captured the imagination of the aristocracy, for whom they were elegant ornamental playthings, symbolizing cosmic and social order. Medieval authors referred to the cosmos as *machina mundi.* Nicole Oresme (d. 1382), medieval mathematician and Bishop of Lisieux, compared God to a divine clockmaker whose activity created and harmoniously regulated the motion of the planets just as he might regulate the inner moral life of the human being. Inspired by the new weight-driven clocks of the fourteenth century, Oresme wrote (1370):

> And these powers are so moderated, tempered, and ordered against their resistances that the movements are made without violence. And except for the lack of violence it is like the situation when a man has made a clock and lets it go and be moved by itself. Thus it was that God let the heavens be moved continually according to the proportions that the moving powers have to their resistances and according to the established order.[5]

In the medieval iconography of the virtue temperance, the figure of Temperantia evolves from a picture of a woman pouring water into a jug of wine, to one holding a pair of compasses (1359) and an hour glass (1350s). By the early 1400s, she is shown adjusting a mechanical timepiece and soon after is surrounded by geared clocks and other measuring and surveying instruments of great intricacy (ca. 1450). Bruegel's "Sin of Sloth" (1557) depicted a giant clock operated by gears, weights, and a verge escapement mechanism, as well as an enormous clock face with a human arm for a dial. His "Virtue of Temperance" depicted compasses, plumb lines, squares, and clocks—instruments used for surveying distances on land, measuring the size of the earth, the distance of the moon, and making other astronomical measurements. Temperance has here clearly become a symbol of measure.[6]

Beginning with manuscript illuminations around 1465, temperance is sometimes replaced by King Solomon, symbol of wisdom, who is now depicted as repairing rather than simply using the clock

223

Figure 22.   Portrait of Jan Fernaguut, by Pieter Pourbus the Elder, from the Musée Communal, Bruges. The covered treadmill machine in the inset dwarfs the merchants standing around it; its inclusion in this portrait of Jan Fernaguut, a successful merchant, is a comment on the power of machines in commerce.

Figure 23. Geared clock (late fifteenth century), Musées Royaux d'Art et Histoire, Brussels. In the sixteenth century, geared mechanical clocks served as symbols of cosmic order; God was the supreme clockmaker. Eventually, God was reduced to a minimal role in the clock model of the universe—he wound up the mechanical cosmos and left it to tick away into eternity.

as a regulatory device. Here the human figure has become the analogue of the divine clockmaker who creates time in equal hours to suit his own needs and intervenes to maintain its order.

For the mechanists, God became a clockmaker and an engineer constructing and directing the world from outside. It was Mersenne's hope and intention to replace *The Imitation of Jesus Christ* by *The Imitation of the Divine Engineer*. The engineer's art gave humanity not only the opportunity to imitate God "in external productions" but also the possibility of dominion over the earth.[7] For Gassendi, God was the external governor and director of the world. For Descartes, the corpuscular world and natural bodies, including the human body, operated according to the same mechanical laws as clocks and other machines:

> And as a clock, composed of wheels and counterweights, observes not the laws of nature when it is ill made, and points out the hours incorrectly, than when it satisfies the desire of the maker in every respect; so likewise if the body of man be considered as a kind of machine, so made up and composed of bones, nerves, muscles, veins, blood, and skin, that although there were in it no mind, it would still exhibit the same motions which it at present manifests involuntarily.[8]

Robert Boyle, however, noted the problematical character of the clock as an autonomous machine for the image of God as a clockmaker. If the clock is truly autonomous, then once it is set in operation it will continue to do so forever and God as clockmaker will have no cause to intervene in its operation: "The world is like a rare clock such as that at Strasbourg, where all things are so skillfully contrived that the engine once being set going—all things proceed according to the artificer's first design, and the motions . . . do not require the peculiar interposing of the artificer, or any intelligent agent employed by him."[9] By the early eighteenth century, the issue of God's intervention and hence the exercise of his providential care in a clocklike mechanical world would come to a head in the debates between Leibniz and Newton. In England, because of the political tensions between order and freedom, the clock metaphor eventually became more convincingly articulated as a balance, symbolic of regulating the balance of power and balance of trade.[10] As the machine technologies and capitalist modes of trade and manufacture, already a part of the growth of medieval guilds, markets, and towns, evolved toward industrialized capitalist society, machines, calculations, and measurements were increasingly integrated into the commercial and industrial life of European society.

The philosophy that the world was a vast machine made of inert particles in ceaseless motion appeared at a time when new and more efficient kinds of machinery were enabling the acceleration of trade and commerce. The development of transportation equipment, navigational techniques, the building of roads and canals, the development of mining technology, the refining of metals and currency, and advances in ballistics machinery were compatible with the image of a mechanical cosmos.[11] These socioeconomic ends could be realized by mining the earth for gold, copper, iron, and coal, by cutting its forests for fuel for the refining of ores, and by constructing ships to be used in transporting the products. The death of the world soul and the removal of nature's spirits helped to support increasing environmental destruction by removing any scruples that might be associated with the view that nature was a living organism. Mechanism substituted a picture of the natural world, which seemed to make it more rational, predictable, and thereby manipulable.

MACHINES AS STRUCTURAL MODELS FOR WESTERN ONTOLOGY AND EPISTEMOLOGY.

The imagery, iconography, and literary metaphor associated with machines extended the experiences of everyday life to the realm of the imagination, where machines became symbols for the ordering of life itself. Out of such symbolic universes evolve conceptual universes as new definitions of reality replace the old.[12] As the machine and clock increased their symbolic power as root metaphors, in response to society's changing needs, wants, and purposes, the symbolic force of the organism declined in plausibility and the organic conceptual framework underwent a fundamental transformation. The images and symbols associated with the machines of everyday life helped to mediate the transition between frameworks.

The emerging mechanical world view was based on assumptions about nature consistent with the certainty of physical laws and the symbolic power of machines. Although many alternative philosophies were available (Aristotelianism, Stoicism, gnosticism, Hermeticism, magic, naturalism, and animism), the dominant European ideology came to be governed by the characteristics and experiential power of the machine. Social values and realities subtly guided the choices and paths to truth and certainty taken by European philosophers. Clocks and other early modern machines in the seventeenth century became underlying models for Western philosophy and science.

Not only were seventeenth-century philosophical assumptions about being and knowledge infused by the fundamental physical structures of machines found in the daily experience of western Europeans, but these presuppositions were completely consistent with another feature of the machine—the possibility of controlling and dominating nature. These underlying assumptions about the nature of reality have today, by retrojection, become guidelines for decision making in technology, industry, and government.

The following assumptions about the structure of being, knowledge, and method make possible the human manipulation and control of nature.

1. Matter is composed of particles (the ontological assumption).
2. The universe is a natural order (the principle of identity).
3. Knowledge and information can be abstracted from the natural world (the assumption of context independence).
4. Problems can be analyzed into parts that can be manipulated by mathematics (the methodological assumption).
5. Sense data are discrete (the epistemological assumption).

Based on these five assumptions about the nature of reality, science since the seventeenth century has been widely considered to be objective, value-free, context-free knowledge of the external world. Additionally, as the twentieth-century philosopher Martin Heidegger argued, Western philosophy since Descartes has been fundamentally concerned with power. "The essence of modern technology lies in enframing"; that is, in the revealing of nature so as to render it a standing reserve, or storehouse. "Physics, indeed as pure theory sets up nature to exhibit itself" and to "entrap" it "as a calculable order of forces."[13]

The new conception of reality developed in the mid-seventeenth century shared a number of assumptions with the clocks, geared mills, and force multiplying machines that had become an important part of daily European economic life. First of all, they shared the ontological assumption that nature is made up of modular components or discrete parts connected in a causal nexus that transmitted motion in a temporal sequence from part to part. Corpuscular and atomic theories revived in the seventeenth century hypothesized a particulate structure to reality, the parts of matter like the parts of machines being dead, passive, and inert. The random motions of the atoms were rearranged to form new objects and forms of being by the action of external forces. Motion was not inherent in the corpuscles, but a primary quality of matter, put into the mundane machine by God. In Descartes' philosophy, motion was initiated at creation and sustained from instant to instant throughout created time; for Newton, new motion in the form of "active principles" (the cause of gravity, fermentation, and electricity) was added periodically to prevent the nonautonomous world machine from running down. For Leibniz, the universal clock was autonomous—it needed no external input once created and set into motion. The ontology of this classical seventeenth-century science modified by energy concepts has become the framework of the Western commonsense view of reality.

The second shared assumption between machines and seventeenth-century science was the law of identity, the idea that A is A, or of identity through change. This assumption of a rational order in nature goes back to the thought of the philosophers Parmenides of Elea (fl. 500 B.C.) and Plato and is the substance of Aristotle's first principle of logic. Broadly speaking, it is the assumption that nature is subject to lawlike behavior and therefore that the domain of science and technology includes those phenomena that can be reduced to orderly predictable rules, regulations, and laws. Events that can be so described can be controlled because of the simple identity of mathematical relationships. Phenomena that "cannot be foreseen or reproduced at will . . . [are] essentially beyond the control of science."[14]

The formal structural dependence of this mathematical method on the features of the mechanical arts was beautifully articulated by Descartes in his *Discourse on Method* (1636): "Most of all I was delighted with mathematics, because of the certainty of its demonstrations and the evidence of its reasoning; but I did not understand its true use, and, believing that it was of service only in the mechanical arts, I was astonished that, seeing how firm and solid was its basis, no loftier edifice had been reared thereupon."[15]

The primary example of the law of identity for Descartes was conservation of the quantity of motion measured by the quantity of matter and its speed, $m|v|$. In the late seventeenth century Newton,

Leibniz, English mathematicians Christopher Wren and John Wallis, and Dutch physicist Christiaan Huygens all contributed to the correction of Descartes' law to accurately describe momentum (mv) as the product of mass and velocity rather than speed, and mechanical energy (mv^2) as the product of the mass and the square of the velocity. Everyday machines were models of ideal machines governed and described by the laws of statics and the relational laws of the conservation of mechanical energy and momentum. The form or structure of these laws, based as they were on the law of identity, was thus a model of the universe. Although the conversion of energy from one form to another and, in particular, the conversion of mechanical motion into heat were not fully understood until the nineteenth century, the seventeenth-century laws of impact were nevertheless, for most natural philosophers, models of the transfer and conservation of motion hypothesized to exist in the ideal world of atoms and corpuscles.

The third assumption, context independence, goes back to Plato's insight that only quantities and context-independent entities can be submitted to mathematical modeling. To the extent that the changing imperfect world of everyday life partakes of the ideal world, it can be described, predicted, and controlled by science just as the physical machine can be controlled by its human operator. Science depends on a rigid, limited, and restrictive structural reality. This limited view of reality is nevertheless very powerful, inasmuch as it allows for the possibility of control whenever phenomena are predictable, regular, and subject to rules and laws. The assumption of order is thus fundamental to the concept of power, and both are integral to the modern scientific world view.

Although Descartes' plan for reducing complexity in the universe to a structured order was comprehensive, he discovered that the very problem that Aristotle had perceived in the method of Plato was inherent in his own scheme. That problem was the intrinsic difficulty, if not impossibility, of successfully abstracting the form or structure of reality from the tangled web of its physical, material, environmental context. Structures are in fact not independent of their contexts, as this third assumption stated, but integrally tied to them. In fact, Descartes was forced to admit, "the application of the laws of motion is difficult, because each body is touched by several others at the same time. . . . The rules presuppose that bodies are perfectly hard and separable from all others . . . and we do not observe this in the world." The enormous complexity of things thus inhibits the analysis in terms of simple elements.[16]

Descartes' method exhibits very precisely the fourth or methodological assumption that problems can be broken down into parts and information can then be manipulated in accordance with a set of mathematical rules and relations. Succinctly stated, his method assumes that a problem can be analyzed into parts, and that the parts can be simplified by abstracting them from the complicating environmental context and then manipulated under the guidance of a set of rules.

His method consisted of four logical precepts:

1. To accept as true only what was so clearly and distinctly presented that there was no occasion to doubt it;
2. To divide every problem into as many parts as needed to resolve it;
3. To begin with objects simple and easy to understand and to rise by degrees to the most complex (abstraction and context independence);
4. To make so general and complete a review that nothing is omitted.

In Descartes' opinion, this method was the key to power over nature, for these methods of reasoning used by the geometricians "caused me to imagine that all those things which fall under the cognizance of man might very likely be mutually related in the same fashion." By following this method, "there can be nothing so remote that we cannot reach to it, or recondite we cannot discover it."[17]

Descartes' method depended on the manipulation of information according to a set of rules: "Commencing with the most simple and general (precepts), and making each truth that I discovered a rule for helping me to find others—not only did I arrive at the solution of many questions which I had hitherto regarded as most difficult but . . . in how far, it was possible to solve them." In the same manner, the operation of a machine depends on the manipulation of its material parts in accordance with a prescribed set of physical operations.

Descartes placed great emphasis on the concept of a plan or form for the ordering of this information, drawing his examples from the practical problem of city planning: "Those ancient cities which, originally mere villages, have become in the process of time great towns, are usually badly constructed in comparison with those which are regularly laid out on a plain by a surveyor who is free to follow his own ideas." He wished his new ideas to "conform to the uniformity of a rational scheme."

In his *De Cive*, written in 1642, Hobbes had advocated the application of this method of analysis to society:

> For everything is best understood by its constitutive causes. For as in a watch, or some such small engine, the matter, figure, and motion of the wheels cannot well be known except it be taken asunder and viewed in parts; so to make a more curious search into the rights of states and duties of subjects, it is necessary, I say, not to take them asunder, but yet that they be so considered as if they were dissolved.[18]

The fifth assumption shared by seventeenth-century science and the technology of machines was the assumption that sense data are atomic. Data are received by the senses as minute particles of information. This assumption about how knowledge is received was articulated most explicitly by Hobbes and the British empiricists, John Locke and David Hume (1711–1776). According to Hobbes, sense data arise from the motions of matter as it affects our sense organs, directly in the case of taste and touch, or indirectly, through a material medium, as in sight, sound, and smell. These sense data can then be manipulated and recombined according to the rules of speech: "But the most noble and profitable invention of all other, was that of speech, consisting of names or appellations and their connection whereby men register their thoughts . . . without which there had been among men neither commonwealth, nor society, nor contract, nor peace."[19] Words are abstractions from reality; sentences or thoughts are connections among words: "The manner how speech serves to the remembrance of the consequence of causes and effects, consists in the imposing of names and the connection of them." Nature cannot be understood unless it is first analyzed into parts from which information can be extracted as sense data: "No man therefore can conceive anything, but he must conceive it in some place and endowed with some determinate magnitude; and which may be divided into parts."

For Hobbes, the mind itself is a special kind of a machine—a calculating machine similar to those constructed by Scottish mathematician John Napier (1550–1617), French philosopher and

Figure 24. Pascal's adding machine (1642). The adding machine, the precursor of the modern computer, was a model for Hobbes' concept of the human brain: to reason was only to add and subtract or to calculate.

mathematician Blaise Pascal (1623–1662), Leibniz, and other seventeenth-century scientists (Fig. 24). To reason is but to add and subtract or to calculate. "When a man reasoneth, he does nothing else but conceive a sum total, from the addition of parcels; or conceive a remainder, from subtraction of one sum from another; which, if it be done by words, is conceiving of the consequence of the names of all the parts, to the name of the whole; or from the names of the whole and one part to the name of the other part." "In sum, in what matter soever there is place for addition and subtraction there is also place for reason; and where these have no place, there reason has nothing at all to do.... For reason ... is nothing but reckoning, that is adding and subtracting."[20] This view is manifested in twentieth-century information theory that according to Heidegger is "already the arrangement whereby all objects are put in such form, as to assure man's domination over the entire earth and even the planets."[21]

The new definition of reality of seventeenth-century philosophy and science was therefore consistent with and analogous to the structure of machines. Machines (1) are made up of parts, (2) give particulate information about the world, (3) are based on order and regularity, (perform operations in an ordered sequence), (4) operate in a limited, precisely defined domain of the total context, and (5) give us power over nature. In turn, the mechanical structure of reality (1) is made up of atomic parts, (2) consists of discrete information bits extracted from the world, (3) is assumed to operate according to laws and rules, (4) is based on context-free abstraction from the changing complex world of appearance, and (5) is defined so as to give us maximum capability for manipulation and control over nature.

Models are abstractions of particular aspects of an assumed reality. Into the abstractive process go human assumptions and values as to the significant features of that reality. "When we speak of A being a model of B, we mean that a theory about some aspects of the behavior of B is also a theory about the same aspects of A."[22] Thus when we say that reality is a model of the machine, we mean that a theory about some aspects of a machine's behavior is also a theory about the same aspects of reality.

Both order and power are integral components of the mechanical view of nature. Both the need for a new social and intellectual order and new values of human and machine power, combined with older intellectual traditions, went into the restructuring of reality around the metaphor of the machine. The new metaphor reintegrated the disparate elements of the self, society, and the cosmos torn asunder by the Protestant Reformation, the rise of commercial capitalism, and the early discoveries of the new science.

The emphasis on God's will and active power associated with the Reformation tended to legitimate human power and activity in worldly affairs and fostered an interest in technological improvement, empirical observation of God's works in the creation, and experimentation to extract and use nature's secrets for human benefit. Focusing on God's intellect and logic, on the other hand, gave an impetus not only to mathematics and the sciences of rational mechanics and dynamics but also to the rational management of nature through predictability and efficient use. The domination of nature depends equally on man as operator deriving from an emphasis on power and on man as manager deriving from the stress on order and rationality as criteria for progress and development. Efficient operation results from the ordered rational arrangement of the components of a system.

The mechanical framework with its associated values of power and control sanctioned the management of both nature and society. The management of natural resources depends on surveying the status of existing resources, and efficiently planning their systematic use and replenishment for the long-term good of those who use them. Attempts at the conservation of resources begun in England and France in the seventeenth century, to be discussed next, are reflections of the managerial mentality associated with the mechanical order and of a new emerging synthesis between the organic and mechanical perspectives.

The organic view of nature, however, did not disappear with the rise of mechanism. While some of the ideas of the mechanists provided a stimulus to the new science, others were criticized and rejected in favor of the older organic theories of nature and society. Thus the theories of Descartes and Gassendi were introduced into England, where they were refined by Boyle, Charleton, Newton, and the virtuosi of the Royal Society. But Descartes and Hobbes were also subjected to criticism by the Cambridge Platonists and the vitalists for whom mechanism could not adequately explain a fundamental observed feature of the natural world—life itself. Both Descartes and Hobbes were more positively received in France, where Descartes stimulated the development of natural philosophy and Hobbes a political philosophy of sovereign rule.[23] The development of mechanism as a world view has already been thoroughly analyzed by others. It is instead the accommodation of organicism to mechanism and the tensions between these two perspectives on which the remaining chapters will focus.

CHAPTER 10

The Management
of Nature

By the early 1660s, the British navy had become distressed over the lack of tall timber to repair the masts and hulls of its ships, reducing its ability to defend the nation. Its commissioners and officers requested that the Royal Society study the state of the king's forest reserves. The English diarist and founding member of the Royal Society, John Evelyn (1620–1706), began an analysis of the destruction caused by wasteful land practices and lack of conservation methods and in 1662 published the results in his *Silva, A Discourse of Forest Trees and the Propagation of Timber in His Majesty's Dominions.* Increasing numbers of shipping vessels were needed for commercial trade, glass-works and iron-works needed charcoal for smelting, and additional forests had been cut for pasture. "Prodigious havoc" had been wreaked through the tendency not only to "cut down, but utterly to extirpate, demolish, and raze . . . all those many goodly woods and forests, which our more prudent ancestors left standing," a devastation that had now reached epidemic proportions.[1]

Since the mid-sixteenth century, England's forest reserves had been declining, in conjunction with rapid increases in population, trade, and the growth of industries that depended on wood. Writing of the shortage of timber in the 1660s, Samuel Pepys (1633–1703) exclaimed in his diary, "God knows where materials can be had."[2] At the time of the Civil War and the Interregnum, widespread destruction had occurred in the forests of the king and the gentry. Forest supervisors had not performed their duties adequately. England, emerging from internal troubles and looking again to the external defense of the nation, found the "wooden walls" for her protection lacking. Something more in the way of wise management and conservation had to be instituted if England's military and commercial superiority were to be maintained "since our forests are undoubtedly the greatest magazines of wealth and glory of this nation, and our oaks the truest oracles of its perpetuity and happiness, as being the only support of the navigation which makes us feared abroad, and flourish at home."[3]

In his *Silva*, Evelyn called for the institution of sound conservation practices that would contribute to steady economic progress. Like many other intellectuals of his generation, Evelyn was a religious moderate of Latitudinarian persuasion and social philosophy. Latitudinarianism, a religious compromise arising after the English Civil War, retained the Anglican's episcopal form of government, but denied its divine origin. Like other moderates, Evelyn accepted the philosophy of self-interest and acquisitiveness, which he perceived to be of primary importance in his society. The goal of Latitudinarian Protestants was to bring these instincts of self-preservation under the moral law of the church, where enlightenment would convert them to the higher service of the public interest. A high proportion of Latitudinarian churchgoers were English merchants, engaged in craft and trade whose prosperous business activities contributed to the progress of the nation. In a society increasingly organized around the dictates of the market, natural philosophers and amateur scientists responded to the need for responsible management in social and commercial affairs. After the Restoration, Latitudinarians accommodated to the Anglican establishment and played an increasingly active role in civic life.[4]

The political events of the Civil War had filled Evelyn with "disgust of the world occasioned by the violence and confusion of the times" so strong that he retreated to a rural homestead where he studied and read. When the Restoration seemed imminent, he reentered active politics with a pamphlet entitled, "An Apology for the Royal Party" (1659). Appointed by King Charles II to a board "to promote trade," he prepared a pamphlet on "Navigation and Commerce; Their Original and Progress" (1674).[5]

Evelyn's interest in conservation of timber represented a new managerial approach to nature. If exploitation were allowed to continue unabated, nature would decay, and human progress would be curtailed. Managerial conservation was an adaptation of the rationalizing tendencies inherent in mechanism applied to the natural environment. It represented an accommodation of the old organic philosophy to the new mechanism—the sell-out of organicism. Already forming out of the older variants in the organic world view was a new alliance with mechanism, which would become fundamental to the history of both conservation and ecology. The organismic, communal orientation of the naturalists, Paracelsian vitalists, and millenarian utopists was thrust aside to make way for efficiency and production in the sustained use of nature for human benefit. A value system oriented to nature as a teacher whose ways must be followed and respected was giving way to a system of human values as the criteria for decision making.

As will be seen, the managerial approach to conservation was basic to the philosophies of Evelyn, naturalist John Ray (1627–1705), William Derham (1657–1735) and the Cambridge Platonists, Henry More (1614–1687), and Ralph Cudworth (1617–1688) in Restoration England, and represented a rudimentary form of a utilitarian ethic later permeating the progressive conservation movement in the United States in the early 1900s and the new reductionist ecology of the 1950s. In these movements, first conservation and then ecology adopted a policy of the responsible management of nature's resources for long-term human benefit and sustained yield. Shortsighted profit making gave way to long-term planning based on scientific study and policy formation. Managerial ecology seeks to maximize energy production, economic yields, and environmental quality through ecosystem modeling, manipulation, and prediction of outcomes. Scientifically trained advisors to government agencies, industries, and universities help to formulate a rational policy for resource use.[6] The roots of this utilitarian approach to ecology are already apparent in the ideas of the natural philosophers of Restoration England.

Among Evelyn's recommendations for improved forest practice was the removal of most iron mills to New England, lest they "ruin Old England." The New Englanders "should hasten thither to supply us with iron for the peace of our days, whilst his Majesty becomes the great sovereign of the ocean and of free commerce." Conservation at home was thus to be purchased at the expense of frontier expansion abroad, a program rationalized by the New Englanders' need to "clear their surfeit of woods" lest they be "hindered in their advance." "It were better to purchase all our iron out of America, than to thus exhaust our woods at home."[7]

Additionally, the forests near existing British iron-works should be managed intelligently. Evelyn's own father had supplied wood to forges and mills by replanting, careful selective cutting, and care of the standing timbers. Another "great ironmaster" from Surrey, Christopher Darrell, had "so ordered his (iron) works that they were a means of preserving even his woods, notwithstanding those insatiable devourers."[8]

To the king, Evelyn recommended enforcement of statutes passed in the time of Queen Elizabeth that had prohibited the cutting of any tree "one foot square" growing within twenty-two miles of London or fourteen miles of navigable rivers or the sea for conversion to charcoal by iron mills. To the gentry who were "slothful" and "deficient" in "providing for their numerous children," Evelyn recommended refurbishment of their lands with new seedlings for the future use and profit of each of their children. Wet lands should be drained by canals and sluices built to regulate the water supply. Other plots should be forested with seedlings in order to recompense and balance any losses. Cattle should be kept out of young plots on commons and private lands for six years until the trees had grown out of their reach, even to the "detriment of a few clamorous and rude commoners."

The earth did not have to be devastated by mining but could instead be treated with respect: "As that of earth, we shall need at present to penetrate no deeper into her bosom, than . . . scarifying the upper mold, and digging convenient pits and trenches, not far from the natural surface, without disturbing the several strata and remoter layers." The soil must be treated with compost, mold, and other vegetable matter.

Evelyn claimed that the gentry, as a result of his recommendations, had set several million seedlings. A 1668 "Act for the Increase and Preservation of Timber within the Forest of Dean," which directed the planting of 11,000 acres with oak was followed in 1698 with 1,022 acres in New Forest. But Baltic oak had to be imported during the Dutch Wars of 1677. Conservation policies were not continued and after the harvest of trees planted during the Restoration, shortages again became acute by the Seven Years' War in 1756. Problems for the navy increased from then until the replacement of wooden vessels by iron.[9]

In France, the neglect of forest conservation and the abuses by forest officials were dealt with by Louis XIV's minister, Jean Baptiste Colbert, who recognized that "France will perish through lack of woods." His report to the king (1661) surveyed forested lands and timber policies in the province of Tours, while his *Memoir on Forests* (1665) recommended the maintenance of timber reserves for future use. The results of these preliminary reports were then incorporated into the French Forest Ordinance of 1669, which reorganized both administrative policy and cropping practices.[10]

Under Colbert's direction, the older method of harvesting timber was altered. The policy (known as *jardinage*) of the sixteenth-century ordinances had been to partition the forest into equal lots and to cut trees from each lot in successive years, leaving some standing trees for reseeding the plot. The disadvantages of this method were an unequal distribution in the density of the second-growth trees, unequal regrowth in adjacent plots due to soil differences, and variation in the size and age of timber cut in successive years on the rotation system.

Under the new method, forests were divided not into equal but into equivalent sections. Coppice woods containing small trees that could be harvested every twenty years for firewood were divided into twenty portions, while stands of large timber for shipbuilding, which could be profitably harvested only every 120 years, were divided into 120 portions, only one of which was to be cut each year. Portions were equivalent with respect to long-term yield, taking into consideration soil variation, exposure, and growth rate, rather than equal as to acreage. The new methods required greater attention to local conditions and types of trees as well as periodic thinning to regulate sustained yield. Forests ceased to be regarded as communal and were instead regulated by the state.

Evelyn devoted another treatise specifically to the discussion of air pollution in London. His *Fumifugium*, like his *Silva* and *Terra*, showed a sophisticated awareness of the important ecological interrelationships between air, water, and the health of living things.[11] The *Fumifugium* was presented in 1661 to Charles II, "who is the very breath of our nostrils," in response to general concern over the increase of "fuliginous and filthy vapors" caused by the burning of "sea coal." Its sulfurous fumes permeated the environs of London, penetrating even into the heart of the royal palace, causing "catarrhs, phthisics, coughs, and consumptions" in the inhabitants, premature deaths in infants, and leaflessness and barrenness in fruit-bearing trees. Air pollution had continued to grow, owing to the increase in "brewers, dyers, lime burners, salt and soap boilers, and . . . other private trades." The pollution had reached such proportions in London that "the sun itself which gives day to all the world besides, is hardly able to penetrate and impart it . . . and the weary traveler, at many miles distance, sooner smells, than sees the city to which he repairs."

Especially deleterious were the effects on human health. On entering London, visitors found that their pores became stopped up, their noses and skin dried out, their eyes stung, their lungs filled with sooty vapors, their humors were apt to "putrify," their spittle became black, and their appetite was lost. This was succeeded by colds, coughs, and catarrhs that remained with them until they returned home.

If one consulted London's weekly bills of mortality, the morbid results of air pollution were all too obvious. According to Evelyn, one half of the people who died in London did so of "phthisical and pulmonic distempers." John Graunt's bills of mortality showed that, over and above the deaths caused by the plague, one in thirty-two died in London as opposed to one in fifty in the country.[12] London was thus the city known for its two kinds of plague. In his sixth observation, Graunt noted that "although seasoned bodies may and do live near as long in London as elsewhere, yet newcomers and children do not, for the smoaks stinks, and close air are less healthful than that of the country: otherwise why do sickly persons remove into the country air?" Seventy-five years later, the situation had not been ameliorated. Corbyn Morris, lamenting the larger number of deaths of children, cited the unhealthiness of the air of London as the primary reason why burials in London exceed christenings. More children under five died in the city than in the country; and more adults of weak constitution would have survived had they been able to make a livelihood in the country rather than in the cities.[13]

To resolve the problem of London's air pollution, Evelyn offered a number of suggestions. First was the planting and reforestation of timber tracts in areas outside of London so that wood could be purchased at reasonable prices. Since Paris, which was equally large, supplied itself with wood, this proposal was not infeasible. Wood could also be brought in from countries to the north by sea. A more immediate solution would be to remove all trades that used coal in their forges and furnaces to new locations outside the city. These would include brewers, dyers, soap and salt boilers, lime burners, and other similar works and would immediately restore the air to the same purity as was presently observed on Sundays when their work ceased.

If the coal-burning trades could not be removed from the city, the heights of the chimneys should be increased, new chimney shapes should be devised to send the smoke higher, experimental work should be conducted on the use of coke instead of coal in smelting, and the building of new furnaces within the city limits should be curtailed. Evelyn's final suggestion was to plant the entire city with the "most fragrant and odoriferous flowers" so that London would be noted instead as the most beautiful and sweetest city in Europe.[14]

NATURE A VEGETABLE.   During the Restoration the Cambridge Platonists, Henry More and Ralph Cudworth, helped to advance Evelyn's managerial approach by adapting the older organic philosophy to the new social and commercial demands of preindustrial capitalism. The basis for their compromise was the reassertion of the essential vegetative constitution of nature, a position of moderation between the dead mechanical cosmos of the mechanists and the animal-like world of the Civil War pantheists (who saw God as active and immanent in every material thing) and enthusiasts (who fanatically gave into all forms of religious passion) (see Chapter 4). From Descartes, they incorporated the concept of "dead stupid" matter and the dualism between matter and spirit, but they attempted to bridge the ontological gap by retaining the world's vital vegetative character. Thus nature's fundamental organic character was maintained, while the unpredictability associated with an animistic world was minimized. The cosmos was reduced to a vegetable—still alive, but not uncontrollably so.

During the English Civil War, More's and Cudworth's Cambridge was a stormy center of divisiveness between the Puritan township and the then Royalist university. More's beliefs in an aristocratic social order, monarchy, the episcopacy, immutable law, and an organic theory of nature implied for him a gradual and natural rather than violent mode of social change. The Paracelsian philosophy of nature associated with the radical sects, chemists, and theosophists was responsible for the disreputable philosophies adhered to by the "palpably mad" enthusiasts, who professed that "God is nothing but the universal matter of the world dressed up in several shapes and forms."[15]

After the Restoration, the ideas of "Latitude-men" such as More and Cudworth became the foundation for a religious compromise between Anglicanism and Puritanism and for a social philosophy of gradual and progressive commercial change. Latitudinarianism represented the restoration of order and social harmony under a new monarch after the failure of the commonwealth to establish a lasting new constitutional form of government.

More's and Cudworth's answer to both Paracelsianism and Hobbism was to reassert the conservative hierarchical theory of society and the cosmos. More's organic cosmos was derived from the Neoplatonic philosophy of a hierarchical chain of being. He recognized four kinds of created spirits existing in a graded hierarchy below God: angels, the souls of humans and animals, and the individual plastic essences of lower vegetative forms of life. All intermediate orders of spirits accepted by some Neoplatonists were rejected.[16]

More and Cudworth both placed great emphasis on the spirit of nature and plastic natures because they wished to restore vital organic life to the dead, inanimate world constructed by the mechanists. Matter, "inert and stupid of itself," must be moved by a spiritual substance. More's spirit of nature was an incorporeal principle that pervaded the universe, directing the parts of matter and their motions and working on matter to give it shape and form through its plastic power.[17] For Cudworth, a similar general plastic nature organized and directed the activity of the entire larger cosmos, ordering the growth and reproduction of plants, herbs, grasses, minerals, and the whole heavens. It was unreasonable, he thought, that each blade of grass should have its own plastic life, but in the case of animals the body of each was formed and conserved by its individual particular plastic nature.[18]

Plastic natures were akin to the incorporeal vegetative souls and seminal forms of the older Aristotelian and Neoplatonic philosophies. But More and Cudworth denied to their plastic natures any functions associated with the sensory or rational portions of the soul such as reason, free will, and sensation. All animistic powers and sensations in the earth, sun, and heavenly bodies were likewise negated so that metaphorically nature could no longer be conceived as a living animal as in pantheistic and enthusiastical philosophies.[19] Only its living vegetative characteristics were retained. More warned, "If there be any sense . . . to be acknowledged in the spirit of nature, a scruple may arise, whether this spirit may not be called the soul of a brute, and so the world itself a brute."[20]

Cudworth drew his ideas on the vegetative character of the general universal plastic nature from an ancient philosophy he referred to as "cosmoplastic stoicism," which "supposes the whole world . . . not to be an animal, but only a great plant or vegetable, having one spermatic form, or plastic natuie, which without any conscious reason or understanding, orders the whole."[21] The ideas of the cosmoplastic philosophers were condemned by him, however, not because of their use of plastic principles but for their atheism.

The vegetative metaphor offered immense advantages for Latitudinarian philosophers such as More and Cudworth who wished to present a moderate rational philosophy of compromise between the excesses of the Civil War enthusiasts with their animistic philosophies of nature and Hobbesian materialists who viewed nature as composed of "dead, stupid matter" forming beings by blind and fortuitous motions. As a cosmoplastic vegetable, nature retained its fundamental organic character. It was alive yet not uncontrollable or unpredictable. The older conception of the world as an animal with sensation and self-activity implied the possibility of unaccountable, irrational actions—one could never be fully certain what an animal might do. Not only did animism imply atheism, but enthusiasm as well, and in both philosophies events could neither be predicted nor controlled. If, on the other hand, one denied sensation and all life to nature, the rest was mechanism—or lifeless matter in motion with atheistic implications.

In More's dualistic philosophy, spirit had the capacity to move, alter, and penetrate matter. In that it could move matter, it was the source of the world's activity; in that it could alter matter, it shaped and rearranged the positions and motions of material particles through its plastic property. Spirit, like matter, was extended and had the capacity either to penetrate and move through matter or for total organic unity with it. It was therefore a principle of organic life within matter, causing motion from within, and unifying and maintaining the organization of its particles.[22] It contrasted with the Cartesian view of motion as external to matter, put into the universe by God and subsequently transmitted from particle to particle. If the role of God in Descartes' system were eliminated, the result was atheism and the blind mechanism attributed to the Hobbists. Endowing the world with a vital vegetative plastic nature avoided the dangers of both enthusiasm and materialism. It allowed for the mediated action and providential care of God in the creation and for human management and stewardship over nature.

More's *Antidote Against Atheism* (1653) dealt with the issues created by the atheistic mechanism of Hobbes, and the underlying problem of man's mechanical power over nature.[23] Like Cicero, More saw evidence of human technological prowess in the use of nature—the lodestone for navigation, timbers for building ships to transport goods, coal for the smelting of ores and working of metals, and quarried stones in the erection of large buildings. God's providence, combined with human reason in the wise use of nature, would prevent runaway technology from taking its toll, as in the case of Daedalus and Icarus. "And the closer we look into the business, we shall discern more evident footsteps of providence in it . . . there could nothing so highly gratify [man's] nature as power of navigation, whereby he, riding on the back of the waves of the sea, views the wonders of the deep and . . . is able to prove the truth of those sagacious suggestions of his own mind."

God sanctioned man's use of nature and "provided him with materials with which he might be able to adorn his present age, and furnish history with the records of egregious exploits both of art and valor." The intricacies and structures of plants, animals, the human body, and the heavenly bodies were examples of the natural order put into the universe by God, and evidence that these organisms, minerals, and metals had been put here for human use. The ox and horse had been provided for pulling carts, ropes and iron for plows, and pulleys and engines for lifting weights. All these could be used to "carry home the fruits of the earth." In developing the technologies related to plants, animals, and metals, humans were carrying out the providential dictates of their creator. God was the wise counsel, superintendent, and overseer of the creation, to whom people were ultimately responsible as his stewards.

MANAGEMENT AND HUMAN STEWARDSHIP.   The organic accommodation to mechanism developed by More and Cudworth, emphasizing the vegetative constitution of the world, was intellectually significant as a background for a managerial philosophy of human stewardship over nature developed in succeeding decades. John Ray and William Derham adapted plastic natures, vital motions, and "ecological" principles to a social philosophy of commercial progress in an ordered designed universe.

In his *Wisdom of God Manifested in the Works of the Creation* (1691), John Ray used plastic natures as an argument against the atheism implicit in the mechanical philosophy of Descartes, Gassendi, and Hobbes. In Ray's opinion, Descartes had erred in his banishment of final causes and in his limitation of God's role to the production of corpuscular matter constrained to obey a few simple laws. The first excluded human ability to discern God's purpose within nature, the second his daily providential care. The Epicurean hypothesis "lately revived" in Christian guise by Gassendi was no better, as it was based solely on space, matter, and declination. As Cicero had aptly shown in his critique of the Epicurean system, there was no sufficient reason why some atoms should swerve and others not, nor what angle of declination they would assume.[24]

The frightening implications of mechanism could be countered by the assertion of a designed universe governed by a providential God as a model for human affairs. In refuting the mechanists and "mechanic theists," Ray turned to plastic natures as evidence for

God's action in the mundane world. One could not adequately account for the growth and form of plants on mere mechanical principles. A vegetative plastic principle was necessary to rule over the entire plant forming the various structure in its roots, stalks, leaves, flowers, and fruit. "I incline to Dr. Cudworth's opinion that God uses for these effects the subordinate ministry of some inferior plastic nature; as, in his world of providence, he doth of angels." Other examples of plastic principles could be found in the vital motion of the heart, the first breaths of the newborn child, and in the monstrosities, sports, and mistakes of nature.[25]

He brought forward the arguments of Latitudinarian divines such as Archbishop John Tillotson (1630–1694) and Bishop Edward Stillingfleet (1635–1699) that a wise God had created a "spacious and well-furnished world." Although not everything in nature was created for the sole benefit of human beings, God had made them "sociable creatures" who could understand, communicate, and experiment. He expected them to make use of metals, stones, timbers, plants, and animals, to advance the sciences, to increase the trade and prosperity of their countries, and in so doing to glorify their creator. When people built beautiful villages, country houses, cities, and castles, God was pleased with their industry.[26]

Moreover, God had seen fit to provide just the right quantity of gold and silver to produce money for the increase of trade. While the love of money was vicious and the root of evil, nevertheless when used properly "it was an admirable contrivance for rewarding and encouraging industry or carrying on trade and commerce certainly, easily, and speedily, for obliging all to employ their various parts and several capacities for the common good, and engaging everyone to communicate the benefit of his particular labor, without any prejudice to himself." Money and the metals were essential to the cultivation of the arts and sciences; where they had not been introduced people were "brutish and savage," devoted like beasts to merely seeking food. Without them, "what a kind of barbarous and sordid life we must necessarily have lived, the Indians in the northern part of America are a clear demonstration." What Ray and his age did not see was the way in which these very metals and money, coupled with alcohol, when introduced destroyed the holistic lifestyle of the American Indians, their interconnection with nature, and their own "natural religion."[27]

In refuting the claims of the atheists that the world was created by chance configurations of atoms, Ray and his follower William Derham, pursuing the lead taken in More's *Antidote Against Atheism*, worked out the argument from design and its teleological implications. The earth as it appeared at the present time, with its great abundance and variety of parts, "must needs be the result of counsel, wisdom, and design." Against those who had argued that the "present earth is nothing else but a heap of rubbish and ruins," he pointed out the advantages to civilization of its present constitution. The earth was filled with sand, gravel, lime, stone, clay, and marble such that there was a variety in the landscape suitable for numerous kinds of plants and animals and the breeding of domestics. The variation in slope of the land and the size of streams and rivers made different areas suitable to various kinds of tillage and herding. Its pleasurable coating of green grass and fragrant flowers with its "many pleasant and nourishing fruits, many liquors, drugs, and good medicines," were marks of its design by a "wise creator and governor."[28]

Ray's philosophy of progress was extended by his follower William Derham to the concept of human stewardship over the whole natural world. Derham's *Physico-Theology* (1713) might today have been called an *ecotheology*. It embodied a number of ecologically sound principles, in a managerial framework of stewardship modeled on man's role as caretaker of God's creation. The manuscript that ultimately resulted in the *Physico-Theology* had been prepared for the Boyle lecture of 1711, an important part of a series of defenses of liberal Protestant social philosophy resting on Newtonian science.

These lectures had been founded according to the wishes of Robert Boyle, who in his will (1691) provided for annual lectures dedicated to proving the Christian religion against "notorious infidels—atheists, theists, pagans, Jews, and Mohammedans," divisions among Christians being excluded. The English Revolution of 1689, like the Civil War, had produced a social and religious disorientation resulting in the need for reassertion of a moderate Latitudinarian religion that would support the continued development of an acquisitive market society.[29]

The natural religion expounded in the Boyle lectures offered arguments for the existence of the deity based on the achievements of natural science and the providential care of God over his creation.

Newtonian natural philosophy and natural religion could be united in the attempt to restore the ordered society necessary for entrepreneurial advance. At the same time, they would provide a bulwark against the socially disruptive views of Hobbists, enthusiasts, and freethinkers. Latitudinarian control over the church had been threatened by the 1689 revolution and the Latitudinarians needed again to codify religious, secular, and social goals.

Although Derham's Latitudinarian affiliation is not as clear as that of the other Boyle lecturers, his ministerial position at Essex was probably obtained with the help of Latitudinarian moderate divines. His *Physico-Theology* accepted the Judeo-Christian ideal of human dominion over nature in the form of a stewardship that supported progress. Man, "the top of the lower world," was given "superiority in the animal world."[30] God was the wise conservator and superintendent of the natural world, who made people in his image as caretakers and stewards on earth. In confirmation, he cited Matthew 25:14: "That these things are the gifts of God, they are so many talents entrusted with us by the infinite Lord of the world, a stewardship, a trust reposed in us; for which we must give an account at the day when our Lord shall call."

God's gifts should be used in secular lives and callings. Religion should be united with the Protestant idea of duty and calling, be it that of "ambassadors of heaven" or "the more secular business of the gentleman, tradesman, mechanic, or only servant." Worldly occupations should be discharged with "diligence," "care," and "fidelity." In Luke 16:2, God had said to the unfaithful steward, "Give an account of thy stewardship, for thou mayest be no longer steward." The idea of man as nature's guardian and caretaker was a managerial interpretation of the doctrine of dominion.

Derham followed Bacon in the idea that we should pry into nature to extract her secrets in order to control nature, even taking over the harshness of Bacon's language: "We can, if need be, ransack the whole globe, penetrate into the bowels of the earth, descend to the bottom of the deep, travel to the farthest regions of this world, to acquire wealth, to increase our knowledge, or even only to please our eye and fancy."[31] There were "many secret, grand functions and operations of nature in the bowels of the earth." Once extracted, these secrets could be used to improve man's lot on earth in order to fulfill the New Testament commandment "Love thy neighbor." Nature contained such an abundance of creatures, minerals, and fossils, and so many of each species, that "there is nothing wanting to the use of man or any other creature of this lower world." Even if every age, or culture, changed its food, clothing, and shelter every day, "still the creation would not be exhausted, still nothing would be wanting for food, nothing for physic, nothing for building and habitation, nothing for cleanliness and refreshment, yea even for recreation and pleasure."

The Creator had supplied the earth with endless bounty for the needs of life on it, but this abundance must be managed properly by God's rational stewards. Derham overestimated the extent of nature's resources, much of the New World being as yet undeveloped. But his unabiding faith in God as the great conserver and supplier of creation's needs caused him to make the optimistic claim that "the munificence of the Creator is such, that there is abundantly enough to supply the wants, the convenience, yea almost the extra agencies of all the creatures, in all places, all ages, and upon all occasions."

The Baconian method was useful also in searching out evidence of the deity from observations of design in nature to fulfill the Protestant commandment, "glorify God." "Let us cast our eyes here and there, let us ransack all the globe, let us with the greatest accuracy inspect every part thereof, search out the inmost secrets of any of the creatures, let us examine them with all our gauges, measure them with our nicest rules, pry into them with all our microscopes and most exquisite instruments, till we find them to bear testimony to their infinite workman."[32]

The inventions through which nature could be utilized to glorify God and improve the human condition included the instruments of astronomy, navigation, and geography and the sciences of geometry, arithmetic, and optics, as well as the technologies useful for building, clothing, feeding, and caring for the health of the human race. God had provided man's soul with an inventive faculty, and hands to carry out his ideas. The present social order, like the present natural order, was the most useful for carrying out God's plan. "There are good political reasons for man's clothing himself . . . many callings and ways of life arise from hence, and (to name no more), the ranks and degrees of men are hereby in some measure rendered visible to others, in the several nations of the earth." The idea that nature could be transformed by the hand of man within a social system where clothing and callings marked one's station in life was part of the ideology of status society with an acquisitive middle class.

God was a caretaker, steward, and wise manager of his entire created world. His design, management, and providential care as the great superintendent of the universe was a model for human management of the earth. Everything created had its own place and purpose in the interdependent balance of nature, even though "there were things of little immediate use to man, in this or any other age." Volcanoes, caverns, and grottos were the "wise contrivances of the Creator, serving to the great uses of the globe and ends of God's government."

Derham made use of not only the principle of ecological interdependence but also the concept of adaptation. All animals had organs marvelously adapted to their habits and to their habitats. Their organs of respiration, motion, and vision were suited to the medium in which they lived. Each lake, pond, hill, and vale had its own group of trees, shrubs, plants, and animals "whose organs of life and action are manifestly adapted to such and such places and things; whose food and physic, and every other convenience of life is to be met with in that very place appointed it."[33]

Another ecological principle was that of population stability. Each valley, forest, or lake was kept in perfect balance so that the number of species in any one place remained constant, and there was sufficient "room, food, and other necessaries." God's wisdom was manifested in "balancing the number of individuals of each species of creatures in that place appointed thereunto." The cause of this adaptation and ecological balance in nature was the great artist who had created the existing frame of the earth by design and conserved and maintained its present constant state.

The optimism expressed by Derham over the fitness of the present constitution of the earth was part of the growing eighteenth-century optimism over the progress of the human race as a rational species in control of its environment. By then Newtonian science had given natural religion a firm basis, and the work of the scientists of the Royal Society had helped to contribute to such areas as natural history, electricity, chemical reactions, medicine, and the practical arts, where Newton had not provided a synthesis. Atheism was refuted far more confidently by the Boyle lecturers of the 1690s and 1700s than by the Cambridge Platonists of the 1650s and 1660s, who lacked the firm Newtonian synthesis and the institutionalization of science as an ideology for organizing other areas of human life. The argument from design and the organic philosophy of John Evelyn, the Cambridge Platonists, and the naturalist theologians represented an adaptation of organicism to the managerial ethos of mechanism and the ideal of progress in an expanding market society. Growth and trade must not be halted due to the depletion of resources, nor should nature be exploited for short-term gain, but if it could be used wisely and understood rationally, nature's abundance would not be exhausted.

Much like its seventeenth-century predecessors, today's managerial ecology subjects nature to rational analysis for long-term planning. By further reducing the vegetative community to an ecosystem, the anthropomorphic connotations of group sharing give way to the physical descriptions and equations associated with quantitative analysis. The thermodynamic laws of energy exchange describe the behavior of physical and biotic components within the mathematical framework of systems theory. The reductionist ecology of Arthur George Tansley, developed in the 1950s, has matured into the "Club of Rome's" computer model, which predicts the "limits to growth" for the entire world system.[34]

But the reductionist model of the new ecologists has its limits. It is difficult, if not impossible, to successfully program contexts and patterns into a computer. Removing components or abstracting data from the environmental context can alter the whole, distorting its behavior. The organismic small-community approach, which relies on human decision makers and participatory democracy rather than on experts, represents an alternative to the managerial ethic that developed out of seventeenth-century mechanism. The "land ethic" of Aldo Leopold (1949), the "declarations of interdependence" of ecology action groups, and the nature religion of Callenbach's Ecotopians (see chapter 3) present a community-oriented ecocentric alternative to the homocentric ethics of ecosystem management.[35]

CHAPTER 11

Women on Nature

Anne Conway and Other Philosophical Feminists

By the late seventeenth century, several reactions to the mechanical philosophies of Descartes, Gassendi, Hobbes, and Boyle had appeared in western Europe. Among these were philosophies that reasserted the fundamental organic unity of nature, such as Cambridge Platonism and vitalism. The Cambridge Platonists, Henry More and Ralph Cudworth, retained the dualistic structure of mind and matter assumed by Descartes and attempted to bridge the gap by the reassertion of plastic natures and the spirit of nature as organic links. The vitalists, however, affirmed the life of all things through a reduction of Cartesian dualism to the monistic unity of matter and spirit. Among the proponents of a vitalist philosophy were Francis Glisson, Jean Baptiste Van Helmont and his son, Francis Mercury Van Helmont, Lady Anne Conway, and Gottfried Wilhelm von Leibniz.

As a philosophy of nature, vitalism in its monistic form was inherently antiexploitative. Its emphasis on the life of all things as gradations of soul, its lack of a separate distinction between matter

253

and spirit, its principle of an immanent activity permeating nature, and its reverence for the nurturing power of the earth endowed it with an ethic of the inherent worth of everything alive. Contained within the conceptual structure of vitalism was a normative constraint. Perhaps it is not an accident to find among its advocates a woman philosopher, Anne Conway, and a wandering scholar-healer, Francis Mercury Van Helmont, both of whom turned to Quakerism as a moral and religious alternative.

There are strong similarities between the dynamics of Paracelsus' four elements, Jean Baptiste Van Helmont's pluralism of living, developing seeds of matter, and the vitalist philosophies of the late seventeenth century. Paracelsus was an important influence on the ideas of the elder Van Helmont, whose philosophy of the insoluble unity of body and soul was in turn transmitted to his son Francis Mercury Van Helmont and to Anne Conway.

Anne Conway, a philosopher whose ideas were praised and respected in her own day, has through scholarly error been almost forgotten in ours. Her only book, *The Principles of the Most Ancient and Modern Philosophy*, was edited and published in 1690 after her death by her friend Francis Mercury Van Helmont. Because her name was withheld from the original Latin title page— the custom regarding female authors in that period—the book was attributed by modern scholars to its editor, Van Helmont. In 1853, the German historian of philosophy, Heinrich Ritter, erroneously based his analysis of the younger Van Helmont's philosophy almost entirely on Conway's book. His discussion became the basis for historian Ludwig Stein's theory (1890) that Van Helmont had transmitted to Leibniz the most fundamental term in his whole philosophy—the *monad*, Leibniz's infinitesimal vital active force.[1] Actually, Van Helmont did use the term, but the book containing his discussion of it was apparently unknown to Ritter. Thus the major textual evidence for attributing Leibniz's appropriation of the term *monad* to Van Helmont, rather than including Anne Conway, was due to inaccurate scholarship. The withholding of Conway's name, as a woman writer, from the Latin edition of her book excluded from recognition her important role in the development of Leibniz's thought.

Francis Mercury Van Helmont, the wandering "scholar gypsy," was introduced to Anne Conway quite fortuitously in 1670. He had come to England in order to deliver to Henry More several letters from Princess Elizabeth of Bohemia and to discuss with him their mutual interest in the Cabala, an esoteric occult and mystical tradition stemming from the Middle Ages. He had only planned to remain in England one month, but through the joint efforts of More and Viscount Edward Conway he was finally persuaded to travel to Ragley to visit the learned Lady Anne Conway, in order to attempt a cure of her incessant and intolerable migraine headaches.[2]

Anne Finch, Viscountess of Conway (1631–1679), as a young woman, had been one of Henry More's most accomplished and brilliant disciples, known to him through her brother, John Finch, a pupil of More at Christ College, Cambridge. An avid reader of philosophy, literature, the classics, mathematics, and astronomy, she was an intelligent, vital conversationalist and had a charming personality. Her home at Ragley Hall in Warwickshire became an intellectual center where lively debates were held with philosophers such as More, Ralph Cudworth, Joseph Glanvill, Benjamin Whichcote, and the younger Van Helmont.[3]

Tormented by headaches, which gradually increased in frequency and severity until they were pronounced incurable after attempts by Europe's most noted physicians, including William Harvey, the noted healer Valentine Greatrakes, and Francis Mercury Van Helmont, Anne Conway nevertheless carried on an active intellectual life. Her book carried on the Cambridge school's interest in spiritualism, Platonism, and cabalism, and bears the influence of Van Helmont. Truer to the Platonic tradition than the writings of either of her colleagues, More or Cudworth, it was far more sweeping in its rejection of Cartesianism and embracement of vitalism.

Van Helmont's intended month in England turned into eight years during which he remained with Lady Conway, unsuccessful in treating her terrible headaches, but introducing stimulating new intellectual avenues for her mind. Henry More likewise spent much time there, experimenting with Van Helmont in the laboratory which the wandering alchemist had set up, and discussing Hebrew and cabalistic texts. Whenever Lady Conway was too ill to do so herself, Van Helmont read to her from a variety of books and pamphlets and reported on the activities of a group of Quakers meeting near Ragley. Under his influence, she began studying the texts of Quakers, Behmenists, Seekers, and Familists—religious sects that

255

had flourished during the period of the Civil War.

More and Conway discussed the philosophy of Jacob Boehme and the Familists in their letters during the years 1667–1670. More was skeptical of the neglect of the power of reason by the Behmenists and Familists, and deplored their tendency toward enthusiasm, but Anne Conway was sympathetic. Because the Quakers were "quiet people," she employed them as servants and also lent her home for their meetings. She made the acquaintance of Quaker leaders—George Fox, George Keith, Isaac Pennington, and Charles Lloyd—and corresponded with William Penn. To the despair of More, who identified the Quakers with the Ranters, Seekers, Familists, and other enthusiasts of the Civil War years, both she and Van Helmont became Quakers—Van Helmont in the spring of 1676 and Conway at least by 1677.

The Quakers, far more than the other Protestant sects, gave both women and men full equality. Quakerism, growing out of discontent with Puritanism, began to spread in southern and eastern England around 1655, carried not only by men but also women preachers, such as Anne Blaykling, Mary Fisher, Dorothy Waugh, Jane Waugh, and Mary Pennington. Women, some of whom left their families behind, became traveling preachers, bearing the Quaker message not only all over England but as far as the Ottoman Empire. Under the leadership of George Fox, separate meeting houses were established, administered, and attended solely by women throughout England.[4]

The Quakers emphasized the inward presence of God, the living Christ within each individual, and the vitality of the living word, as opposed to the deadness of tradition and inertness of the written word. They distinguished between the historical figure of Christ, who had died, and the voice of the living, vital Christ within.

Van Helmont, George Keith, and Anne Conway saw much in common between the Cabala and the Quaker doctrines of the "inner light," "Christ within," and the Christian trinity. The three collaborated over a four-year period on a treatise entitled the "Two Hundred Queries ... Concerning the Doctrine of the Revolution of Humane Souls" (1684). In subsequent years, this book became the bone of contention between the Quakers and Keith and Van Helmont, because of their emphasis on the transmigration of souls and the reality of the historical figure of Christ.[5]

During his stay at Ragley, Van Helmont also wrote his *Cabbalistical Dialogue*, which was included in Knorr von Rosenroth's *Kabbalah Denudata* in 1677. Anne Conway's manuscript, written some time during this period, probably in 1672 or 1673 bears Van Helmont's influence. After her death in February 1679, her coffin was inscribed with the epitaph "Quaker Lady." Van Helmont returned to the Continent, taking the manuscript with him, where in 1690 it was published in Holland in Latin translation and in 1692 in English.[6]

In March 1696, Van Helmont arrived in Hanover, where he remained for several months, meeting with Leibniz each morning at nine for philosophical discussion. According to Leibniz, Van Helmont took the desk, while Leibniz became the pupil, interrupting frequently to ask for greater clarification. Van Helmont recounted to Leibniz the history of the "extraordinary woman," the Countess of "Kennaway," and his own relationship with Henry More and John Locke. From him, Leibniz learned of Anne Conway's metaphysics and her studies of the works of Plato, Plotinus, and the Cabala.[7] In a 1697 letter to English divine Thomas Burnet (1635–1715) Leibniz, having read her book, went so far as to state:

> My philosophical views approach somewhat closely those of the late Countess of Conway, and hold a middle position between Plato and Democritus, because I hold that all things take place mechanically as Democritus and Descartes contend against the views of Henry More and his followers, and hold too, nevertheless, that everything takes place according to a living principle and according to final causes—all things are full of life and consciousness, contrary to the views of the Atomists.[8]

Leibniz spoke subsequently with praise and approval of both Lady Anne Conway and Van Helmont, although the latter he often found puzzling and quixotic. In the *New Essays Concerning Human Understanding*, begun in 1697 and published posthumously in 1765, Leibniz referred to both as explicating the doctrine of vitalism better than their Renaissance predecessors, writing that he saw:

> ... how it is necessary to explain rationally those who have lodged life and perception in all things, as Cardan, Campanella, and better than they, the late Countess of Connaway, a Platonist, and our friend, the late M. François Mercure Van Helmont (although elsewhere bristling with unintelligible paradoxes), with his friend the late Mr. Henry More.[9]

The elements of Conway's system were a significant influence in the important period of Leibniz's thought, leading up to the writing of his "Monadology" (1714).

ANNE CONWAY'S MONISTIC VITALISM.   Whereas the Cartesians and the Cambridge Platonists, More and Cudworth, were dualists, Anne Conway, like Van Helmont, was a monist. In her philosophy, there was no essential difference between spirit and body and, moreover, the two were interconvertible. She distinguished her views sharply from those of Descartes and also from More and Cudworth on these points. Body was condensed spirit and spirit was subtle, volatile body. Body and spirit were not contrary entities, the first impenetrable and divisible, the other penetrable and indivisible, as More had held. Matter was not dead, "stupid," and devoid of life, as Descartes and the Cambridge Platonists had thought. For Lady Conway, an intimate bond and organic unity existed between the two. Body and soul were of the same substance and nature, but soul was more excellent in such respects as swiftness, penetrability, and life.[10]

If, as More asserted, spirit was the principle of motion in dead, unorganized matter, and if spirit could see, hear, and sense of itself, then it would have no need for body or sense organs. But since the soul felt pain and grief when the body was cut or wounded, the two must be united and of one substance. Otherwise the soul, as an independent substance, could simply move away from the suffering of a damaged body and thereby be insensitive to it.[11]

Her break from Descartes and the other Cambridge Platonists was sharpest on the issue of dualism. She insisted that her philosophy was not Cartesianism in a new form, as she perceived that of her friends to have been, but fundamentally anti-Cartesian:

> For first, as touching the Cartesian Philosophy, this says that every body is a mere dead mass, not only void of all kind of life and sense, but utterly uncapable thereof to all eternity; this grand error also is to be imputed to all those who affirm body and spirit to be contrary things, and inconvertible one into another, so as to deny a body all life and sense.[12]

Body and spirit were interconvertible because they were of the same substance and differed only as to mode. The distinctions made between the attributes of matter as impenetrable and extended, and spirit as penetrable and unextended, could not to be assigned respectively to two separate substances. Body was simply the grosser part of a thing and spirit the subtler. The penetration of spirits within a body caused it to swell and puff up, an alteration that might or might not be visible to the senses. Just as spirit and body could interpenetrate, so a less gross body or spirit could penetrate a more gross one. Penetrability like other properties of objects (heat, weight, and solidity), was relative. The dualists had "not yet proved that body and spirit are distinct substances."[13]

Matter and spirit were united as two different aspects of the same substance. Division into parts, ordinarily attributed to bodies, was equally an attribute of spirit. Just as bodies were composed of lesser bodies, the human spirit was composed of several spirits under one governing spirit. Conversely, motion and figure, which were supposed to be attributes of extended matter, applied equally to spirit, for spirit was even more movable and figurable than body.[14]

This and other ideas in her treatise she extracted from works on the Cabala included in the *Kabbalah Denudata,* such as the *Philosophia Kabbalistica Dissertatio* and the *Adumbratio Kabbalae Christianae.* In the *Kabbalah Denudata*, spirit was the capacity to enlarge or contract itself by sending out light from a center. Matter was a "naked center or a point wanting eradiation."[15] Interest in cabalistic literature was keen among the members of the Cambridge school, and both More and Cudworth had at times viewed Descartes as the restorer of the true philosophy of Moses. One of More's works least appreciated by modern scholars, his *Conjectura Cabbalistica* (1653), written before he had read the *Zohar* and admitted to be the product of his own imagination, was nevertheless an important influence on John Milton. More subsequently repudiated the Cabala in a treatise in the *Kabbalah Denudata*, entitled "The Fundamentals of Philosophy." But the Cabala was an important source of validation to those philosophers who wished to restore life and spirit to the dead world of the mechanists. Cudworth, More, and Conway all used it to argue that the ancient wisdom that perceived a total unity and vitality in the universe was the true knowledge, whereas the dead mechanical world of the moderns was a distortion emphasizing only the atomistic aspect of old gnosis.[16]

Like More and Cudworth, Lady Conway differed from Descartes on the subject of cosmic and animal mechanism. Although Descartes had discovered many mechanical laws, nature was not a machine, but a living body: "But yet in nature, and her operations, they [natural operations] are far more than merely mechanical . . . like a clock, wherein there is not vital principle of motion; but a living body, having life and sense, which body is far more sublime than a mere mechanism, or mechanical motion."[17] Likewise, animals were not machines, composed of "mere fabric" or "dead matter," but had spirits within them "having knowledge, sense, and love, and divers other faculties and properties of a spirit."[18]

An individual atom of dead matter, the building block of the mechanists, could never, if isolated, do anything to develop or perfect itself, for an atom had no internal motion and no capacity for sensation. Having no sight, taste, or hearing from within, it could receive nothing from without.

Like other organicists of the period, Conway based her system of creation not on the machine but on the great, hierarchical chain of being, modified to incorporate an evolution or transmutation to higher forms, based on the acquisition of goodness and perfection. Conway denied that any created essences could reach God's essence, which was infinitely perfect, but within the creation there was an ascension up the scale of being. Dust and sand were capable of successive transmutation to stones, earth, grass, sheep, horses, humans, and the noblest spirits, so that after a long period of time they could achieve the perfections common to the highest creatures; that is, "feeling, sense, and knowledge, love, joy, and fruition, and all kind of power and virtue."[19]

Creation was like a ladder whose steps were species placed in finite, rather than infinite, distances from one another. Hence,

> . . . stones are changed into metals, and one metal into another; but lest some should say these are only naked bodies and have no spirit, we shall observe the same not only in vegetables, but also in animals, like as barley and wheat are convertible the one into the other, and are in very deed often so changed. . . . And in animals worms are changed into flies and beasts, and fishes that feed on beasts, and fishes of a different kind do change them into their own nature, and species.[20]

This, she believed, was consistent with the biblical account that the waters brought forth birds and fishes and the earth, beasts and creeping things at the command of the Creator.

The transmutation of spirits into new bodies after death was effected by the soul's plastic nature, a concept obtained from More and Cudworth, hypothesizing a force capable of forming matter into new shapes:

> And when the said brutish spirit returns again into some body, and has now dominion over that body, so that its plastic faculty has the liberty of forming a body, after its own idea and inclination (which before in the humane body, it had not); it necessarily follows, that the body, which this vital spirit forms, will be brutal, and not humane.... Because its plastic faculty is governed of its imagination, which it doth most strongly imagine to itself, or conceive its own proper image; which therefore the external body is necessarily forced to assume.[21]

Leibniz, differing from Conway and Van Helmont on this point, not only argued against transmigration or metempsychosis in animals, but also against the idea of plastic natures. Plastic natures could not move, alter, or change the direction of a body, all motion being consonant with the system of preestablished harmony. In a letter of 1710, he called plastic natures an outmoded theory.[22]

Anne Conway radically opposed Hobbes and Spinoza, both of whom had reduced nature to a monistic materialism that denied any distinction between God and his creation. Like Conway, they accepted the interconvertibility of all things, but their materialism admitted no distinction between lower and higher forms and saw God as interconvertible with corporeal species. To Conway, sense and knowledge were far more noble than the Hobbesian reduction to the mechanical motion of corpuscles would allow. Vital action occurred by intrinsic presence, a more subtle kind of penetration than mechanical action. Since the whole of creation was alive, every motion within it was vital—"a motion of life" or "vital virtue." All bodies had "not only quantity and figure, but life also." They were "not only locally and mechanically but vitally movable" and could "receive and transmit vital action." Old and New Testament texts supported her view that "all things have life, and do really live in some degree or measure."[23]

In much of her discussion of the essential spiritual vitality of the whole world, Anne Conway's thought converged with that of Leibniz, and she was for this reason held in high esteem by him. Like Leibniz, who believed that in each portion of matter there was a whole world of creatures, each one containing within it also an entire world, Anne Conway wrote that "in every creature, whether

261

the same be a spirit or a body, there is an infinity of creatures, each whereof contains an infinity, and again each of these, and so *ad infinitum*."[24]

Like Leibniz, who wrote that there was nothing dead or fallow in the universe, Conway asked, "How can it be, that any dead thing should proceed from him, or be created by him, such as is mere body or matter.... It is truly said of one that God made not death, and it is true, that he made no dead thing: For how can a dead thing depend of him who is life and charity?" Death was not annihilation, but "a change from one kind of and degree of life to another." Dead body could not receive goodness nor perfect itself in any way; changes in motion or shape would not help it to attain life or improve itself intrinsically. This idea was echoed in Leibniz's statement that "Every possible thing has the right to aspire to existence in proportion to the amount of perfection it contains in germ."[25]

Like Leibniz, who stressed the interconnectedness of all spirits (or minds) in a "kind of fellowship with God," so that the totality composed the City of God, Lady Conway based her system on the interdependence of all creatures under God in a "certain society or fellowship ... whereby they mutually subsist one by another, so that one cannot live without another." Each creature had a "central or governing spirit" having dominion over the other spirits which composed it. "The unity of spirits that compose or make up this center or governing spirit, is more firm and tenacious than that of all the other spirits; which are, as it were, the angels or ministering spirits of their prince or captain." Akin to this was Leibniz's dominant monad unifying the simple monads.[26]

But unlike Leibniz, who held to a system of preestablished harmony to solve the problem of the dualism between the body and the spirit, and unlike More and Cudworth, who used plastic natures to unify the two worlds, Conway followed the *Kabbalah Denudata* and the ancient system of the Hebrews. She argued that the soul was of one nature and substance with the body, "although it is many degrees more excellent in regard of life and spirituality, as also in swiftness of motion, and penetrability, and divers other perfections." Between the two extremes of gross and subtle bodies were "middle spirits," which either joined body and soul or, if absent, dissolved its unity. Similarly, Jesus Christ functioned as a middle nature or medium uniting the soul of man to God.[27]

Anne Conway's vitalism was an influential reaction against the ideas of the mechanists. She was well versed in and sharply critical of the ideas of her adversaries, Descartes, Hobbes, and the Dutch philosopher Benedictus Spinoza (1632–1677), as well as her teachers and friends, More and Cudworth. Ritter, mistaking the work of Conway for that of Van Helmont, saw the author of the *Principles* as carrying out a wide-ranging battle against the Cartesian philosophy of dualism and against the basis of mechanical physics in general.[28] Of the two vitalists, Conway had a far more rigorous and critical, and less enthusiastic, mind than Van Helmont. For these reasons, as well as for the compatibility of her ideas with his own, Leibniz found her work more "extraordinary" and less "paradoxical" than that of Van Helmont.

Yet Anne Conway's philosophy ultimately did not go beyond the limits of the categories of substance philosophy within which she worked. Her monistic resolution of the mind-body problem, although more parsimonious than the dualism of Descartes, was simply a reduction of all of reality to the idealist category of spirit. By denying the validity of body as an explanatory category, her philosophical framework was unable to provide a satisfactory description of empirical phenomena. Unlike Leibniz, whose system of preestablished harmony and "well-founded phenomena" obeying mechanical laws also fell short of a solution, she did not even address herself to the issue of bodies and their interactions.

Furthermore her assumption of the transmigration of souls, and the concepts of "middle natures," plastic natures, and vital virtues that composed the core of her vitalism were based neither on rigid logical consistency nor on firm empirical evidence, a problem that continued to weaken the case for vitalists and holists of the nineteenth and twentieth centuries, such as German biologist Hans Driesh, French philosopher Henri Bergson, and South African statesman Jan Christiaan Smuts. Like other protagonists in the mechanist-vitalist debates that have continued ever since the rise of mechanism, her embrace of vitalism was based on metatheoretical commitments. Her philosophy falls within a post-Cartesian scientific tradition that operates on the assumption that the living and nonliving constitute two fundamental categories of reality.[29] Her commitment to spirit as the solution to the dualistic dilemma derived not only from the logic of philosophical alternatives, but also from psychological needs connected to her physical health and her adoption of Quakerism as a spiritual refuge friendly to women.

THE MONADS OF CONWAY, VAN HELMONT, AND LEIBNIZ. By September, after Van Helmont's March 1696 arrival in Hanover, one finds in Leibniz's writings the first use of the term *monad* to characterize his concept of "individual substance." In long hours of conversation with Leibniz and the Electress Sophie, Van Helmont spoke about his own ideas, those of Anne Conway, and of Knorr von Rosenroth's *Kabbalah Denudata*, conversations which Leibniz found "very instructive," in contrast to Van Helmont's books, which were more enigmatic.[30]

Prior to 1696, Leibniz had used the terms *entelechie, formes substantielles, unité substantielle, point metaphysical,* and *forces primitives* interchangeably to mean "individual substance." But in 1696, the disparate elements of his metaphysics coalesced when he began using the concept of the monad to represent an independent individual—a substance endowed with perception and activity—existing in a state of accommodation and consensus with other substances.

A theory expounded by Dühring in 1869 held that Leibniz had based his whole system on Bruno's book, *De Maximo et Minimo* and had borrowed the term *monad* from his *De Monade.* The influence of Bruno on Leibniz was traced back to Nicholas of Cusa, from whom Bruno had obtained the term. But Ludwig Stein pointed out that Leibniz had merely mentioned Bruno's name in 1666, 1682, and 1690. In 1691, he had referred to Bruno's *De Monade* but does not seem to have known his whole doctrine. Although Leibniz's philosophy of individual substance and immaterial force was well developed by that year, he did not mention Bruno's work or appropriate the term *monad* until 1696, five years after the time of Bruno's most likely influence on the doctrine of the monad. He was in fact so little familiar with the writings of Bruno that in 1708, writing to La Croze he misspelled the title of Bruno's *Lo Spaccio* as "Specchio."[31]

The term *monad*, which had appeared in the writings of Anne Conway, Van Helmont, and the *Kabbalah Denudata*, probably came to Leibniz's attention during his conversations with Van Helmont. In her *Principles*, Conway had written,

> But as was said before, God cannot do that which is contrary to his wisdom and goodness, or any of his attributes. (Mathematical division of things is never made in minima; but things may be physically divided into their least parts; as when concrete matter is so far divided that it departs into physical monads, as it was in the first state of its materiality. Concerning the production of matter, see *Kab. denud. Tom.* I, Part 2, p. 310 following; and *Tom.* 2. the last Tract p. 28, Numb. 4,5. Then it is again fit to resume its activity, and become a spirit, as it happens in our meats.) [32]

The sections from the *Kabbalah Denudata* to which she referred stated that matter had been made from a coalition of spiritual monads in a state of inactivity, or stupor, out of which is created the material world. In slipping downward, these material monads retain some of the original light, such that if excited in a certain manner they can emit radiations peculiar to the matter and seeds of the various classes of animals, plants, and the inanimate. Matter consists of singular monads, deprived of their own motion, but disposed toward it through the capacity for light and irradiation. [33]

These indivisible monads as the basis of all life were close to Leibniz's own metaphysical position on individual substance. Through his own acquaintance with Knorr von Rosenroth, he knew the *Kabbalah Denudata*, in which had appeared "The Cabbalistical Dialogue" of Van Helmont. In it, Van Helmont had stated,

> For these are our positions: (1) That the creator first brings into being a spiritual nature. (2) And that either arbitrarily (when he please;) or continually, as he continually understands, generates, etc. (3) That some of these spirits for some certain cause or reason, are slipt down from the state of knowing, of penetrating or of moving into a state of impenetration. (4) That these monads or single beings now become spiritless or dull, did cling or come together after various manners. (5) That this coalation or clinging together, so long as it remains such is called *matter*. (6) That out of this matter, all things material do consist, which yet shall in time return again to a more loosened and free state. [34]

Elaborating on the nature of the matter produced, he wrote, "matter is made by a coalition or clinging together of spiritual degenerate dull monads or single beings and that this coalition is called creation." Another example of his use of the term *monad* was as a single being in a state of death: "After . . . a spirit is immediately created, it does for certain assignable causes . . . descend into that state of death, that it admits of the qualities and name of matter, being now a natural monad or single being, and a very atom."

As Stein pointed out, the monad first appears in Leibniz's "Letter to Fardella" of September 3 and 13, 1696, shortly after Van Helmont's visit: "All substance seems to me to be wonderfully fruitful in its operations. But I do not hold substance, that is a monad, to be produced from substance."[35]

Then in the 1698 essay "On Nature Itself," Leibniz characterized the monad in terms of internal force and consensus of actions:

> What we can establish about the external . . . actions or creatures may better be explained elsewhere; in fact, I have already partly explained it—the intercourse of substance or of monads, namely, arises not from an influence but from a consensus originating in their preformation by God, so that each one is adjusted to the outside while it follows the internal force and laws of its own nature. It is also in this that the union of soul and body consists.[36]

At a later point in the same essay, he defined the monad more explicitly as containing appetite and perception: "Neither is it [body] to be taken for a simple modification but for something which perseveres and is constitutive and substantial. This I customarily call a *monad*, which contains perception and appetite, as it were."[37]

Then in the year 1714, the doctrine of the monadology was set out in detail in his well-known essays "On the Principles of Nature and of Grace" and the "Monadology." Here he expounded a vitalistic metaphysics holding that the world was really alive, unlike the constructed mechanical world of well-founded phenomena. Organic life was divisible to infinity, still retaining its organic living character. "There is a world of creatures, living beings, animals, entelechies, souls, in the smallest particle of matter." "Each part of matter can be thought of as a garden full of plants or as a pond full of fish. But each branch of the plant, each member of the animal, each drop of its humors, is also such a garden or such a pond."[39]

Unlike the dialectically related organisms of the Renaissance naturalists, Leibniz's vital substances were noninteractive. Although in one sense every monad was related to every other monad through a pre-established harmony, in another sense there were no relations between monads at all. And although the level of activity of each vital substance was characterized by the heightening or diminution of its own internal perception, no monad actually perceived another monad.

Life and death, like activity and passivity, were reciprocal and interconvertible states of substance. Sleep, like death, was a diminution of perception in which the soul was like a simple monad. If perceptions were not activated, we would be continually in a state of stupor, like the naked monads. "When there is a large number of small perceptions with nothing to distinguish them we are stupified. . . . Death can produce this state in animals for a time." These ideas are very close in language and in content to the younger Van Helmont's view that matter is to spirit, as a dead man is to a living man—the same in substance, but dull, blind, resting and "in privation" and the passages from the *Kabbalah Denudata* referred to by Conway.[40]

It seems clear, therefore, that Leibniz appropriated the term *monad* from Conway and Van Helmont, its origins stemming initially from the Cabala. The influential role that Anne Conway's ideas played in his decision to use this concept has hitherto not been recognized, due to a series of scholarly errors originating from Heinrich Ritter's assumption that Van Helmont was the author rather than the editor of her *Principles*.

The basic elements that went into Leibniz's concept of the monad had been well developed by 1686, the crucial year of synthesis in which the main tenets of his philosophy were laid out in the *Discourse on Metaphysics,* the *Correspondence with Arnauld,* and the "Brief Demonstration of a Notable Error of Descartes." By then he had set out the concept of an individual substance whose essence was perception and activity, the animation of matter, the concept of the organic continuity of life, the idea of preestablished harmony, and the metaphor of each soul mirroring the universe from its own point of view. He had read the work of the Cartesians, medieval Scholastics, Dutch microscopists such as Anton Van Leeuwenhoek and Jan Swammerdam, and the Cambridge Platonists. During the decade 1686–1696, he refined many of these fundamental ideas and developed his system of dynamics in more detail. In addition he read and incorporated into his philosophy ideas from Chinese philosophy, the Jewish philosopher Maimonides (1135–1204), and the Cabala.[41] The writings of Francis Mercury Van Helmont and Anne Conway served to confirm and buttress his vitalistic view of nature and to stimulate the coalescence of his ideas into a "monadology."

Despite its philosophical weaknesses, vitalism represented an important reaction to Cartesian mechanism and dualism. At a time when mechanism was turning all of nature into something dead, inanimate, and void of sensation, thereby creating a subtle justification for the domination and control of nature, the vitalists along with the Cambridge Platonists raised voices of protest. They perceived the dangers in the reduction of matter to dead, inert atoms the motion of which stemmed from externally imposed forces rather than from the immanent spontaneity of vital principles. The older organic view of nature, however, was dying, along with an inherent value system that paid recognition to the life and worth of all things, the concept of cyclical renewal, and the binding of nature into a close-knit holistic unity. In the light of our current ecological crisis, which stems in part from the loss of this organic value system, we might regret that the mechanists did not take their vitalistic critics more seriously.

W OMEN AND THE "NEW PHILOSOPHY."   The almost total neglect by historians of philosophy of the work of Anne Conway raises a question about a cluster of women who studied and contributed to the philosophy, science, and educational literature of the seventeenth and eighteenth centuries. Do they not also deserve more detailed study and evaluation than has been accorded them? Besides Anne Conway, other women with great intellectual gifts whom Leibniz took seriously as students of philosophy included Sophie, the Electress of Hanover; her daughter Sophia Charlotte, queen of Prussia after 1701; the latter's ward, Princess Caroline (1683–1737), later queen of Great Britain, in answer to whose questions the entire Leibniz-Clarke correspondence of 1716 was directed; and Lady Damaris Masham (1658–1708), daughter of Ralph Cudworth, who educated her, a friend and student of John Locke, and a theological writer with whom Leibniz carried on an extensive correspondence. One of the most brilliant women of the eighteenth century, Madame Gabrielle Émelie du Châtelet (1706–

1749), was a principal expounder of Leibniz's system.[42] An expanding group of educated women began to participate in the philosophical and intellectual life of the period.

By the late seventeenth century, upper-class English women were noticing and reacting to the economic and educational advances men had made, while their own opportunities had been by comparison significantly constricted. They argued that differences in male and female achievement stemmed not from female intellectual inferiority, but from differences in childrearing practices, educational opportunities, and social position. Hannah Wooley, writing in 1655, Bathsua Makin, writing in 1673, and Mary Astell, writing in 1694, deplored women's lack of education and advocated the study of philosophy, foreign languages, medical care, household accounts, and writing. Their ideal went far beyond the emphasis on morals, Christian virtue, chastity, and the reading of the scriptures that had characterized women's education in the Renaissance.[43]

Translations were made of Henry Cornelius Agrippa's 1525 essay *On the Nobility and the Excellency of the Female Sex*, and François Poulain de la Barre's French treatise, *The Woman as Good as the Man* (written in 1673), which argued for the equality of the sexes. Agrippa's treatise had been presented to Margaret of Austria in 1509. Although not printed until 1525, it was subsequently reprinted many times before its English translations in 1652 and 1670. Agrippa marshalled numerous arguments to make a case for the superiority of women over men. Eve, whose name meant life, was created last in the chain of creatures and was therefore more perfect. Her body was more beautiful, her face unspoiled by a beard. As a mother, the woman contributed more in material and intellect to the embryo than the man. A female could conceive without a man: witness the Virgin Mary. Whereas Mary was the best human being the world has ever known, Judas, a man, was the worst known sinner. Since Jesus the Redeemer was a man, it was a man, Adam, who had committed the original sin. Great women had excelled in the past and were only prevented from achievement by the monopoly and tyranny of men in education.[44]

While learned ladies had always been present among the educated nobility, and women had contributed to science and mathematics from earliest times, the "scientific lady" was a product of the Scientific Revolution. Leading the way towards recognition of women as students of the new philosophy was Margaret Cavendish, Duchess of Newcastle, a member of the famous Newcastle circle which in the mid-seventeenth century had played a major role in the formation of the mechanical philosophy. A feminist who between 1653 and 1671 wrote some fourteen scientific books about atoms, matter and motion, butterflies, fleas, magnifying glasses, distant worlds, and infinity, her ideas and theories are often inconsistent, contradictory, and eclectic, which is attributable at least in part to her lack of formal education—a lack she herself deplored. She was acutely aware of the problems of leisured ladies who were made "like birds in cages to hop up and down in their houses." "We are shut out of all power and authority by reason, we are never employed either in civil nor marshall affairs, our counsels are despised and laughed at, the best of our actions are trodden down with scorn by the overweaning conceit men have of themselves and through a despisement of us."[45] An epistle in her book *Poems and Fancies* (1653), written to Mistress Toppe, a now unknown lady, lamented the "truth" that "our sex hath so much waste time, having but little employments, which makes our thoughts run wildly about, having nothing to fix them upon, which wild thoughts do not only produce unprofitable, but indiscreet actions, winding up the thread of our lives in snarls."[46]

Another epistle in the same book, addressed "To All Writing Ladies," noted that in different ages different types of spirits rule and have power; sometimes they are masculine, sometimes feminine. The present age had produced many feminine writers, rulers, actors, and preachers and was perhaps a feminine reign. "Let us take advantage, and make the best of our time . . . in any thing that [might] bring honour to our sex."[47]

Her preface to *Poems and Fancies*, requested the support of her own sex for a work "belonging most properly to themselves." "All I desire," she said, "is fame . . . but I imagine I shall be censured by my own sex and men will cast a smile of scorn upon my book, because they think thereby, women encroach too much upon their prerogatives; for they hold books as their crown, and the sword as their scepter, by which they rule and govern."[48]

Poems and Fancies begins with Nature calling a council consisting of the female principles, Motion, Figure, Matter, and Life, to advise her on creating the world. Life, Figure, and Motion all agree that Death is the "great enemy" who does not obey Nature's laws, undoes Form, and corrupts Matter.

> First Matter she brought the Materials in,
> And Motion cut, and carv'd out everything.
> And Figure she did draw the Formes and Plots,
> And Life divided all out into Lots.
> And Nature she survey'd, directed all,
> With the foure Elements built the World's Ball.

Though Death finally submits, he continues his attempt to obstruct and hinder Nature in all her efforts. Nevertheless, she creates a world made up of atoms—square, round, long, sharp, and so on— which form the vegetables, minerals, and animals of the everyday world. By their combinations atoms make heat and cold, life and death, and cause illnesses such as dropsy, consumption, and colic.

> Small Atomes of themselves a World may make,
> A being subtle, and of every shape:
> And as they dance about, fit places finde,
> Such formes as best agree, make every kinde.

While many of her early writings emphasized the empirical methodology of the telescope and the microscope, her later works shifted toward a rationalist critique of empiricism and developed a materialist ontology. In the *Grounds of Natural Philosophy* (1668), she took the materialist stance that motion was inherent in matter and that nature was self-knowing and perceptive. A "first cause" was nonsensical because the immaterial could not have material motion. Matter might be motionless but all motion must be attended by matter. Corporeal bodies could not have incorporeal perceptions; thoughts were corporeal motions united by conjunction.[49]

In an attempt to gain recognition for her achievements, Margaret Cavendish insisted on a visit to the all-male scientific society, the Royal Society of London, where scientific experiments and instruments were displayed for her surveillance. Samuel Pepys, the London gossip and journalist, "did not like her at all," but John Evelyn was "pleased with her "fanciful habit, garb, and discourse." Excluded from membership in the Royal Society because of her sex, she invented her own scientific community in *The Blazing World*

(1666), which would bring her the fame and recognition for which she hungered. "I am not covetous, but as ambitious as ever any of my sex is, or can be; which though I cannot be Henry the Fifth or Charles II, yet I endeavor to be Margaret the First." The sole survivor of a shipwreck, in which all the men have been killed, a lady resembling the Duchess finds herself on an island where she marries the Emperor and becomes Margaret I. She founds schools and societies and receives scientific instruction from beast-men who walk upright. Bear-men and bird-men are her experimental philosophers, who bring telescopes and microscopes for her investigations. Fish-men and worm-men answer her questions about the sea and earth, while the ape-men, her chemists, give an account of transmutations. Fox-men are her politicians, and spider- and lice-men teach her mathematics. Thus the Duchess, in her fantasies, poems, and many prefaces to her voluminous writings presented one of the earliest explicitly feminist perspectives on science.[50]

Men likewise participated in teaching science to women, but often with a more condescending tone. Bernard Fontenelle in his *Conversations on the Plurality of Worlds* (1686) attempted to educate women in the intricacies of the Cartesian universe through the device of a bright and beautiful young lady who is instructed in the mysteries of vortex motion and other worlds. His Marchioness of G "without tincture of learning understands what is said to her, and without confusion rightly apprehends...." Although a woman might not have the application of the mind "for scientific discovery, she can approach it as many do a romance or novel when they would retain the plot." On six successive evenings the marchioness walks with her philosopher lover through formal French gardens as he instructs her in scientific matters. Throughout the discourse the Marchioness is praised for her agreeableness and acuity in understanding the philosopher's teachings. "And yet were her company but half so agreeable, I am persuaded all the world would run mad after wisdom...."[51]

Significantly, in this new age when God had become an engineer and mathematician, nature in Fontenelle's fantasia had become a housewife. "Nature is a great housewife, she always makes use of what costs least let the difference be ever so inconsiderable and yet this frugality is accompanied with an extraordinary magnificence, which shines through all her works; that is, she is magnificent in the design, but frugal in the execution."

Fontenelle's book was translated into English in 1688 by playwright Aphra Behn. Although she approved of the idea of instructing women in science, Behn found the Marchioness somewhat less than convincing, because for a student her comments vascillated between silly and excessively profound.

WOMEN AS AN AUDIENCE FOR SCIENCE. As capitalist forms of economic life displaced women from household production, creating more leisure time, noble and bourgeois ladies became audiences for the new scientific discoveries. One step in the long process of institutionalizing and integrating classical physics and philosophy into Western consciousness was scientific popularization. A tradition of textbooks for the lay person, popular lectures, museums, and itinerant scientific demonstrators, as well as the foundation of professional societies for the more learned, began to emerge by the late seventeenth century. Part of this effort was directed specifically toward the female.

The attempt by bourgeois entrepreneurs to capture a wider female audience lay behind the publication of scientific serials for city and country ladies of all social levels. Pamphleteer John Dunton began including in 1691 a "Ladies Day" section in his twice-weekly serial, the *Athenian Mercury*. The annual *Ladies Diary* begun in 1704 by John Tipper taught science and posed difficult mathematical and astronomical questions for its audience to solve. The semiweekly *Free Thinker,* published from 1718 to 1721 and later bound into permanent three-volume sets, featured articles on natural history for the "philosophical girl" who "did not aspire to masculine virtues" but was above female capriciousness.[52]

Whereas Fontenelle had explained the Cartesian vortex heavens to his female audience, eighteenth century popularizers appropriated his method and format to educate women in Newtonian science and to enhance their abilities as salon conversationalists. Francesco Algarotti (1718–1764) consulted with Madame du Châtelet, translator of the French edition of Isaac Newton's *Principia Mathematica*, in preparing his *Newtonianism for the Ladies* (1737), and then to her diappointment dedicated it to Fontenelle. In six dialogues, a learned lady was instructed in Newtonian physics and optics. The theory of light and color as produced by prisms and lenses and the use and construction of telescopes and microscopes prepared her and her readers for their own experimentation. An English translation of Algarotti's book was made two years later by Elizabeth Carter at the age of twenty-two—*Sir Isaac Newton's Philosophy Explained for the Use of the Ladies.*[53]

In the mid-eighteenth century, scientific entrepreneurs Benjamin Martin and James Ferguson manufactured telescopes and microscopes, wrote scientific books for well-informed middle-class ladies and gentlemen, and invited women to attend popular lectures. In Martin's *Young Gentlemen and Lady's Philosophy* (1750s) and Ferguson's *Easy Introduction to Astronomy, for Young Gentlemen and Ladies* (1768) the authors' device was an Oxford-educated brother who returned home to teach his eager sister the secrets of the new science, with the expectation that the newly converted sister would purchase her own scientific instruments. Even children were recruited into the ever-expanding scientific market by John Newberry's best-seller, *The Newtonian System of Philosophy Adapted to the Capacities of Young Gentlemen and Ladies . . . Being the Substance of Six Lectures Read to the Lilliputian Society, by Tom Telescope* (1766). Tom Telescope interrupts the frivolous games of the Lilliputian boys and girls with lectures on astronomy, natural history, air pressure experiments, and geography.

Because the Scientific Revolution itself was not isolated from social, economic, and intellectual changes, its meaning for women was directly tied to these developments. The women who contributed to the intellectual life of late seventeenth-century England were not only reacting against the constriction of women's roles into the domestic sphere and to the slow pace of women's educational advancement, but, as in the Reformation, were responding to opportunities for women made possible within the more radical and traditionally more egalitarian religious sects. Caught up by the excitement of the "new science," educated women, along with men, became an eager audience for the new ideas. And the dissemination of the new learning facilitated its spread into other spheres of human life, where it was becoming institutionalized as a problem-solving methodology and as the dominant conceptual framework, reshaping the image of the cosmos and society. The contributions of Leibniz and Newton both helped to elaborate the new synthesis and to expose the underlying tensions brought about by the mechanization of the world view.

CHAPTER 12

Leibniz and Newton

The world in which we live today was bequeathed to us by Isaac Newton and Gottfried Wilhelm von Leibniz. Twentieth-century advances in relativity and quantum theory notwithstanding, our Western commonsense reality is the world of classical physics. The legacy left by Newton was the brilliant synthesis of Galilean terrestrial mechanics and Copernican-Keplerian astronomy; that of Leibniz was dynamics— the foundation for the general law of conservation of energy. Both contributions are fundamental in generality; they describe and extend over the entire universe. Classical physics and its philosophy structure our consciousness to believe in a world composed of atomic parts, of inert bodies moving with uniform velocity unless forced by another body to deviate from their straight line paths, of objects seen by reflected light of varying frequencies, and of matter in motion responsible for all the rich variations in colors, sounds, smells, tastes, and touches we cherish as human beings. In our daily lives, most of us accept these teachings as givens, without much critical reflection on their origins or associated values. To Newton, Leibniz, and their followers, however, the situation was not so straightforward. They saw their mechanics, their philosophies, and their own beliefs about God and nature as deeply divergent from each other.

The problem that the mechanization of the world raised for the generation after Descartes and Hobbes was the very issue of the "death of nature." If the ultimate principles were matter and motion, as they were for the first generation of mechanists, or even matter, motion, void space, and force as they became for Newton, this left unresolved the central issue of explaining the motion of life forms in a dead cosmos. Like many others, Newton was not satisfied with Descartes' dualistic solution, which reduced the human being to a ghost-in-the-machine whose mind could change the direction but not initiate bodily motion, and categorized animals as mere beast machines. Hobbes' monistic materialism, which further reduced the will and mind to material motion, raised the specter of atheism. Nor, like the Cambridge Platonists who had schooled him, could Newton entertain the pantheistic assumption that God was immanent in matter, together with its associated radical intellectual and social implications. He specifically argued against this position in his queries to the Latin edition of his *Opticks* in 1706: "And yet we are not to consider the world as the body of God, or the several parts thereof, as the parts of God."[1] God was neither a living animal-writ-large nor the soul of the world.

Yet as the most powerful synthesis of the new mechanical philosophy, Newton's *Philosophiae Naturalis Principia Mathematica* (*The Mathematical Principles of Natural Philosophy,* 1687) epitomized the dead world resulting from mechanism. Throughout the complex evolution of his thought, Newton clung tenaciously to the distinguishing feature of mechanism—the dualism between the passivity of matter and the externality of force and activity. But he refined this ontology in significant ways. The *Principia* and *Opticks* transformed the mechanical philosophy into a mechanical science, counterposing a fourfold ontology of matter, motion, force, and void space to the simpler plenum of matter in motion postulated by Descartes and Hobbes.

For Descartes, matter had been inert and passive: bodies continued in a state of rest or motion in a straight line unless acted on by another moving body, and change in motion resulted from contact between bodies. Newton departed from the strict passivity this earlier mechanical philosophy had assigned to matter by associating with it a complex, overlapping set of passive forces, while nevertheless maintaining the basic assumption that "matter is a passive principle and cannot move itself."[2] By its *vis insita* (innate force) a body continued in its state of rest or uniform motion, a state that could be altered only with difficulty. The *vis inertiae* (force of inertia) was the force of corporeal matter by which a body resisted an externally impressed force. The innate *vis conservans* (conserving force) maintained a body's forward direction by a succession of impulses.[3]

Like Descartes, Newton viewed changes in motion as external in origin, rather than as the internal activity central to organicism. His *vis impressa* (impressed force) was an external impressed force acting on the body so as to change its state of motion or rest. Likewise external to matter were various active principles such as gravity, fermentation, and cohesion necessary for explaining changes and activity not produced by impact. Gravitational force, unlike the impressed contact forces, acted at a distance, attracting all particles of matter toward each other according to the inverse square of the distance between them ($1/r^2$).

The mathematization of the world picture presented in the *Principia,* based on the dualism between the passivity of matter and the externality of force, epitomized the success of the mechanical analysis of nature. Mechanism eliminated from the description of nature concepts of spatial hierarchy, value, purpose, harmony, quality, and form central to the older organic description of nature, leaving material and efficient causes—matter and force. Motion was not an organic process but a temporary state of a body's existence relative to the motion or rest of other bodies. The mathematizing tendencies in Newtonian thought which emphasized not the process of change, but resistance to change, the conservation of a body's motion, and the planets and satellites as ideal spheres and point sources of gravitational force were manifestations of the mechanical philosophers' concern with geometrical idealization, stability, structure, being, and identity, rather than organic flux, change, becoming, and process. In mechanism the primacy of process was thus superseded by the stability of structure.

Completely consistent with this restructuring of the cosmos as passive matter and external force was the division of matter into atomic parts separated by void space. The book of nature was no longer written in symbols, signs, and signatures, but in corpuscular characters. The atomic analysis of matter ultimately became an exemplar for the atomic division of data, problems, and events on a global scale.

During the two decades following the publication of the *Principia,* Newton contemplated an atomic view of material particles distributed throughout void space, rearrangeable into new configurations by the action of external forces. The 1706 edition of the *Opticks* elaborated the structure of matter as hard atoms:

> God in the beginning formed matter in solid, massy, hard, impenetrable moveable particles, of such sizes and figures and with such other properties and in such proportion to space as most conduced to the end for which he form'd them; and that these primitive particles being solids are incomparably harder than any porous bodies compounded of them; even so very hard as never to wear or break in pieces; no ordinary power being able to divide what God himself made one in the first creation. . . . And therefore, that nature may be lasting, the changes of corporeal things are to be placed only in the various separations and new associations and motion of these permanent particles.[4]

These immutable unobservable atoms of which bodies and light were composed varied individually in size, shape, and weight, but their matter was homogeneous and their primary properties invariant. Extension, shape, solidity, and inertial mass were the primary or universal qualities possessed by all bodies. Secondary qualities (color, taste, sound, smell, and touch) unique to individual bodies were produced by the separation, association, and motion of the ultimate atomic constituents. All observable compound bodies were of several orders of composition. Thus primary atoms of different shapes and sizes united together to form first composition particles such as gold or silver. The first composition particles combined to form gold of the second and higher compositions subject to chemical reactions and transmutations from which the first composition particles could be recovered.[5]

Newton's speculations on atomic structure as presented in the 1713 edition of the *Principia* and the queries to the 1706 and 1717 editions of the *Opticks* became a foundation for eighteenth-century experimental philosophers who wished to complete the task of reducing known phenomena to simple laws which, like the law of gravitation, would quantify other mechanical, chemical, electrical, and thermal observations. Additionally, the laws of Newton's mechanical "system of the world" predicting the ordered motions of both terrestrial and celestial bodies served as a cosmological exemplar for political and economic order in English society. Published during the Restoration period following the turmoil of the English Civil War, the *Principia* aided the Latitudinarian cause for order and moderation in religious and political affairs.[6] Moreover, its conceptual framework, emphasizing external force and passive matter divided into rearrangeable components, could provide a subtle sanction for the domination and manipulation of nature necessary to progressive economic development. If eventually the religious framework providing for God's constant care and for the attainment of human grace were removed, as it was in the eighteenth century, the possibilities for intellectual arrogance toward nature would be strengthened.

Leibniz likewise developed a mechanical philosophy of nature as one component of his thought. His world of corporeal phenomena, governed by efficient causes and mechanical laws imposed initially by a rational creator, like Newtonian mechanics, held implications for the rational management of nature from which human progress would result.

Leibniz's dynamics, developed during the years 1686–1695, defined the "force" of a body in motion to be the product of its quantity of matter and the distance through which it fell under acceleration. This living force, or *vis viva*, (mv^2 or mass times velocity squared; now, as $1/2\ mv^2$, called kinetic energy), was conserved in all elastic impacts. In semielastic and inelastic collisions it was temporarily stored in the small parts of the body's matter and therefore not lost to the universe.[7]

For Leibniz, "force" was the foundation for an understanding of both the phenomenal and spiritual universes. Primitive active force, an activity or striving toward a future state, (later defined as the essence of his monad) was a true substance, while derivative force (mv^2) observed in impacts between corporeal bodies, was not fully real, but was grounded in primitive force and subject to the laws of nature. Corporeal objects were not substances, but collections of confused minds (monads), perceived to be extended bodies. The properties of these ostensibly extended bodies—size, shape, inertia, impenetrability, and motion—were "well-founded" in the states of existence of the monads which constituted them. Leibniz thus assigned extension, which for Descartes was a substance, to the world of well-founded phenomena (*phenomena bene fundata*), arguing that extension and motion were merely attributes of phenomenal bodies, while force, on the other hand, was real. Inertness or passivity, an essential property of matter for Descartes and Newton, was for Leibniz simply an expression of the limitation placed on the monad because of the accommodation of its life to the unfolding lives and activities of all the other monads. Mechanical phenomena obeyed the laws of efficient causation, whereas monads or true substances were governed by final causes.[8]

Nature was manageable through rational understanding and efficient action. Despite the fact that "certain parts of [the earth] grow up wild again or again suffer destruction and deterioration," nevertheless eventually the entire globe would be brought into cultivation and assume a garden-like character: "It is thus that even now a great part of our earth has received cultivation and will receive more and more." Leibniz's optimism over the progress of human civilization and the internal development of the universe as a whole guided the philosophies of the eighteenth-century Enlightenment. "In addition to the general beauty and perfection of the works of God", he wrote, "we must recognize a certain perpetual and very free progress of the whole universe, such that it advances always to still greater improvement."[9] Some commentators have found in Leibnizian philosophy, with its emphasis on self-contained independence, internal development, and progress, a justification for laissez-faire capitalism.

Leibniz applied his interest in a universal logical language and mathematical method to practical inventions which would foster the capitalist spirit. His design for a calculating machine which he called a "living bank-clerk" would, he believed, be useful in business, surveying, military affairs, and astronomy. He worked on a new kind of pump that could be used to remove water from the Harz mines in Germany. He designed "catadoptic tubes" of mirrors and perspective lenses to improve the science of optics and a submarine to aid in navigating through storms, dangerous seas, and naval combats.

While the mechanical analysis of nature was an important component in the systems of both Leibniz and Newton, their views about God's role in the mechanical universe and the fundamental nature of reality were very different. For Leibniz, only the phenomenal world was mechanical; the real world of substance was organic. The conflict between Leibniz and Newton in their famous debates of 1716 was in actuality a conflict over the concepts of God, matter, and nature underlying the organic and mechanical traditions. The issue that stimulated the debate was the character of God's role in a clocklike mechanical universe that operated according to the mathematical laws of nature. Was God a rational creator who constructed a perfectly operating and well-maintained machine, or did the machine require his intervention and care to avoid decay and ultimate breakdown?[10] Leibniz considered the necessity of God's intervention in the machine of the universe to be a limitation on his wisdom and foresight. Newton and his spokesperson theologian Samuel Clarke (1675–1729) argued that God's glory and power were manifested in his providential care and interposition. The world as Newton and Clarke viewed it could have been a different world, for it depended on the free exercise of God's will to continually sustain its existence.

For Leibniz, however, a world created by God's will without his logic might have resulted in an ill-constructed, inferior world. The existing world must be consistent with the principle of noncontradiction; its beings must exhibit nothing mutually destructive or incompatible. Because God operated rationally within the laws of logic in creating it, this world was the best of all possible worlds. Yet another principle, that of sufficient reason, was needed to explain the existence of this particular world and no other. This second principle was necessary in proceeding from the laws of logic to actual existence. By God's sufficient reason, his logic was united with his creative power.

The distinction between these concepts of God stressing either his reason or his power was related to the issue of the organic immanence of divine law *within* nature versus God's imposition of natural laws *on* the creation and to the difference between the older view that nature was an intelligent organism and the newer scientific view of the world as a machine. In the Greek organic analogy, the natural world rationally ordered its own movements according to immanent laws, whereas the mechanical view held the world to be devoid of life and intelligence—hence its motions were regular and imposed from outside in the form of natural laws. Thomas Aquinas, fusing the Greek idea of the rationality inherent in nature with the Christian concept of a creator, had held a quasi-immanentist view of natural law.[11]

Whereas immanent law goes back to the Stoic view of nature, imposed law is a product of the Judeo-Christian world view. Many Protestants during the Reformation adopted this Judaic conception of the imposition of God's will and law on the creation. The Judeo-Christian stress on the immediacy of God's will and power was closely related to the sense of power over nature that was growing as active daily life became increasingly organized around the power of machine technology.

The view of the mechanists—Gassendi, Boyle, and Newton—and of mainstream science was based on the theory that *things* rather than relations are the ultimate reality, relations being externally imposed by God in the form of natural laws. For Leibniz, on the other hand, activity and relations were internal and followed from a doctrine of natural law as quasi-immanent. The relationship between the inherent activity of the monad and the immanence of divine law is summed up in his essay "On Nature Itself" (1698): "Primary matter is merely passive but not a complete substance; there must be added to it a soul . . . or primitive force of action which is itself the *inherent law impressed upon it by divine command*."[12] Although God initially impresses the laws of nature on the universe, these laws are manifested in the inherent internal development of the simultaneous states of all the monads and in their intrinsic mutual relations.

Leibniz sharply delineated the difference between his own interpretation of nature and that of Boyle's follower, mechanist Christopher Sturm, who held a doctrine of the external imposition of law and of external relations:

> He admits . . . that motions now taking place result by virtue of an eternal law once established by God, which law, he calls a volition and command. . . . I ask whether this volition and command, or if you prefer, this divine law, once established, had bestowed upon things only an extrinsic denomination or whether it has truly conferred upon them . . . an *internal law* from which their actions and passions follow, even if this law is mostly not understood by the creatures in which it inheres.

For Leibniz, the world of substance was really organic; every being in the universe, from living animals down to the simple monad, was alive or composed of living parts. "Thus there is nothing fallow, sterile, or dead in the universe; no chaos, no confusions, save in appearance." Monads as individual vital substances were characterized by an internal principle of change or striving; each has a perception which is raised or diminished. The monads act only from within, as their own internal lives or perceptions unfold; they are all created together in the beginning and annihilated together in the end, but cannot die or be born naturally. They are impermeable to natural influence, for "they have no windows through which anything can enter or depart."[13] Thus change occurs as the result of an internal immanent principle rather than from the action of an external imposed force, as in mechanism.

Leibniz stressed the idea of a life and perception permeating all things. The main distinction between his philosophy and that of the mechanists lay in the idea that substance was life, not dead matter. He criticized the "advocates of the new philosophy" for "maintain[ing] the inertness and deadness of things." As he had once written to Jansenist theologian Antoine Arnauld (1612–1694): "All matter must be full of animated, or at least living, substances."[14]

Like the elements of Paracelsus and the seeds of the elder Van Helmont, the actions of the Leibnizian monads exist in a continual state of mutual harmony. Relations among the internal states of monadic lives constitute the laws of nature; the overall result is a consensus of individual actions producing that which is best for the whole and the maximization of perfection in the world. Each monad mirrors the universe in its own way, its life unfolding simultaneously with the lives of all other monads in a preestablished harmony.[15]

Leibniz's dynamic vitalism was thus in direct opposition to the "death of nature." In this vitalistic component of his later thought we find an organic orientation which, like the vitalism of his predecessors in its reverence for the pervasive life of the cosmos, can be construed as antiexploitative; normative constraints are contained within the framework itself and imposed by it. His principle of self-contained internal development central to the organic world view sharply contrasted with the mechanistic theory that change is reactive—the product of external influences on a passive entity.

Like Leibniz, Newton was deeply concerned about the problem

of organic life raised by the mechanization of the cosmos. Although his system of the world was to become the mechanistic model for the future, he privately believed it to be at best a partial truth. The mechanical laws of passive matter as set out in the *Principia* were insufficient to explain the laws and causes of vital life and violent actions. Newton addressed this issue in his published queries to the *Opticks* and in various unpublished drafts of queries prepared for the 1706 and 1717 editions.

In Query 23 of the 1706 edition (Query 31, in the 1717 edition), he wrote:

> The *vis inertiae* is a passive principle by which bodies persist in their motion or rest, receive motion in proportion to the force impressing it, and resist as much as they are resisted. By this principle alone there never could have been any motion in the world. Some other principle was necessary for putting bodies into motion and now they are in motion some other principle is necessary for conserving the motion.[16]

But in unpublished variants to the queries, he was even more adamant about the inadequacies of the laws of the motion of passive matter for explaining the origin of new and vital motions:

> If you think that the *vis inertiae* is sufficient for conserving motion, pray tell me the experiments from whence you gather thy conclusion. Do you learn by any experiment that the beating of heart gives no new motion to the blood, that the explosion of gunpowder gives no new motion to a bullet or that a man by his will can give no new motion to his body? Do you learn by experiment that the beating of your heart takes away as much motion from something else as it gives to the blood or that explosion takes away as much motion from something else as it gives to a bullet or that a man by his will takes away as much motion from something else as he gives to his body? If so, tell me your experiments; if not your opinion is precarious. Reasoning without experience is very slippery.[17]

A mode for God's continued action and providential care was essential in a clocklike mechanical universe seemingly governed only by the laws of passive matter. God's recruitment of new motion and his renewal of the activity of the cosmos was necessary because decay in the world system was evident; periodic repair of the frame of nature and the continual replenishment of its vital motions were needed. Newton, deeply troubled over the inadequacy of his own analysis of the laws of mechanics for explaining life and will, looked to older traditions for answers to these fundamental questions. His clandestine interest in alchemy and in the secrets of the ancient kingdoms had been in part a search for clues to more general laws governing living as well as mechanical systems.

In an unpublished manuscript of about 1674, "Of Nature's Obvious Laws and Processes in Vegetation," possibly inspired by the interest in vegetative principles among the Cambridge Platonists, he inquired about the action of a latent vegetative spirit produced in fermentation.[18] "No spirit," he wrote, "searches bodies so subtly, piercingly, and quickly as does the vegetable spirit." "The earth resembles a great animal or rather inanimate vegetable," which "draws in etheral breath for its daily refreshment and vital ferment and transpires again with gross exhalations. And according to the condition of all other things living, ought to have its times of beginning, youth, old age, and perishing." The vegetative spirit produced in fermentation is "nature's universal agent, her secret fire, the only ferment and principle of all vegetation."

Fermentations and mineral dissolutions on the earth continually produced a large quantity of gentle air that rose and buoyed up the clouds, ascending into the ethereal regions. There the air crowded the ether, forcing it to descend towards the earth, causing gravitation and creating a circulation "very agreeable to nature's proceedings." The ether was the carrier of the vital vegetative spirit, and bodies breathed in both of them together:

> Note that tis more probable the ether is but a vehicle to some more active spirit. The bodies may be concreted of both together; they may imbibe ether as well as air in generation, and in that ether the spirit is entangled.

The life of all matter depended on a gentle heat in order to produce life; its withdrawal resulted in death. The continual source of new life was therefore to be found in fresh fermentation. While the mechanical changes of gross corpuscles accounted for the sensible qualities of things, the subtle secret workings of nature took place by means of the vegetative spirit produced in fermentation—"an exceeding subtle and unimaginably small portion of matter diffused through the mass, which, if it were separated, there would remain but a dead and inactive earth."

Why did Newton attribute such importance to the concept of fermentation? Fermentation had had a long and clear historical connection with motion and activity and could be viewed as a source of violent change. From a political standpoint a ferment carried the connotation of agitation—the inflaming and fomenting of passions and tumult. A ferment could "work up to foam and threat the government." In alchemy and chemistry changes in the properties of metals were thought to be produced by a ferment operating within them. The action of yeast on dough and the brewing of beer produced an internal commotion and effervescence. All were examples of new motions generated in both living and nonliving things.[19]

Newton, at work on his queries to the *Opticks* in the early 1700s, still presumed these violent motions resulting from fermentation to be operative in cosmic chemical processes. The fermentation of sulfurous steams with minerals deep within the "bowels of the earth ... if pent up in subterraneous caverns burst the caverns with a great shaking of the earth," generating tempests and hurricanes, landslides and boiling seas. In the air fermentation caused lightning, thunder, and fiery meteors.[20]

But fermentation was not only an important cause of violent cosmic motions resulting from chemical reactions, it was also a cause of the life motions of animals and vegetables. It was responsible for "the beating of the heart by means of respiration," and of "perpetual motion and heat." Without fermentation as an active principle, "all putrefaction, generation, vegetation, and life would cease."

The draft queries also show how thoroughly Newton was convinced of the pervasiveness of vital life in animal, vegetable and mineral matter. "We cannot say that all nature is not alive," he wrote in one draft, and, in another, "All matter duly formed is attended with signs of life."[21] In the drafts he also discussed the human will as a clear example of the "recruitment" of new motion not explained by the laws of impressed force or by Descartes' principle of the conservation of motion:

Matter is a passive principle and cannot move itself. It continues in its state of moving or resting unless disturbed. It receives motion proportional to the force impressing it. And resists as much as it is resisted. There are passive laws and to affirm that there are no other is to speak against experience for we find in ourselves a power of moving our bodies by our thought. Life and will are active principles by which we move our bodies and thence arise other laws of motion unknown to us.[22]

Newton's answer to the problem of the revitalization of the cosmos was to replenish its motion through "active principles" such as gravity and fermentation:

> Seeing . . . the variety of motion which we find in the world is always decreasing, there is a necessity of conserving and recruiting it by active principles, such as are the *cause of gravity,* by which planets and comets keep their motions in their orbs, and bodies acquire great motion in falling; and *the cause of fermentation,* by which the heart and blood of animals are kept in perpetual motion and heat . . . for we meet with very little motion in the world, besides what is owing to these active principles.[23]

Without these active principles, Newton warned, "the bodies of the earth, planets, comets, sun, and all things in them would grow cold and freeze, and become inactive masses . . . and the planets and comets would not remain in their orbs."[24] For Newton, fermentation thus furnished an antidote to the "death of nature" implicit in a mechanical universe, a universe founded on passivity and having an inherent tendency towards decay, decline, and eventual death. Unsatisfied with the mechanistic analysis of phenomena, he, like Leibniz, was searching for the causes and laws that would unify biological processes, just as his gravitational theory had synthesized physical interactions.

Both Newton and Leibniz, however, are known today chiefly for their contributions to mathematics and mechanics. Moreover, a cultural research program extending from the seventeenth century to the present day has resulted in mechanical models of the self, society, and the cosmos. Thus the human body and the human psyche are treated as reactive, conditionable entities, and the human brain as a computer. The body politic has become a pluralism of atomized interest groups. The mechanistic cosmos has been extended to encompass chemical, electrical, thermodynamic, and cellular phenomena. During the three centuries in which the mechanical world view became the philosophical ideology of Western culture, industrialization coupled with the exploitation of natural resources began to fundamentally alter the character and quality of human life. Through popular scientific education, through commonsense empirical philosophy and natural religion, and through the spread of scientific, rationalizing tendencies to manufacturing, government bureaucracies, and medical and legal systems, the mechanical science, method, and philosophy created in the seventeenth century have gradually become institutionalized as a form of life in the Western world.

Between 1500 and 1700 an incredible transformation took place. A "natural" point of view about the world in which bodies did not move unless activated, either by an inherent organic mover or a "contrary to nature" superimposed "force," was replaced by a non-natural non-experiential "law" that bodies move uniformly unless hindered. The "natural" perception of a geocentric earth in a finite cosmos was superseded by the "non-natural" commonsense "fact" of a heliocentric infinite universe. A subsistence economy in which resources, goods, money, or labor were exchanged for commodities was replaced in many areas by the open-ended accumulation of profits in an international market. Living animate nature died, while dead inanimate money was endowed with life. Increasingly capital and the market would assume the organic attributes of growth, strength, activity, pregnancy, weakness, decay, and collapse obscuring and mystifying the new underlying social relations of production and reproduction that make economic growth and progress possible. Nature, women, blacks, and wage laborers were set on a path toward a new status as "natural" and human resources for the modern world system. Perhaps the ultimate irony in these transformations was the new name given them: rationality.[25]

In 1500 the parts of the cosmos were bound together as a living organism; by 1700 the dominant metaphor had become the machine. Although machines and the cosmic *machina mundi* had been parts of the ancient and medieval worlds, the organic conception of nature had been sufficiently integrative as a framework to override changes and discrepancies within it. Similarly, although the mechanistic analysis of reality has dominated the Western world since the seventeenth century, the organismic perspective has by no means disappeared. It has remained as an important underlying tension, surfacing in such variations as the Romantic reaction to the Enlightenment, American transcendentalism, the ideas of the German *Naturphilosophen,* the early philosophy of Karl Marx, the nineteenth-century vitalists, and the work of Wilhelm Reich. The basic tenets of the organic view of nature have reappeared in the twentieth century in the theory of holism of Jan Christiaan Smuts, the process philosophy of Alfred North Whitehead, the ecology movements of the 1930s and 1970s, alternative analyses in nuclear physics (the "bootstrap" model), and developmental theories in psychology. Some philosophers have argued that the two frameworks are fundamentally incommensurable. Although such a perception of the dichotomy is too extreme, as the fusions between the two perspectives discussed in previous chapters have shown, a reassessment of the values and constraints historically associated with the organic world view may be essential for a viable future.[26]

Epilogue

The mechanistic view of nature, developed by the seventeenth-century natural philosophers and based on a Western mathematical tradition going back to Plato, is still dominant in science today. This view assumes that nature can be divided into parts and that the parts can be rearranged to create other species of being. "Facts" or information bits can be extracted from the environmental context and rearranged according to a set of rules based on logical and mathematical operations. The results can then be tested and verified by resubmitting them to nature, the ultimate judge of their validity. Mathematical formalism provides the criterion for rationality and certainty, nature the criterion for empirical validity and acceptance or rejection of the theory.

The work of historians and philosophers of science notwithstanding, it is widely assumed by the scientific community that modern science is objective, value-free, and context-free knowledge of the external world. To the extent to which the sciences can be reduced to this mechanistic mathematical model, the more legitimate they become as sciences. Thus the reductionist hierarchy of the validity of the sciences first proposed in the nineteenth century by French positivist philosopher August Comte is still widely assumed by intellectuals, the most mathematical and highly theoretical sciences occupying the most revered position.

The mechanistic approach to nature is as fundamental to the twentieth-century revolution in physics as it was to classical Newtonian science, culminating in the nineteenth-century unification of mechanics, thermodynamics, and electromagnetic theory. Twentieth-century physics still views the world in terms of fundamental particles—electrons, protons, neutrons, mesons, muons, pions, taus, thetas, sigmas, pis, and so on. The search for the ultimate unifying particle, the quark, continues to engage the efforts of the best theoretical physicists.

Mathematical formalism isolates the elements of a given quantum mechanical problem, places them in a latticelike matrix, and rearranges them through a mathematical function called an *operator*. Systems theory extracts possibly relevant information bits from the environmental context and stores them in a computer memory for later use. But since it cannot store an infinite number of "facts," it must select a finite number of potentially relevant pieces of data according to a theory or set of rules governing the selection process. For any given solution, this mechanistic approach very likely excludes some potentially relevant factors.

Systems theorists claim for themselves a holistic outlook, because they believe that they are taking into account the ways in which all the parts in a given system affect the whole. Yet the formalism of the calculus of probabilities excludes the possibility of mathematizing the gestalt—that is, the ways in which each part at any given instant take their meaning from the whole. The more open, adaptive, organic, and complex the system, the less successful is the formalism. It is most successful when applied to closed, artificial, precisely defined, relatively simple systems. Mechanistic assumptions about nature push us increasingly in the direction of artificial environments, mechanized control over more and more aspects of human life, and a loss of the quality of life itself.

In the social sphere, the mechanistic model helps to guide technological and industrial development. In *The Technological Society,* Jacques Ellul discussed the techniques of economics and the mechanistic organization of specialties inherent in and entailed by the machines and mathematical methods themselves. The calculating machine, punch card machine, microfilm, and computer transform statistical methods and administrative organization into specialized agencies centered around one or more statistical categories.

Econometric models and stochastics are used to operate on statistical data in order to analyze, compare, and predict. In social applications, attempts to predict public reaction through the calculus of probabilities may make a public informed of its conformation to a trend act in the inverse manner.

> But the public, by so reacting falls under the influence of a new prediction which is completely determinable. . . . It must be assumed, however, that one remains within the framework of rational behavior. The system works all the better when it deals with people who are better integrated into the mass . . . whose consciousness is partially paralyzed, who lend themselves willingly to statistical observations and systematization.[1]

Such attempts to reduce human behavior to statistical probabilities and to condition it by such psychological techniques as those developed by B. F. Skinner are manifestations of the pervasiveness of the mechanistic mode of thought developed by the seventeenth-century scientists.

Holism was proposed as a philosophical alternative to mechanism by J. C. Smuts in his book *Holism and Evolution* (1926), in which he attempted to define the essential characteristics of holism and to differentiate it from nineteenth-century mechanism. He attempts to show that

> Taking a plant or animal as a type of whole, we notice the fundamental holistic characters as a unity of parts which is so close and intense as to be more than a sum of its parts; which not only gives a particular conformation or structure to the parts but so relates and determines them in their synthesis that their functions are altered; the synthesis affects and determines the parts so that they function toward the "whole"; and the whole and the parts therefore reciprocally influence and determine each other and appear more or less to merge their individual characters.[2]

Smuts saw a continuum of relationships among parts from simple physical mixtures and chemical compounds to organisms and minds in which the unity among parts was affected and changed by the synthesis. "Holism is a process of creative synthesis; the resulting wholes are not static, but dynamic, evolutionary, creative. . . . The explanation of nature can therefore not be purely mechanical; and the mechanistic concept of nature has its place and justification only in the wider setting of holism."

The most important example of holism today is provided by the science of ecology. Although ecology is a relatively new science, its philosophy of nature, holism, is not. Historically, holistic presuppositions about nature have been assumed by communities of people who have succeeded in living in equilibrium with their environments. The idea of cyclical processes, of the interconnectedness of all things, and the assumption that nature is active and alive are fundamental to the history of human thought. No element of an interlocking cycle can be removed without the collapse of the cycle. The parts themselves thus take their meaning from the whole. Each particular part is defined by and dependent on the total context. The cycle itself is a dynamic interactive relationship of all its parts, and process is a dialectical relation between part and whole. Ecology necessarily must consider the complexities and the totality. It cannot isolate the parts into simplified systems that can be studied in a laboratory, because such isolation distorts the whole.

External forces and stresses on a balanced ecosystem, whether natural or man made, can make some parts of the cycle act faster than the systems' own natural oscillations. Depending on the strength of the external disturbance, the metabolic and reproductive reaction rates of the slowest parts of the cycle, and the complexity of the system, it may or may not be able to absorb the stresses without collapsing.[3] At various times in history, civilizations which have put too much external stress on their environments have caused long-term or irrevocable alterations.

By pointing up the essential role of every part of an ecosystem, that if one part is removed the system is weakened and loses stability, ecology has moved in the direction of the leveling of value hierarchies. Each part contributes equal value to the healthy functioning of the whole. All living things, as integral parts of a viable ecosystem, thus have rights. The necessity of protecting the ecosystem from collapse due to the extinction of vital members was one argument for the passage of the Endangered Species Act of 1973. The movement toward egalitarianism manifested in the democratic revolutions of the eighteenth century, the extension of citizens' rights to blacks, and finally, voting rights to women was thus carried a step further. Endangered species became equal to the Army Corps of Engineers: the sail darter had to have a legal hearing before the Tellico Dam could be approved, the Furbish lousewort could block construction of the Dickey-Lincoln Dam in Maine, the red-cockaded woodpecker must be considered in Texas timber management, and the El Segundo Blue Butterfly in California airport expansion.

The conjunction of conservation and ecology movements with women's rights and liberation has moved in the direction of reversing both the subjugation of nature and women. In the late nineteenth and early twentieth centuries, the strong feminist movement in the United States begun in 1842 pressed for women's suffrage first in the individual states and then in the nation. Women activists also formed conservation committees in the many women's organizations that were part of the Federation of Women's Clubs established in 1890. They supported the preservationist movement for national, state, and city parks and wilderness areas led by John Muir and Frederick Law Olmsted, eventually splitting away from the managerial, utilitarian wing headed by Gifford Pinchot and Theodore Roosevelt.[4]

Today the conjunction of the women's movement with the ecology movement again brings the issue of liberation into focus. Mainstream women's groups such as the League of Women Voters took an early lead in studying and pressing for clean air and water legislation. Socialist-feminist and "science for the people" groups worked toward revolutionizing economic structures in a direction that would equalize female and male work options and reform a capitalist system that creates profits at the expense of nature and working people.

The March 1979 accident at the Three-Mile Island nuclear reactor near Harrisburg, Pennsylvania, epitomized the problems of the "death of nature" that have become apparent since the Scientific Revolution. The manipulation of nuclear processes in an effort to control and harness nature through technology backfired into disaster. The long-range economic interests and public image of the power company and the reactor's designer were set above the immediate safety of the people and the health of the earth. The hidden effects of radioactive emissions, which by concentrating in the food chain could lead to an increase in cancers over the next several years, were initially downplayed by those charged with responsibility for regulating atomic power.

Three-Mile Island is a recent symbol of the earth's sickness caused by radioactive wastes, pesticides, plastics, photochemical smog, and fluorocarbons. The pollution "of her purest streams" has been supported since the Scientific Revolution by an ideology of "power over nature," an ontology of interchangeable atomic and human parts, and a methodology of "penetration" into her innermost secrets. The sick earth, "yea dead, yea putrified," can probably in the long run be restored to health only by a reversal of mainstream values and a revolution in economic priorities. In this sense, the world must once again be turned upside down.

As natural resources and energy supplies diminish in the future, it will become essential to examine alternatives of all kinds so that, by adopting new social styles, the quality of the environment can be sustained. Decentralization, nonhierarchical forms of organization, recycling of wastes, simpler living styles involving less-polluting "soft" technologies, and labor-intensive rather than capital-intensive economic methods are possibilities only beginning to be explored.[5] The future distribution of energy and resources among communities should be based on the integration of human and natural ecosystems. Such a restructuring of priorities may be crucial if people and nature are to survive.

Notes

INTRODUCTION

1. James Murry, et al., eds., *The Oxford English Dictionary* (Oxford, England Clarendon Press, 1933), vol. 7, pp. 41–42, 194–95; vol. 6, pp. 284–85.

CHAPTER 1:
NATURE AS FEMALE

1. On the tensions between technology and the pastoral ideal in American culture, see Leo Marx, *The Machine in the Garden* (New York: Oxford University Press, 1964). On the domination of nature as female, see Annette Kolodny, *The Lay of the Land* (Chapel Hill: University of North Carolina Press, 1975); Rosemary Radford Ruether, "Women, Ecology, and the Domination of Nature," *The Ecumenist* 14 (1975): 1–5; William Leiss, *The Domination of Nature* (New York: Braziller, 1972). On the roots of the ecological crisis, see Donald Hughes, *Ecology in Ancient Civilizations* (Albuquerque: University of New Mexico Press, 1976); Lynn White, Jr., *Medieval Technology and Social Change* (New York: Oxford University Press, 1966); and L. White, Jr., "Historical Roots of Our Ecologic Crisis," in White, Jr. *Machina ex Deo* (Cambridge, Mass.: M. I. T. Press, 1968), pp. 75–94; Reijer Hooykaas, *Religion and the Rise of Modern Science* (Grand Rapids, Mich.: Eerdmans, 1972); Christopher Derrick, *The Delicate Creation: Towards a Theology of the Environment* (Old Greenwich, Conn.: Devin-Adair, 1972). On traditional rituals in the mining of ores and in metallurgy, see Mircea Eliade, *The Forge and the Crucible,* trans. Stephan Corrin (New York: Harper & Row, 1962), pp. 42, 53–70, 74, 79–96. On the divergence between attitudes and practices toward the environment, see Yi-Fu Tuan, "Our Treatment of the Environment in Ideal and Actuality," *American Scientist* (May-June 1970): 246–49.

2. Stanley Cavell, "Must We Mean What We Say?" in Colin Lyas, ed., *Philosophy and Linguistics* (London: Macmillan, 1971), pp. 131–65; see esp. 148, 165. On frameworks and values see Charles Taylor, "Neutrality in Political Science," in Alan Ryan, ed., *The Philosophy of Social Explanation* (London: Oxford University Press, 1973) pp. 139–70, see esp. pp. 144–46, 154–55.

3. On the Elizabethan view of nature and Shakespeare's *King Lear,* see John Danby, *Shakespeare's Doctrine of Nature: A Study of King Lear* (London: Faber and Faber, 1949), pp. 20–21, 26, 28, 133, 126.

4. Richard Hooker, *Of the Laws of Ecclesiastical Polity,* ed. W. Speed Hill (Cambridge, Mass.: Harvard University Press, 1977, first published, 1594): Bk. I, Chap. 3.3, p. 66, Chap. 3.4, p. 68; Chap. 8, pp. 3–5. Danby, pp. 24–27. Eustace M. W., Tillyard, *The Elizabethan World Picture* (New York: Random House Vintage, 1959 [?]), p. 46.

5. Danby, p. 133. The interpretation of Cordelia is Danby's.

6. The discussion of the pastoral tradition draws on Joseph W. Meeker, *The Comedy of Survival: Studies in Literary Ecology* (New York: Scribner's, 1972), pp. 81–89, and L. Marx, *The Machine in the Garden,* pp. 26, 28–29, 36–43. Also relevant are Walter W. Greg, *Pastoral Poetry and Pastoral Drama* (London: Bullen, 1906); William Empson, *Some Versions of the Pastoral* (London: Chatto & Windus, 1950); Michael Putnam, *Virgil's Pastoral Art: Studies in the Eclogues* (Princeton, N.J.: Princeton University Press, 1970); Bruno

Snell, "Arcadia: The Discovery of a Spiritual Landscape," in *The Discovery of the Mind: The Greek Origins of European Thought,* trans. T. G. Rosenmeyer (Oxford: Blackwell, 1953); Hallett Smith, "Pastoral Poetry," *Elizabethan Poetry: A Study in Conventions, Meanings, and Expression* (Cambridge, Mass.: Harvard University Press, 1952); Erwin Panofsky, "Et in Arcadia Ego: Poussin and the Elegaic Tradition," in *Meaning in the Visual Arts: Papers in and on Art History* (Garden City, N.Y.: Doubleday, 1957).

7. Quoted in Meeker, p. 82.

8. Quoted in Meeker, pp. 82–83.

9. Plato, *The Timaeus* (written ca. 360 B.C.), in *The Dialogues of Plato,* trans. Benjamin Jowett (New York: Random House, 1937), vol. 2, pp. 14, 16, 17, 18, 21. For a commentary, see Francis MacDonald Cornford, *Plato's Cosmology* (New York: Liberal Arts Press, 1937). On Plato's views on women, see Anne Dickason, "Anatomy and Destiny: The Role of Biology in Plato's Views of Women," *The Philosophical Forum* 5 (1973–74): 45–53.

10. Bernard Silvestris, *De Mundi Universitate* (written, ca. 1136), ed. Carl Sigmund Barach and Johann Wrobel (Innsbruck: Wagner, 1876); B. Silvestris, *Cosmographia,* trans. Winthrop Wetherbee (New York: Columbia University Press, 1973), pp. 65–127; George D. Economou, *The Goddess Natura in Medieval Literature* (Cambridge, Mass.: Harvard University Press, 1972), pp. 54, 63.

11. Alain of Lille, *De Planctu Naturae* (written ca. 1202), in Thomas Wright, ed., *The Anglo-Latin Satirical Poets and Epigrammatists* (Wiesbaden: Kraus Reprint, 1964), vol. 2, pp. 441, 467; Economou, pp. 73, 76, 77, 82. English translation: Alain of Lille, *The Complaint of Nature,* trans. Douglas Moffat (New York: Henry Holt, 1908), see esp. pp. 3, 4, 11, 15, 41, 44. "I marvel," then I said, "wherefore certain parts of thy tunic, which should be like the connection of marriage, suffer division in that part of their texture where the fancies of art give the image of man." "Now from what we have touched on previously," she answered, "thou canst deduce what the fig-

ured gap and rent mystically show. For since, as we have said before, many men have taken arms against their mother in evil and violence, they thereupon, in fixing between them and her a vast gulf of dissension, lay on me the hands of outrage, and themselves tear apart my garments piece by piece, and, as far as in them lies, force me, stripped of dress, whom they ought to clothe with reverential honor, to come to shame like a harlot. This tunic, then, is made with this rent, since by the unlawful assaults of man alone the garments of my modesty suffer disgrace and division" [p. 41].

12. Aristotle, *Metaphysics* (written ca. 335–322 B.C.), in Richard McKeon, ed., *The Basic Works of Aristotle* (New York: Random House, 1971), p. 755, line 1015ᵃ15; see also *Physics,* in *Basic Works,* p. 237, line 193ᵃ28; pp. 240–1, lines 194ᵇ16–195ᵃ14.

13. Aristotle, *De Generatione Animalium* (written ca. 335–322 B.C.), trans. Arthur Platt (Oxford, England: Clarendon Press, 1910), Bk. I, Chap. 19, lines 729ᵇ13, 730ᵃ1–3. See also lines 739ᵇ21–27, 727ᵇ31, 730ᵃ15, 730ᵇ10–25. For a discussion, see Maryanne Cline Horowitz, "Aristotle and Woman," *Journal of the History of Biology* 9, no. 2 (Fall 1976): 183–213; Joseph Needham, *A History of Embryology* (Cambridge, England: The University Press, 1934); Anthony Preus, "Science and Philosophy in Aristotle's Generation of Animals," *Journal of the History of Biology* 3 (Spring 1970): 1–52. Arthur William Meyer, *The Rise of Embryology* (Stanford, Cal.: Stanford University Press, 1939), Chap. 2, pp. 17–27. Caroline Whitbeck, "Theories of Sex Difference," *The Philosophical Forum* 5 (Fall-Winter 1973–74): 54–80, esp. 55–57.

14. Nicolaus Copernicus, *On the Revolutions of the Heavenly Spheres,* trans. A. M. Duncan (New York: Barnes & Noble, 1976; first Latin ed. 1543), Bk. I, Chap. 10, p. 50. See also Frank Sherwood Taylor, *The Alchemists* (New York: Schumann, 1949), Chap. 2; Walter Pagel and Maryanne Winder, "The Higher Elements and Prime Matter in Renaissance Naturalism and in Paracelsus," *Ambix* 21 (1974): 93–127 (see p. 95).

15. Elaine Pagels, "What Became of God

the Mother? Conflicting Images of God in Early Christianity," *Signs* 2, no. 2 (Winter 1976): 297. See also pp. 293–303 for a discussion of women and gnosticism. The discussion of gnosticism draws on Hans Jonas, *The Gnostic Religion* (Boston: Beacon Press, 1958), Chaps. 2-4, 7, 8, and Kurt Seligmann, *Magic, Supernaturalism, and Religion* (New York: Pantheon Books, 1948), pp. 60–66.

16. Quoted in Seligmann, p. 85. See also p. 80.

17. *Ibid.*, pp. 128–29.

18. The following discussion of gnosticism draws on Walter Pagel and Maryanne Winder, "The Eightness of Adam and Related Gnostic Ideas in the Paracelsian Corpus," *Ambix* 16 (1969): 119–39; Pagel and Winder, "The Higher Elements," pp. 94, 96–97, and Walter Pagel, "Das Rätsel der Acht Mütter im Paracelsischen Corpus," *Sudhoff's Archive für Geschichte der Medizin* 59 (1975): 254–66.

19. Theophrastus Paracelsus, *Selected Writings,* ed. J. Jacobi (Princeton, N.J.: Princeton University Press, 1951), p. 27. On the gnostic sources of Paracelsus' philosophy, see Walter Pagel, "Paracelsus and the Neoplatonic and Gnostic Tradition," *Ambix* 8 (1960): 125–66; W. Pagel, "Paracelsus, Traditionalism and Medieval Sources," in Lloyd G. Stevenson and Robert D. Multhauf, eds., *Medicine, Science, and Culture* (Baltimore, Md.: Johns Hopkins Press, 1968), pp. 57–75; W. Pagel, *Paracelsus* (New York: Karger, 1958), pp. 204–17. Paracelsus' attitude toward women, however, remains ambiguous; for example, "How can one be an enemy of woman—whatever she may be? The world is peopled with her fruits, and that is why God lets her live so long, however loathsome she may be" (Jacobi, ed., p. 26); "He who contemplates woman should see in her the material womb of man; she is man's world, from which he is born" (*Ibid.*).

20. Ralph Cudworth, *The True Intellectual System of the Universe* (New York: Gould and Newman, 1838; first published 1678), vol. 1, p. 404.

21. Carl G. Jung, *Alchemical Studies* in Herbert Read and others, eds., *Collected Works* (Princeton, N.J.: Princeton University Press, 1953), vol. 13, pp. 211–44; F.S. Taylor, Chap. 11; K. Seligmann, p. 98.

22. Thomas Vaughan, "Anima Magica Abscondita," in Arthur E. Waite, ed., *The Works of Thomas Vaughan* (London: Theosophical Publishing House, 1919; first published 1650), p. 94.

23. M. Tullius Cicero, *Of the Nature of the Gods,* ed. T. Francklin (London, 1775), Bk. II, Chap. 8, p. 96. See also Chaps. 53, 60.

24. Lucius Seneca, *Physical Science in the Time of Nero; Being a Translation of the Quaestiones Naturales of Seneca,* (written ca. A.D. 65), trans. John Clarke (London: Macmillan, 1910). Quotations in order are taken from Bk. VI, Chap. 16, pp. 244–45; Bk. III, Chap. 15, pp. 126, 127. On the Stoic conception of nature see Eduard Zeller, *The Stoics, Epicureans, and Sceptics* (London: Longmans Green, 1870), pp. 134–94.

25. Translated and quoted in F. M. Cornford, *Plato's Cosmology,* p. 330.

26. Gabriel Harvey, *Pleasant and Pitthy Familiar Discourse of the Earthquake in Aprill Last,* quoted in Walter M. Kendrick, "Earth of Flesh, Flesh of Earth: Mother Earth in the *Faerie Queen,*" *Renaissance Quarterly* 27 (1974): 548–53, see p. 544. On the earth's veins and bowels, see Georg Agricola, *De Re Metallica,* 1556, trans. Herbert C. Hoover and Lou H. Hoover (New York: Dover, 1950; first published, 1556); p. 1. See also excerpts of Agricola, *De Ortu et Causis Subterraneorum* (first published 1546), in Kirtley F. Mather, ed., *Source Book in Geology* (New York: McGraw-Hill, 1939), p. 7; Athanasius Kircher, *Mundus Subterraneous* (Amsterdam, 1678) in Mather, ed., pp. 17–19.

27. Bernardino Telesio, *De Rerum Natura Iuxta Propria Principia,* (Naples, 1587; first published, 1565). Excerpts translated in Arturo B. Fallico and Herman Shapiro, eds. and trans., *Renaissance Philosophy* (New York: Modern Library, 1967), vol. 1, pp. 308–9.

28. Giordano Bruno, *The Expulsion of the Triumphant Beast* (first published 1584), trans. and ed. Arthur D. Imerti (New Brunswick, N. J.: Rutgers University Press, 1964), p. 72.

29. Marco Antonio della Frata et Montalbano, *Pratica Minerale Trattrato* (Bologna, 1678), p. 2; as quoted in Frank Dawson Adams, *The Birth and Development of the Geological Sciences* (New York: Dover, 1938), p. 306.

30. Telesio, in Fallico and Shapiro, p. 309.

31. Aristotle, *Meteorologica* (written, ca. 335–322 B.C.), trans. E. W. Webster, in W. D. Ross, ed., *The Works of Aristotle Translated into English* (Oxford, England: Clarendon Press, 1923), vol. 3, line 339a. See Adams, p. 84; examples cited include Konrad von Megenberg, *Das Buch der Natur* (Augsberg: J. Bämler, 1475); Gregorius Reisch, *Margarita Philosophica* (Strasbourg, 1504); Hieronymus Savonarola, *Compendium Totius Philosophiae tam Naturalis quam Moralis* (Venice, 1542): Giorgio Camillo Maffei, *Scala Naturale* (Venice, 1563); Andreas Baccius, *De Gemmis et Lapidibus Pretiosis* (Frankfurt, 1603); Leonardus Camillus, *Speculum Lapidum* (Venice, 1502), Bk. I.

32. Adams, pp. 84–90, 290, 292; examples cited include Thomas Sherley, *A Philosophical Essay Declaring the Probable Causes Whence Stones Are Produced in the Greater World* (London: 1672), pp. 23–128; M.J.C. Schwiegger, *De Ortu Papidum,* dissertation, University of Wittenburg, 1665; Jerome Cardan, *De Subtilitate* (Nuremberg, 1550); Bernard Palissy, *Discours Admirables de la Nature des Eau et Founteines tant Naturelles qu'Artificelles, des Metaus, des Sels et Salines* (Paris, 1580), pp. 122, 134; John Webster, *Metallographa, or An History of Metals* (London, 1671), p. 70; Robert Boyle, *The Skeptical Chemist* (London: Everyman's Library, 1967), pp. 194, 202; Edward Jorden, *A Discourse of Natural Bathes and Mineral Waters* (London, 1669), p. 51.

33. Adams, pp. 90–4; examples include Agricola, *De Ortu et Causis Subterraneorum;* Giorgio Baglivi, *De Vegetatione Lapidum in Opera Omnia* (London, 1714), p. 497.

34. Paracelsus, ed. Jacobi, p. 25; Pagel, "Das Rätsel der Acht Mütter," p. 255; Paracelsus, in Karl Sudhoff and Wilhelm Mattiessen, eds., *Sämtliche Werke* (Munich: Barth, 1922–1933), Abt. I, vol. 13, p. 88.

35. Basil Valentine, "The Fifth Key," from *The Practica, with Twelve Keys, and an Appendix Thereto Concerning the Great Stone of the Ancient Sages,* reprinted in *The Hermetic Museum Restored and Enlarged* . . . (Frankfurt, 1678), trans. Arthur Edward Waite (New York: Samuel Weiser, 1974; first published 1893), vol. I, p. 333.

36. Henry More, "An Antidote Against Atheism" (first published 1653) in *A Collection of Several Philosophical Writings of Dr. Henry More* (London, 1712), p. 65.

37. Adams, pp. 94, 102–36; Cardan, *De Subtilitate,* Bk. VII; D. C. Goodman, "The Saltish Seed: Crystals and Life in the Seventeenth Century," *Abstracts of Papers Presented to the XVth International Congress of the History of Science,* (Edinburgh, 1977), p. 167.

38. Smohalla (Columbia Basin Tribes), voicing fundamental cause of all great periods of Indian unrest (mid-1800s), quoted in Alfonso Ortiz and Margaret Ortiz, eds., *To Carry Forth the Vine* (New York: Columbia University Press, 1978). On Indian, animistic, and Western belief systems, see Calvin Martin, *Keepers of the Game* (Berkeley and Los Angeles: University of California Press, 1978); Henri Frankfurt and H. A. Frankfurt, *The Intellectual Adventure of Ancient Man* (Chicago: University of Chicago Press, 1977); Lynn White, Jr., "Historical Roots of Our Ecologic Crisis" (cited in note 1); Garrett Hardin, "The Tragedy of the Commons," *Science* 162 (1968): 1243–48.

39. Peter Martyr, *The History of Truayle in the West and East Indies and other countryes lying either way toward the fruitful and ryche Moluceeas, gathered in part and done into Englyshe by Richard Eden, Newly set in order augmented and finished by Richard Willes Third Decade* (London, 1577), quoted in Adams, p. 287. This viewpoint also appeared in Johann Rudolf Glauber's *Operis Mineralis* (Amsterdam, 1652),

Pt. II and in the 18th century in Johann Joachim Becker (1635–1682), *Natur Kündigung der Metallen* (Frankfurt, 1705) and Emmanuel König, *Regnum Minerale* (Basel, 1703). Discussed in Adams, pp. 286–89, 293.

40. Albaro Alonzo Barba, *The Art of Metals,* trans. Earl of Sandwich (London, 1669; first published 1640), p. 49; as quoted in Adams, p. 294. R. Boyle, *The Skeptical Chemist* (p. 191), concurred in the belief that stones grew from the roofs of caves, and that water could be transformed into minerals. This occurrence "is very notable because from thence we deduce that earth, by a metallic plastik principle latent in it, may be in processe of time changed into a metal."

41. Pliny, *Natural History* (written ca. A.D. 23–79), trans. J. Bostock and H. T. Riley (London: Bohn, 1858), vol. 6, Bk. 33, Chap. 1, pp. 68–69. Subsequent quotations are from pp. 69, 71, 70 and 205–206.

42. Publius Ovid, *Metamorphoses* (written A.D. 7), trans. Rolfe Humphries (Bloomington: Indiana University Press, 1955), Bk. I, p. 6, line 100. Subsequent quotations are from lines 101–11, 137–43, 155–62. See also Kendrick, "Earth of Flesh," p. 539.

43. Seneca, *Natural Questions,* trans. John Clarke, Bk. V, Chap. 15, pp. 207–8.

44. See Erik Erikson on the conflict of values in the mining regions surrounding the "transition from the agrarian preoccupation with mud, soil, and fertility to the miner's preoccupation with rock, dirt, and the chances of a haul; and beyond this, the mercantile aim of amassing metal and money, shiny and yet 'dirty' and all subject to a new adventurous and boundless avarice which the church tried to crush with all her might—and at the same time to monoplize." E. Erikson, *Young Man Luther* (New York: Norton, 1958), pp. 53–61. On the revival of mining in the fifteenth century after a long lapse from the fall of Rome through the Middle Ages, see William Barclay Parsons, *Engineers and Engineering in the Renaissance* (Cambridge, Mass: M.I.T. Press, 1968), pp. 177–200.

45. Adams, pp. 172–73. Niavis [Dr. Paul Schneevogel], *Judicum Jovis* (Leipzig: Kachelhofen, n.d.).

46. Henry Cornelius Agrippa, *De Incertitude et Vanitate Omnium Scientiarum et Artum* (Antwerp, 1530). English translation: *The Vanity of Arts and Sciences* (London, 1694), pp. 81, 82.

47. Georg Agricola, *De Re Metallica,* pp. 6–7, 8.

48. Richard Trexler, "Measures Against Water Pollution in Fifteenth-Century Florence," *Viator* 5 (1974): 455–67, see pp. 463, 466–67. Other articles on attempts to deal with water pollution in medieval Europe include Lynn Thorndike, "Sanitation, Baths, and Street Cleaning in the Middle Ages and Renaissance," *Speculum* 3 (1928): 192–203, and Ernest Sabine, "City Cleaning in Medieval London," *Speculum* 12 (January 1937): 19–43.

49. Agricola, *De Re Metallica,* Hoover trans., quotations on pp. 12, 13, 14, 12, 17, 16.

50. Johannes Mathesius, *Bergpostilla, oder Sarepta, darinn von allerley Bergkwerck und Metallen, was ir ey enschaft und natur* (Nürnberg, 1578).

51. Quoted in Kendrick, "Earth of Flesh," p. 538. Edmund Spenser, *The Faerie Queen,* ed. John Upton, 2 vols. (London, 1758; first published 1590/5), Bk. II, Canto 7, verse 17. On Spenser's participation in the sixteenth-century debates on mining, his use of Agricola, and the description of Mammon's forge, see Kendrick, pp. 537-41. *Faerie Queen,* vol. 1, Bk. II, Canto VII, verse 15: "At the well-head the purest streames arise;/But mucky filth his braunching armes annoyes,/And with uncomely weedes the gentle wave accloyes." Mammon's forge is described in vol. I, Bk. II, Canto VII, verses 35–36. On Spenser's association of lust with mining see Kendrick, pp. 544–45.

52. John Milton, *Paradise Lost* (first published 1667), ed. Scott Elledge (New York: Norton, 1975), Bk. I, lines 684–90.

53. John Donne, *Poems of John Donne,* ed. Herbert Grierson (London: Oxford University Press, 1957), p. 35. For an analysis of "Love's Alchemie," see Clay Hunt, *Donne's*

Poetry (New Haven, Conn.: Yale University Press, 1954), pp. 33-41.

54. Donne, *Poems,* Elegie XVIII, "Love's Progress," p. 104, lines 27–36, lines 91–94; Elegie XIX, p. 107, lines 25–30.

CHAPTER 2:
FARM, FEN, AND FOREST

1. The following discussion of manorial farm ecology is based on B. H. Slicher Van Bath, *The Agrarian History of Western Europe, A.D. 500–1850* (London: Arnold, 1963), pp. 71–73; J. H. Clapham, ed., *The Cambridge Economic History of Europe,* vol. 1, *The Agrarian Life of the Middle Ages* (Cambridge, England: Cambridge University Press, 1941), esp. Chaps. 7, 8; Gottfried Pfeiffer, "The Quality of Peasant Living in Central Europe," in William L. Thomas, Jr., ed., *Man's Role in Changing the Face of the Earth* (Chicago; University of Chicago Press, 1970), vol. 1, pp. 240–77; E. Estyn Evans, "The Ecology of Peasant Life in Western Europe," in Thomas, ed., *Man's Role,* pp. 217–39. On the plow and communal organization, see Lynn White, Jr., *Medieval Technology and Social Change* (New York: Oxford University Press, 1960), Chaps. 2–3; Jerome Blum, "The European Village as Community: Origins and Functions," *Agricultural History* 45 (1971): 157–78; J. Blum, "The Internal Structure and Polity of the European Village Community from the Fifteenth to the Nineteenth Century," *The Journal of Modern History* 43 (1971): 541–76; Theodore Shanin, "Peasantry: Delineation of a Sociological Concept and a Field of Study," *European Journal of Sociology* 12 (1971): 289–300; John Duncan Powell, "On Defining Peasants and Peasant Society," *Peasant Studies Newsletter* 1, no. 3 (July 1972): 94–99; David Sabean, "Markets, Uprisings and Leadership in Peasant Societies: Western Europe, 1381–1789," *Peasant Studies Newsletter* 2, no. 3 (July 1973): 17–19; T. J. Byres and C. A. Curwen, "Summary of Eric Hobsbawm, 'Peasants and Politics,'" *Peasant Studies Newsletter* 1, no. 3 (July 1972): 109–14.

2. On the energy base of preindustrial

economies, see Lewis Mumford, *Technics and Civilization* (New York: Harcourt Brace Jovanovich, 1934), Chaps. 3–4; Werner Sombart, *Der Moderne Kapitalismus,* 3 vols. (Munich: Duncker & Humbolt, 1928), esp. vols. 1–2.; Marc Bloch, "The Advent and Triumph of the Water Mill," in *Land and Work in Medieval Europe: Selected Papers* (Berkeley: University of California Press, 1967), pp. 136–67; Bradford Blaine, "The Enigmatic Watermill," in Bert S. Hall and Delno C. West, eds., *On Pre-Modern Technology and Science* (Malibu, Cal.: Undena, 1976), pp. 163–76; William H. Te Brake, "Air Pollution and Fuel Crises in Preindustrial London," *Technology and Culture* 16 (1975): 337–59.

3. Blum, "The European Village," pp. 160–61. On the plow, the three field system, and communal organization, see White, Jr., *Medieval Technology and Social Change,* Chaps. 2–3.

4. Robert Brenner, "Agrarian Class Structure and Economic Development in Pre–Industrial Europe," *Past and Present,* no. 70 (February 1976): 30–75; Te Brake, pp. 354–55.

5. Te Brake, pp. 355–6. Emmanuel LeRoy Ladurie, "Zero Population Growth: Population and Subsistence in Sixteenth Century Rural France," *Peasant Studies Newsletter* 1, no. 2 (April 1972): 60–65; Karl F. Helleiner, "The Population of Europe from the Black Death to the Eve of the Vital Revolution," in J. H. Clapham, ed., *The Cambridge Economic History of Europe,* vol. 4, *The Economy of Expanding Europe in the 16th and 17th Centuries* (Cambridge, England: Cambridge University Press, 1967), pp. 1–95; J. F. D. Shrewsbury, *A History of the Bubonic Plague in the British Isles* (Cambridge, England: Cambridge University Press, 1970).

6. The discussion of peasant-landlord conflict draws on Brenner, pp. 50–1; 56–7; Douglass C. North and Robert Paul Thomas, "The Rise and Fall of the Manorial System: A Theoretical Model," *The Journal of Economic History* 31, no. 4 (December 1971): 777–803, see pp. 780–81; Blum,

"The European Village," pp. 165, 167–68; David Sabean, "Family and Land Tenure: A Case Study of Conflict in the German Peasants' War (1525)," *Peasant Studies Newsletter* 3, no. 1 (January 1974): 1–15, see pp. 3, 4; Pfeiffer, pp. 266–67.

7. Brenner, p. 71. Relevant studies of the French peasantry include Emmanuel Le-Roy Ladurie, *Les Paysans de Languedoc,* 2 vols. (Paris: École Pratique des Hautes Études et Mouton, 1966), and LeRoy Ladurie, "Zero Population Growth," pp. 60–65. Pierre Goubert, "The French Peasantry of the Seventeenth Century: A Regional Example," in Trevor Aston, ed., *Crisis in Europe, 1560–1660* (New York: Basic Books, 1965); Marc Bloch, *Les Characteres Originaux de l'Histoire Rurale Francaise,* 2 vols. (Paris: Colin, 1956); Roland Mousnier, *Peasant Uprisings in Seventeenth Century France, Russia and China,* trans. Brian Pearce (New York: Harper & Row, 1970); Boris Porchnev, *Les Soulevements Populaires en France de 1623 à 1648* (Paris: S.E.V.P.E.N., 1963). On southern Europe, see Fernand Braudel, *The Mediterranean and the Mediterranean World in the Age of Philip II,* 2 vols. (London: Collins, 1973), vol. 2, pp. 734–56.

8. On the rise of capitalism in Europe, see Immanuel Wallerstein, *The Modern World System* (New York: Academic Press, 1975). The discussion of the Netherlands is based on Jan de Vries, *The Dutch Rural Economy in the Golden Age, 1500–1700* (New Haven, Conn.: Yale University Press, 1974), pp. 34, 55, 62–68, 119–21, 136, 142–47, 149–51, 153–54; B. H. Slicher Van Bath, "The Rise of Intensive Husbandry in the Low Countries," in J. S. Bromley and others, eds., *Britain and the Netherlands* (London: Chatto and Windus, 1960), vol. 1, pp. 130–153; Slicher Van Bath, *Agrarian History,* pp. 278–79. See also Violet Barber, *Capitalism in Amsterdam in the Seventeenth Century* (Ann Arbor: University of Michigan Press, 1963) and Herman Van Der Wee, *The Growth of the Antwerp Market and the European Economy* (The Hague: Nijhoff, 1963), vol. 2, esp. pp. 289–322.

9. Brenner, p. 63. The following discussion of English farming is based on Mildred Campbell, *The English Yeoman under Elizabeth and the Early Stuarts* (New Haven, Conn.: Yale University Press, 1942), esp. pp. 32, 33, 61; Albert J. Schmidt, *The Yeoman in Tudor and Stuart England* (Washington, D.C.: Folger Shakespeare Library, 1961), p. 4; Joan Thirsk, ed., *The Agrarian History of England and Wales,* vol. 4, *1500–1640* (Cambridge, England: Cambridge University Press, 1967), see esp. Chap. 3; B. A. Holderness, *Pre-Industrial England: Economy and Society, 1500–1750* (London: Dent, 1976), Chap. 3.

10. John Taylor, *Brood of Cormorants* (published 1622), quoted in Schmidt, p. 8.

11. Quoted in W. E. Tate, *The English Village and the Enclosure Movements* (London: Gollancz, 1967), p. 63. On the enclosure movements, see also G. Slater, *The English Peasantry and the Enclosure of Common Fields* (London: Constable, 1907).

12. Gervase Markham, *The English Husbandman* (London, 1613), quoted in Schmidt, p. 12.

13. Olivier de Serres, *Theatre d'Agriculture,* (Paris, 1804; first published 1600), preface; Walter Blith, *The English Improver Improved,* (London, 1652), "Epistle Dedicatory," pp. 4–5; as quoted in Paul H. Johnstone, "In Praise of Husbandry," *Agricultural History* 2 (1937): 84, 86, 88.

14. On Dutch hydraulic technology, see J. Van Veen, *Dredge, Drain, Reclaim* (The Hague: Nijhoff, 1949), pp. 25–33, 39–48; Slicher Van Bath, *Agrarian History,* pp. 201–2. L. E. Harris, "Land Drainage and Reclamation," in Charles Singer and others, eds., *A History of Technology* (New York: Oxford University Press, 1957), pp. 300–323; S. J. Foekema Andrea, "Embanking and Drainage Authorities in the Netherlands During the Middle Ages," *Speculum* 27 (1952): 158–67.

15. The following account of the English fens is drawn from H. C. Darby, *The Draining of the Fens* (Cambridge, England: Cambridge University Press, 1956), pp. 27–155; Joan Thirsk, *English Peasant Farm-*

ing: *The Agrarian History of Lincolnshire from Tudor to Recent Times* (London: Routledge & Kegan Paul, 1957), pp. 6–7, 111–29.

16. Thirsk, *English Peasant Farming,* pp. 111, 121, 123.

17. *Ibid.,* pp. 9, 125, 127, 129; Darby, p. 53.

18. Quotations in the following paragraphs are quoted in Darby, pp. 46, 48, 49 (text and Note 2), 51, 52, 69, 55, 58, 90, 68.

19. The following discussion draws on Robert G. Albion, *Forests and Sea Power: The Timber Problem of the Royal Navy, 1652–1852* (Cambridge, Mass.: Harvard University Press, 1926): pp. 97, 118, 121; H. C. Darby, "The Clearing of the Woodland in Europe," in W. L. Thomas, Jr., ed., *Man's Role,* vol. 1, pp. 183–216; W. G. Hoskins, *The Making of the English Landscape* (London: Hodder and Stoughton, 1955), Chap. 1 6; F. V. Emery, "England Circa 1600," in H. C. Darby, ed., *The New Historical Geography of England* (Cambridge, England: Cambridge University Press, 1973), pp. 248–301; H. C. Darby, "The Age of the Improver, 1600–1800" in *New Historical Geography,* pp. 302–88; Charles Petit-Dutaillis, "The Forest," in *Studies and Notes Supplementary to Stubbes Constitutional History* (Manchester, England: Manchester University Press, 1915), vol. 2, pp. 147–304.

20. Quoted in Te Brake, p. 345.

21. Quoted in G. G. Coulton, *Social Life in Britain from the Conquest to the Reformation* (Cambridge, England: Cambridge University Press, 1918), p. 333.

22. Slicher Van Bath, pp. 72–73; Te Brake, p. 157; John U. Nef, *The Rise of the British Coal Industry* (Hamden, Conn.: Archon, 1966), vol. 1, pp. 156–64.

23. Darby, "Age of the Improver," in *New Historical Geography,* p. 304; Emery, "England Circa 1600," in *New Historical Geography,* p. 250; Te Brake, p. 356.

24. Andrew B. Appleby, "Common Land and Peasant Unrest in Sixteenth Century England: A Comparative Note," *Peasant Studies Newsletter* 4, no. 3 (July 1975): 20.

25. Albion, p. 119; J. U. Nef, "The Progress of Technology and the Growth of Large-Scale Industry in Great Britain, 1540–1640," *Economic History Review* 5 (1934): 3–24.

26. Eugene F. Rice, *The Foundations of Early Modern Europe, 1450–1559* (New York: Norton, 1970), pp. 45–6. See also Pierre Jeanin, *Merchants of the Sixteenth Century* (New York: Harper & Row, 1972).

27. Nef, "Progress of Technology," *passim.* T. S. Ashton, *The Industrial Revolution 1760–1830* (New York: Oxford University Press), p. 29.

28. Albion, pp. 99, 120; Frederic Chapin Lane, *Venetian Ships and Shipbuilders of the Renaissance* (Baltimore, Md.: Johns Hopkins University Press, 1934), Chaps. 8, 9, 12.

29. Albion, pp. 117–23. The relevant timber laws were 35 Hen. VIII, c. 17; I Eliz. c. 15.

30. Nef, "The Progress of Technology," pp. 10, 15; J. U. Nef, "Coal Mining and Utilization," in Charles Singer and others, eds., *A History of Technology* (New York: Oxford University Press, 1957), vol. 3, pp. 72–88, see p. 77; J. U. Nef, *The Rise of the British Coal Industry,* vol. 1, pp. 19–22, 190–223.

31. Nef, *Rise of the British Coal Industry,* vol. 1, p. 247, plate.

32. Albion, pp. 97, 124–125.

CHAPTER 3:
ORGANIC SOCIETY AND UTOPIA

1. John of Salisbury, *The Statesman's Book,* [*Selections*] *From the Policraticus,* 1159, trans. John Dickinson (New York: Knopf, 1927). Relevant discussions of organic theories of society include Otto Gierke, *Political Theories of the Middle Ages,* trans. Frederic William Maitland (Cambridge, England: Cambridge University Press, 1900); Harry Elmer Barnes, "Representative Biological Theories of Society." *Sociological Review* 17 (1925): 121–30, 183–95, 294–300; F. W. Coker, "Organismic Theories of the State," in *Studies in History, Economics, and Public Law, Columbia Studies in the Social Sciences* 38, no. 2, (1910): 1–209; H. J. McCloskey,

"The State as an Organism, as a Person, and as an End in Itself," *Philosophical Review* 72 (July 1963): 306–25; T. D. Weldon, *States and Morals: A Study in Political Conflicts* (New York: McGraw-Hill, 1947); Ferdinand Tönnies, *Community and Society,* trans. Charles P. Loomis (East Lansing: Michigan State University Press, 1957); Robert Redfield, *The Little Community and Peasant Society and Culture* (Chicago: University of Chicago Press, 1973); critiques of Redfield emphasize variations in social stratification within different peasant communities. Emile Durkheim, *The Division of Labor in Society* (New York: Free Press, 1933), Chaps. 2, 7. Durkheim used the terms *organic* and *mechanical solidarity* in a different sense than is meant here. *Mechanical solidarity* referred to the lack of differentiation in a society while *organic solidarity* resulted from the differentiation brought about by the division of labor and characterized modern society.

2. Salisbury, *Policraticus,* trans. Dickinson, discussion and quotations from pp. 65, 243, 67, 95, 247, 259, 39, 245–47, 251.

3. Aristotle, *Politics,* trans. Benjamin Jowett, in Richard McKeon, ed., *The Basic Works of Aristotle* (New York: Random House, 1941), see esp. Bk. I, Chap. 3, lines 1253ª 19–27; Bk. VII, Chap. 9; E. Barker, *The Political Thought of Plato and Aristotle* (London: Methuen, 1906), pp. 276–81. On Aquinas see George H. Sabine, *A History of Political Theory* (New York: Holt, 1937), pp. 247–51.

4. On Nicolas of Cusa see Gierke, pp. 23–24. On organic analogies in Elizabethan England see James Emerson Phillips, Jr., *The State in Shakespeare's Greek and Roman Plays* (New York: Octagon Books, 1972), esp. pp. 3, 8–10, 66, 68, 70–71, 155; sources cited include Edward Forset, *A Comparative Discourse of the Bodies Natural and Politique . . .* (London, 1606), pp. 35–36; Thomas Floyd, *The Picture of a Perfit Commonwealth* (London, 1600), p. 19; William Fulbecke, *The Pandectes of the Law of Nations* (London, 1602); William Shakespeare, *Coriolanus,* act 1, sc. 1, lines 99–158; William Shakespeare, *Henry V,* act 1, sc. 2, lines 183–213.

5. Lawrence Stone, *The Family, Sex, and Marriage in England: 1500–1800* (New York: Harper & Row, 1977), pp. 85–93, 116–118. On mobility, see L. Stone, "Social Mobility in England, 1500–1700," *Past and Present,* no. 33 (April 1966): 16–55; Alan Everitt, "Social Mobility in Early Modern England," *Past and Present,* no. 33 (April 1966), pp. 56–73; David Herlihy, "Three Patterns of Social Mobility in Medieval Society," *The Journal of Interdisciplinary History* 3 (Spring 1973): 623–48. On the decline of the aristocracy, see L. Stone, *The Crisis of the Aristocracy, 1558–1641* (New York: Oxford University Press, 1967) and Davis Britton, *The French Nobility in Crisis, 1560–1640* (Stanford, Cal.: Stanford University Press 1969). On the rise of the merchant class, see Pierre Jeanin, *Merchants of the Sixteenth Century* (New York: Harper & Row, 1972) and John U. Nef, *Industry and Government in France and England, 1540–1640* (Philadephia: American Philosophical Society, 1940).

6. C. B. Macpherson, *The Political Theory of Possessive Individualism: Hobbes to Locke* (Oxford, England: Oxford University Press, 1962), p. 49; Stone, pp. 123–35.

7. Jean Bodin, *Six Books of the Commonwealth,* (first published 1576), abridged and trans. M. J. Tooley (New York: Macmillan, 1955), Bk. I, Chaps. 2, 6, 7, 8, and Bk. VI, Chap. 5; William Archibald Dunning, *A History of Political Theories from Luther to Montesquieu* (New York: Macmillan, 1931), pp. 81–82, 94–96.

8. James I, *The Trew Law of Free Monarchies,* (first published 1616), in *The Political Works of James I,* (Cambridge, Mass.: Harvard University Press, 1918), pp. 54, 64–65; Dunning, pp. 215–16.

9. Donald Worster, *Nature's Economy: The Roots of Ecology* (San Francisco: Sierra Club Books, 1977), Chap. 11, 15; Weldon, pp. 30–34; Daniel Gasman, *The Scientific Origins of National Socialism* (New York: Elsevier, 1971).

10. Discussion of the communal model is based on Jerome Blum, "The European Vil-

lage as Community: Origins and Functions," *Agricultural History* 45, no. 3 (1971): 158–60, 164–65; 171–74; O. Gierke, *Political Theories,* pp. 37–60; O. Gierke, *The Development of Political Theory,* trans. Bernard Freyd (New York: Norton 1939), pp. 143–63; John Duncan Powell, "On Defining Peasants and Peasant Society," *Peasant Studies Newsletter* 1, no. 3 (July 1972): 94–99; T. J. Byres and C. A. Curwen, "Summary of Eric Hobsbawm, 'Peasants and Politics' " *Peasant Studies Newsletter* 1, no. 3 (July 1972): 109–14.

11. On the breakup of the village community in rural Europe, see Jerome Blum, "The Internal Structure and Polity of the European Village Community from the Fifteenth to the Nineteenth Century," *The Journal of Modern History* 43, no. 4 (December 1971): 541–76, esp. p. 569, and J. Blum, *The End of the Old Order in Rural Europe* (Princeton, N. J.: Princeton University Press, 1978), esp. Chaps. 10–15.

12. John Locke, "An Essay Concerning the True Original Extent and End of Civil Government," (first published 1690), *Two Treatises of Government,* ed. Peter Laslett (Cambridge, England: Cambridge University Press, 1960), Second treatise, Chap. V, secs. 32, 28, 37, 35. Subsequent quotations from secs. 46, 48. See also Macpherson, pp. 201–2.

13. Norman Cohn, *The Pursuit of the Millennium: Revolutionary Millenarians and Mystical Anarchists of the Middle Ages* (New York: Oxford University Press, 1970); Arthur Leslie Morton, *The Everlasting Gospel: A Study in the Sources of William Blake* (New York: Haskell House, 1966); Christopher Hill, *The World Turned Upside Down* (New York: Viking Press, 1972); Eric Hobsbawm, *Primitive Rebels* (New York: Norton, 1959), pp. 58–59.

14. Tommaso Campanella, *La Città del Sole* (written 1602), (Rome: Colombo Editore, 1953; first published 1623). Citations refer to Thomas Halliday's English translation, "The City of the Sun," in Henry Morley, ed., *Ideal Commonwealths* (New York: Colonial Press, 1901), pp. 141–79. Halliday's English translation is slightly abridged. A complete translation in French by Jules Rosset appears in *Oeuvres Choisies de Campanella, precedées d'une Notice par Mme. Louise Colet* (Paris, 1844); Johann Valentin Andreä, *Christianopolis: An Ideal State of the Seventeenth Century,* ed. and trans. Felix Emil Held (Urbana: University of Illinois Press, 1914; first published 1619), pp. 133–280. Important commentaries on these utopias include Marie Louise Berneri, *Journey Through Utopia* (Boston: Beacon Press, 1950), Chap. 2; Eleanor Blodgett, "Bacon's *New Atlantis* and Campanella's *Civitas Solis:* A Study in Relationships," *PMLA* 46 (1931): 763–80; Nell Eurich, *Science in Utopia: A Mighty Design* (Cambridge, Mass.: Harvard University Press, 1967); Felix Emil Held, "Introductory Essay" in Andreä, *Christianopolis,* pp. 3–128; Lewis Mumford, *The Story of Utopias* (New York: Boni and Liveright, 1922), Chaps. 4, 5.

15. On Campanella's philosophy of nature, see Tommaso Campanella, *De Sensu Rerum et Magica,* (written 1591) (Frankfurt, 1620); Bernardino Telesio, *De Rerum Natura,* 1565 (Naples, 1587; first published 1565); selections in Arturo B. Fallico and Herman Shapiro, eds. and trans., *Renaissance Philosophy* (New York: Modern Library, 1967), vol. 1, on Telesio see pp. 302–38, on Campanella see pp. 338–79; Leon Blanchet, *Campanella* (New York: Franklin, 1920), pp. 30, 33–36, 38, 40; D. P. Walker, *Spiritual and Demonic Magic from Ficino to Campanella* (New York: Kraus Reprints, 1958), pp. 203–36; Frances Yates, *Giordano Bruno and the Hermetic Tradition* (New York: Vintage Books, 1969), pp. 360–97; and F. Yates, *The Art of Memory* (Chicago: University of Chicago Press, 1966); Bernardino M. Bonansea, *Tommaso Campanella* (Washington, D. C.: Catholic University of America Press, 1969), p. 29.

16. John Warwick Montgomery, *Cross and Crucible: Johann Valentin Andreae, 1586–1654), Phoenix of the Theologians* (The Hague: Nijhoff, 1973), vol. 1, pp. 63, 68; Held, "Introduction" to Andreä, *Christianopolis,* p. 13; Eurich, *Science in Utopia,* pp.

147–52; Richard Foster Jones, *Ancients and Moderns* (Berkeley: University of California Press, 1965), pp. 88, 131–32.

17. Campanella, "City of the Sun," p. 178.
18. *Ibid.*, p. 169.
19. *Ibid.*
20. Yates, *Bruno,* Chap. 4, pp. 62–83; D. P. Walker, *Spiritual and Demonic Magic from Ficino to Campanella* (reprint ed., Nedeln/Lichtenstein, Germany: Kraus Reprint, 1969; first published 1958), Chaps. 1 and 2, pp. 3–59.
21 Andreä, *Christianopolis,* p. 143.
22. *Ibid.*, p. 231.
23. *Ibid.*, p. 161.
24. *Ibid.*, p. 157.
25. *Ibid.*, p. 160.
26. *Ibid.*, quotes on pp. 154, 157.
27. Campanella, "City of the Sun," p. 158.
28. *Ibid.*, pp. 158–59.
29. *Ibid.*, p. 150, 147; Andreä, *Christianopolis,* p. 161.
30. Andreä, p. 157, 161.
31. Campanella, "City of the Sun," p. 153.
32. *Ibid.*, p. 157.
33. *Ibid.*, p. 155.
34. Andreä, *Christianopolis,* p. 210.
35. *Ibid.*, p. 260.
36. Held, "Introduction," p. 27; Sheldon Wolin, *Politics and Vision* (Boston: Little, Brown, 1960), pp. 189, 191, 177.
37. Andreä, *Christianopolis,* pp. 165, 170, 145–47.
38. Held, "Introduction," pp. 27, 106.
39. Campanella, "City of the Sun," p. 171.
40. *Ibid.*, pp. 147, 156.
41. Holmes Rolston, III, "Is There an Ecological Ethics?" *Ethics* 85 (1975): 93–109.
42. Ernest Callenbach, *Ecotopia* (Berkeley: Banyan Tree Books, 1975).

CHAPTER 4:
THE WORLD AN ORGANISM
1. For a recent critique of the mechanical, reductionist, scientific methodology as applied to the environmental deterioration of the four ancient elements, see Barry Commoner, *The Closing Circle* (New York: Knopf, 1972), esp. Chaps. 1–6, 10. For a critique of holism, see D. C. Phillips, *Holistic Thought in Social Science* (Stanford, Cal.: Stanford University Press, 1976). On

the historical roots of ecology, their origins in organismic philosophies, and the systems approach of the new ecology, see Donald Worster, *Nature's Economy* (San Francisco: Sierra Club Books, 1977), esp. Chaps. 3, 4, 14. On romanticism and American preservationism, see Roderick Nash, *Wilderness and the American Mind* (New Haven, Conn.: Yale University Press, 1967), Chaps. 3–8. For an analysis of the philosophical differences and points of overlap between the organic and mechanical philosophies of nature, see Steven Pepper, *World Hypotheses* (Berkeley: University of California Press, 1970), esp. Chaps. 9, 11. A more recent discussion arguing for the incommensurability of these world systems is William Overton and Hayne Reese, "Models of Development: Methodological Implications," in John R. Nesselroade and Hayne Reese, eds., *Life Span Developmental Psychology* (New York: Academic Press, 1973), pp. 65–86.
2. On Greek and Renaissance organismic ideas, see R. J. Collingwood, *The Idea of Nature* (Oxford, England: Clarendon Press, 1945); Ernst Cassirer, *Individuum und Kosmos in der Philosophie der Renaissance* (Leipzig: Teubner, 1927); Eduard Zeller, *The Stoics, Epicureans, and Sceptics* (London: Longmans Green, 1870), pp. 134–94. Rudolph Aller, "Microcosmus from Anaximandros to Paracelsus," *Traditio 2* (1944): 319–407. On the sixteenth-century world view see Eustace M. W. Tillyard, *The Elizabethan World Picture* (New York: Random House, 1959 [?]), Chaps. 1, 2, 4. Arthur O. Lovejoy, *The Great Chain of Being* (Cambridge, Mass.: Harvard University Press, 1936).
3. Giovanni Battista della Porta, *Magiae Naturalis* (Naples, 1558). English trans.: G. B. della Porta, *Natural Magic,* facsimile edition, ed. Derek J. Price (New York: Basic Books, 1957; first published 1658), p. 13.
4. Theophrast von Hohenheim Paracelsus, "Philosophia ad Atheniensis," in *Sämtliche Werke,* eds. Karl Sudhoff and Wilhelm Matthiessen (Munich: Barth, 1922–1933), Abt. I, vol. 13, p. 409. Eng. trans.: Paracelsus, "The Philosophy Addressed to the Ath-

enians," in *The Hermetic and Alchemical Writings* ed. and trans. Arthur E. Waite (London: Elliott, 1894), vol. 2, p. 268; Peter J. French, *John Dee: The World of an Elizabethan Magus* (London: Routledge & Kegan Paul, 1972), p. 93

5. Della Porta, 1658 facsimile, p. 14.'

6. Bernardino Telesio, *De Rerum Natura Iuxta Propia Principia* (Naples, 1587; first-published 1565). Quotations from excerpts trans. by Arturo B. Fallico and Herman Shapiro in *Renaissance Philosophy*, vol. 1, *The Italian Philosophers* (New York: Modern Library, 1967), p. 309.

7. Tommaso Campanella, *De Sensu Rerum et Magia* (written 1591) (Frankfurt, 1620) Quotations from excerpts trans. Fallico and Shapiro, p. 340.

8. Gene A. Brucker, *Renaissance Florence* (New York: Wiley, 1969), pp. 68–78, 87, 124, 228–9; Frances A. Yates, *Giordano Bruno and the Hermetic Tradition* (New York: Vintage, 1964), Chaps. 2–5. Yates stresses the synthesis of Neoplatonism and Hermeticism in the natural magic tradition. The Hermetic texts are collected in the *Corpus Hermeticum*, ed. A. D. Nock and trans. A.-J. Festugière, 4 vols. (Paris, 1945–54). On Renaissance Neoplatonism and Hermeticism see A.-J. Festugière, *La Révélation d'Hermès Trismégiste*, 4 vols. (Paris, 1950–54); Paul Oskar Kristeller, *The Philosophy of Marsilio Ficino*, trans. Virginia Conant (New York: Columbia University Press, 1943); P. O. Kristeller, *Renaissance Thought: The Classic, Scholastic, and Humanist Strains* (New York: Harper & Row, 1961); Frances Yates, "The Hermetic Tradition in Renaissance Science," in Charles Singer, ed. *Art, Science, and History in the Renaissance* (Baltimore, Md.: Johns Hopkins University Press, 1968), pp. 255–74.

9. Henry Cornelius Agrippa, *De Occulta Philosophia Libri Tres* (Antwerp, 1531). English trans.: Agrippa, *Natural Magic*, trans. Willis F. Whitehead (reprint ed., New York: Weiser, 1971; first published, 1651), pp. 62–63.

10. *Ibid.*, p. 120.

11. Della Porta, 1658 facsimile, p. 8.

12. Thomas Vaughan (Eugenius Phila-

lethes), *Works,* ed., A. E. Waite, (New York: University Books, 1968), p. 79.

13. Agrippa, trans. Whitehead, p. 69. See also Charles G. Nauert, *Agrippa and the Crisis of Renaissance Thought* (Urbana: University of Illinois Press, 1965), p. 267. Also useful is C. G. Nauert, "Magic and Scepticism in Agrippa's Thought," *Journal of the History of Ideas 18* (1957): 161–82.

14. Vaughan, p. 78.

15. Agrippa, trans. Whitehead, quotations in order, on pp. 70, 71, 72, 78.

16. Della Porta, 1658 facsimile, p. 8.

17. Agrippa, *De Occulta Philosophia,* Bk. II, quoted in Yates, *Bruno,* p. 136.

18. Della Porta, 1658 facsimile, quotations in order on pp. 2, 33, 63, 64, 73.

19. On the critique of Aristotle, see Walter Pagel, "The Reaction to Aristotle in Seventeenth Century Biological Thought," in E. Ashworth Underwood, ed., *Science, Medicine and History* (London: Oxford University Press, 1953), vol. 1, pp. 489–509. On the divergence between the traditional scholastic Aristotelian approach and the criticisms of the Italian naturalists, Telesio, Bruno, and Campanella, see Charles Schmitt, "Towards a Reassessment of Renaissance Aristotelianism," *History of Science* 11 (1973): 159–73, esp. p. 164. On Telesio's reaction against Aristotle see Allen Debus, *The Chemical Dream of the Renaissance* (Cambridge, England: Heffer, 1968), pp. 8–13, 24–25. On the complex subject of Aristotelianism and its influence in the late Renaissance, see Neal Ward Gilbert, "Renaissance Aristotelianism and its Fate: Some Observations and Problems," in John P. Anton, ed., *Naturalism and Historical Understanding, Essays on the Philosophy of John Herman Randall, Jr.* (Albany, N.Y.: State University of New York Press, 1967), pp. 42–52; Charles Schmitt, *Critical Survey and Bibliography of Studies on Renaissance Aristotelianism, 1958–1969* (Padua: Editrice Antenore, 1971), and C. Schmitt "John Case on Art and Nature," *Annals of Science* 33 (1976): 543–59; John Herman Randall, Jr., *The School of Padua and the Emergence of Modern Science* (Padua: Editrice Antenore, 1961); Edward F. Cranz, *A Bibliography of Aristotle Edi-*

tions, 1501–1600 (Baden Baden: Koerner, 1971).

20. Bernardino Telesio, *De Rerum Natura,* trans. Fallico and Shapiro, quotations in order on pp. 316, 318, 315, 319, 312. On Telesio's life and thought, see F. Fiorentino, *Bernardino Telesio,* 2 vols. (Florence, 1872–1874). Useful commentaries on Telesio include Neil Van Deusen, "The Place of Telesio in the History of Philosophy," *Philosophical Review* 44 (1935): 317–34 and N. Van Deusen, *Telesio: First of the Moderns* (New York, 1930). That the concept of the dialectic was fundamental to Renaissance magical and organic philosophies I owe to discussions with David Kubrin and to his *How Sir Isaac Newton Helped Restore Law 'n' Order to the West* (San Francisco: Kubrin, 1972).

21. Tommaso Campanella, *De Sensu Rerum et Magia,* trans. Fallico and Shapiro, quotations on pp. 341–42, 345. Campanella's philosophy of nature was also developed in his *Philosophia Sensibus Demonstrata* (Naples, 1591). See also T. Campanella, . . . *Epilogo Magno* (Rome: Reale Accademia d'Italia, 1939). Relevant commentaries on Campanella's life and thought include Leon Blanchet, *Campanella* (Paris: Alcan, 1920); Bernardo M. Bonansea, *Tommaso Campanella* (Washington, D.C.: Catholic University of America Press, 1969); Luigi Amabile, *Fra Tommaso Campanella,* 3 vols. (Naples: Morano, 1882).

22. Giordano Bruno, *Spaccio de la Bestia Trionfante* (Paris, 1584). English trans., *The Expulsion of the Triumphant Beast,* trans. and ed. Arthur D. Imerti (New Brunswick, N.J.: Rutgers University Press, 1964), quotations on pp. 75–76, 90, 91. See also Yates, *Bruno,* Chaps. 11–20; Paul Henri Michel, *The Cosmology of Giordano Bruno,* trans. R. E. W. Maddison (Ithaca, N.Y.: Cornell University Press, 1973), pp. 111–112; Edward Gosselin and Lawrence Lerner, "Galileo and the Long Shadow of Bruno," *Archives Internationales d'Histoire des Sciences,* 25 (December 1975), pp. 223–46; E. Gosselin and L. Lerner "Giordano Bruno," *Scientific American* 228 (1973): 84–94; Antoinette Mann Paterson,

The Infinite Worlds of Giordano Bruno (Springfield, Ill.: Thomas, 1970); Dorothy Singer, *Giordano Bruno: His Life and Thought* (New York: Schumann, 1950); Robert Westman, "Magical Reform and Astronomical Reform: The Yates Thesis Reconsidered," in R. Westman and J. E. McGuire, *Hermeticism and the Scientific Revolution* (Los Angeles: University of California Press, 1977), pp. 3–91; Hélène Védrine, *La Conception de la Nature chez Giordano Bruno* (Paris: Vrin, 1967).

23. Irving Louis Horowitz, *The Renaissance Philosophy of Giordano Bruno* (New York: Ross, 1952), p. 29, 31, 37–38, 41, 27.

24. Quotes in Horowitz, p. 27.

25. Giordano Bruno, *De Monade, Numéro et Figura,* written ca. 1590 (Frankfurt, 1614); also in *Opere Latine,* ed. F. Fiorentino and others, 3 vols. (Naples, 1897–1901), vols. 1–2, pp. 323–473. For a discussion see Harold Höffding, *A History of Modern Philosophy,* trans. B. E. Meyer (London: Macmillan, 1915), vol. 1, pp. 130, 138–39.

26. Rosario Villari, "Naples: The Insurrection in Naples of 1585," in Eric Cochrane, ed. *The Late Italian Renaissance, 1525–1630* (New York: Macmillan, 1970), pp. 305–06. The following discussion draws on Villari, pp. 311, 313–14, 323; Leon Blanchet, *Campanella* (New York: Franklin, 1920), pp. 33–40; I. Frith, *Life of Giordano Bruno the Nolan* (London: Trubner, 1887), pp. 12, 13, 28–30; Singer, *Bruno;* Yates, *Bruno.*

27. Paracelsus, "Archidoxis," (first published 1570), in *Sämtliche Werke,* B. Aschner, ed. (Jena: Fischer, 1930), vol. 3, pp. 1–87. See esp. pp. 9, 10, 12, 13. English trans.: Paracelsus, *His Archidoxis Comprised in Ten Books Disclosing the Genuine Way of Making Quintessences, Arcanums, Magisteries, Elixers, etc.* trans. J. H. Oxon (London, 1660). Relevant commentaries on the philosophy of Paracelsus include Kurt Goldammer, *Paracelsus: Natur und Offenbarung* (Hanover: Oppermann, 1953); Walter Pagel, "Paracelsus and the Neoplatonic and Gnostic Tradition," *Ambix* 8 (1960): 125–66; W. Pagel, "Paracelsus: Traditionalism and Medieval Sources," in Lloyd G.

Stevenson and Robert P. Multhauf, eds., *Medicine, Science, and Culture* (Baltimore, Md.: Johns Hopkins University Press, 1968), pp. 57–64; Oswei Temkin, "The Elusiveness of Paracelsus," *Bulletin of the History of Medicine* 26 (1952): 201–17. On Paracelsus' theory of matter and the elements, see W. Pagel, *Paracelsus* (Basel: Karger, 1958), pp. 38, 82–95, and W. Pagel, "The Prime Matter of Paracelsus," *Ambix* 9 (October, 1961): 117–35.

28. Paracelsus, "Philosophia ad Athenienses," Sudhoff, ed., *Sämtliche Werke*, Abt. I, vol. 13, pp. 389–423. The following discussion is based on pp. 390, 410–11, 403, 407, 395–96, 398. Two English translations exist: Paracelsus, *Three Books of Philosophy Written to the Athenians in Philosophy Reformed and Improved*, trans. H. Pinnell (London, 1657), and Paracelsus "The Philosophy Addressed to the Athenians," in Waite, ed., *Hermetic and Alchemical Writings*, vol. 2, pp. 247–81. On the question of authenticity, see Pagel, *Paracelsus*, pp. 89–91, notes 237–38 and W. Pagel, "Recent Paracelsian Studies," *History of Science* 12 (1974): 200–11. Allen Debus treats the text as if it were Paracelsus' own in his book *The Chemical Philosophy: Paracelsian Science and Medicine in the Sixteenth and Seventeenth Centuries*, 2 vols. (New York: Science History Publications, 1977), vol. 1, pp. 55–57.

29. Paracelsus, "Philosophia ad Athenienses," ed. Sudhoff, vol. 13, p. 406; trans. Waite, vol. 2, p. 266.

30. Paracelsus, "Philosophia ad Athenienses," ed. Sudhoff, vol. 13, p. 409; trans. Waite, vol. 2, p. 269.

31. Owen Hannaway, *The Chemists and the Word: The Didactic Origins of Chemistry* (Baltimore, Md.: Johns Hopkins University Press, 1975), pp. 26, 30–31, 32. See also D. P. Walker, "The Astral Body in Renaissance Medicine," *Journal of the Warburg and Courtald Institutes* 21 (1958): 119–33; W. Pagel, *Paracelsus*, pp. 122–23.

32. Paracelsus, "Credo," *Selected Writings*, ed. Jolande Jacobi (Princeton, N.J.: Princeton University Press, 1951), pp. 5, 109.

33. Pagel, *Paracelsus*, pp. 5–30, esp. 17,

23; Jacobi, ed., p. 4; Waite, ed., vol. 1, p. 20.

34. On Paracelsus' influence in chemistry and medicine, see Allen Debus, *The Chemical Philosophy;* A. Debus, *The English Paracelsians* (New York: Watts, 1965) and A. Debus, *The Chemical Dream of the Renaissance* (Cambridge, England: Heffer, 1968). On the political ideas associated with Paracelsianism, see P. M. Rattansi, "Paracelsus and the Puritan Revolution," *Ambix* 11 (1963): 24–32. On popular magic, see Keith Thomas, *Religion and the Decline of Magic* (New York: Scribner's, 1971), pp. 228–29.

35. Johann Baptista Van Helmont, *Works*, trans. J. Chandler (London: Lodowick, Lloyd, 1664), quotations on pp. 32, 30, 29, 32, 35, 29, 30. On Van Helmont's philosophy, see W. Pagel, "The Religious and Philosophical Aspects of Van Helmont's Science and Medicine," *Supplements to the Bulletin of the History of Medicine*, no. 2 (1944), p. 20; W. Pagel "The Spectre of Van Helmont," in Mikulas Teich and Robert Young, eds., *Changing Perspectives: Essays in Honor of Joseph Needham* (London: Heineman, 1973), pp. 100–109; Debus, *Chemical Philosophy*, vol. 2, pp. 295–344.

36. Rattansi, "Paracelsus and the Puritan Revolution," pp. 28, 30; D. Kubrin, *How Sir Isaac Newton Helped;* D. Kubrin, "The Social Origins of Mechanical Thinking in the West: Magic, Dialectics, and Class Struggle in the Background of Newton's Natural Philosophy" (forthcoming); Christopher Hill, *The World Turned Upside Down: Radical Ideas During the English Revolution* (New York: Viking Press, 1972).

37. Gerrard Winstanley, "A Declaration to the Powers of England, and to all the Powers of the World, shewing the cause why the common people of England have begun and give consent to digge up, manure and sowe corn upon George-Hill, in Surrey; by those that have subscribed, and thousands more that gives consent," (written 1649), in *Works*, ed. George H. Sabine (Ithaca, N.Y.: Cornell University Press, 1941), p. 251; Hill, pp. 86, 104.

38. The following quotations are quoted in

Hill, pp. 92–93, 179, 114, 151, 165, 177.
39. Hill, pp. 251–52, 254, 256–57.
40. Keith Thomas, "Women in the Civil War Sects," *Past and Present,* no. 13 (April 1958): 42–62, see pp. 44–45, 47, 50–51.
41. Robert Fludd, *Philosophia Moysaica* (Govdae: Rammazenius, 1638); Eng. trans.: R. Fludd, *Mosaicall Philosophy* (London, 1659), pp. 59–60, 79–80, 91–92. See also, Allen G. Debus, *The English Paracelsians* (New York: Watts, 1965), pp. 105 ff.; A. Debus, *Chemical Philosophy,* vol. 1, pp. 205–93; W. Pagel, "Religious Motives in the Medical Biology of the XVIIth Century," *Bulletin of the Institute of the History of Medicine* [Johns Hopkins University] 3 (February 1935): 270; Kubrin, *How Sir Isaac Newton Helped,* p. 15.

CHAPTER 5:
NATURE AS DISORDER: WOMEN AND WITCHES

1. Thomas Elyot, *The Boke Named the Governour,* ed. Foster Watson (London: Dent, 1907; first published, 1531), pp. 3, 4; William Shakespeare, *Troilus and Cressida,* act 1, sc. 3, lines 75–137; Richard Hooker, *Of the Laws of Ecclesiastical Polity,* ed. W. Speed Hill (Cambridge, Mass.: Harvard University Press, 1977; first published 1594), Bk. I; James Emerson Phillips *The State in Shakespeare's Greek and Roman Plays* (New York: Octagon, 1972), pp. 5, 6, 78. Eustace M. W. Tillyard, *The Elizabethan World Picture* (New York: Random House Vintage, [1959?]), pp. 1–16.
2. Bernard de Fontenelle, *Week's Conversation on the Plurality of Worlds,* trans. William Gardiner (London, 1737; first published 1686), pp. 17–18.
3. Johannes Kepler, "Letter to Herwart von Hohenburg," February 10, 1605, trans. and quoted in Gerald Holton, "Johannes Kepler's Universe: Its Physics and Metaphysics," in *Thematic Origins of Scientific Thought: Kepler to Einstein* (Cambridge, Mass.: Harvard University Press, 1973) p. 72. On Copernicus, Tycho Brahe, and the celestial spheres, see Thomas Kuhn, *The Copernican Revolution* (Chicago: University of Chicago Press, 1970); William H. Donahue, "The Dissolution of the Celestial Spheres: 1595–1650," unpublished doctoral dissertation, University of Cambridge, 1972. On the breakup of the organic cosmos, see Marjorie Hope Nicolson, *The Breaking of the Circle* (New York: Cambridge University Press, 1960). On the breakup of hierarchies in the cosmos and the body, see S. F. Mason, "The Scientific Revolution and the Protestant Reformation," *Annals of Science* 9 (1953): 64–81. On the history of the circulation of the blood, see Gweneth Whitteridge, *William Harvey and the Circulation of the Blood* (New York: Elsevier, 1971) and Tibor Doby, *Discoveries of Blood Circulation* (New York: Schumann, 1963), Chaps. 5–8.
4. John Donne, "An Anatomie of the World: The First Anniversary," in *The Poems of John Donne,* ed. Herbert Grierson (London: Oxford University Press, 1957), lines 56–58. On Godfrey Goodman's *The Fall of Man* (1616) and George Hakewell's reply in his *Apologia* (1627), see Victor Harris, *All Coherence Gone* (Chicago: University of Chicago Press, 1949), and Richard Foster Jones, *Ancients and Moderns* (Berkeley: University of California Press, 1965), Chap. 2.
5. Niccolò Machiavelli, *The Prince* and the *Discourses* (written 1513) (New York: Random House, Modern Library, 1950), Chap. 25, p. 91, 94; Chap. 13, p. 64.
6. On the symbolism of Shakespeare's *The Tempest,* see Leo Marx, *The Machine in the Garden* (New York: Oxford University Press, 1964), pp. 46–57. On picaresque literature see Joseph Meeker, *The Comedy of Survival* (New York: Scribner's, 1972), pp. 92–97. Lazarillo de Tormes, *The Life of Lazarillo de Tormes, His Fortunes and Adversities,* trans. from the 1554 edition (London: Black, 1908); Hans Jakob Christoffel von Grimmelhausen, *Simplicius Simplicissmus* trans. Hellmuth Weissenbaum and Lesley Macdonald (London: Calder, 1964; first published 1667).
7. See Roderick Nash, *Wilderness and the American Mind* (New Haven, Conn.: Yale University Press, 1967), Chap. 1, pp. 8–22;

John Passmore, *Man's Responsibility For Nature* (New York: Scribner's, 1974), Chap. 1.

8. Quotations appear in Richard Ashcraft, "Leviathan Triumphant: Thomas Hobbes and the Politics of Wild Men," in Edward Dudley and Maximillian E. Novak, eds., *The Wild Man Within* (Pittsburgh, Penn.: University of Pittsburgh Press, 1973), pp. 147, 152, notes 66, and Alden T. Vaughan, "English Policy and the Massacre of 1622," *William and Mary Quarterly* 35 (January 1978): 57–84, see pp. 78–79. On images of the American Indian, see also Roy Harvey Pearce, *The Savages of America* (Baltimore, Md.: Johns Hopkins University Press, 1953).

9. Robert Burton, *Anatomy of Melancholy,* 3 vols. (New York: Dutton, 1932; first published 1621), vol. 3, quotations on pp. 56, 55.

10. Keith Thomas, *Religion and the Decline of Magic* (New York: Scribner's, 1971), p. 569, note 2; Louis B. Wright, "The Popular Controversy over Women," in *Middle-Class Culture in Elizabethan England* (Chapel Hill: University of North Carolina Press, 1935), pp. 466, 473, 490; Anon. "Aristotle's Masterpiece," in *The Works of Aristotle the Famous Philosopher,* 1684 (reprint ed., New York: Arno Press, 1974; first published 1684 [?]), pp. 16, 17, 18.

11. Donne, *Poems,* ed. Grierson, p. 211.

12. Wright, p. 470; Hannelore Sachs, *The Renaissance Woman* (New York: McGraw-Hill, 1971), pp. 46, 94, 102; Natalie Z. Davis, "Women on Top: Sexual Inversion and Disorder in Early Modern Europe," in *Society and Culture in Early Modern France* (Stanford, Cal.: Stanford University Press, 1977), pp. 124–51; see p. 136. On the legends and iconography surrounding Aristotle and Phyllis see Jane Campbell Hutchinson, "The Housebook Master and the Folly of the Wise Man," *Art Bulletin* 48, no. 1 (March 1966): 73–78; Maryanne Cline Horowitz, "Aristotle and Woman," *Journal of the History of Biology* 9, no. 2 (Fall 1976): 183–213, see pp. 189–91; George Sarton, "Aristotle and Phyllis," *Isis* 14, no. 1 (May 1930): 8–19.

13. Heinrich Institor and Jacob Sprenger, *The Malleus Maleficarum* (New York: Bloom, 1970; first published 1486). Grillot de Givry, *Witchcraft, Magic, and Alchemy* (New Hyde Park, N.Y.: University Books, 1958), quotation on Ziarnko on p. 76, see also pp. 77–84.

14. Henri Boguet, *An Examen of Witches,* trans., John Rodker (Bungay, England: Clay, 1929; first published ca. 1590), pp. 29, 31–33, 125, 128; Ronald Seth, *Stories of Great Witch Trials* (London: Barker, 1967), p. 98; Institor and Sprenger in *The Malleus Maleficarum,* argued that the supposed sexual lust of women was a direct threat to the sexuality of the male. Witches were accused of "inclining the minds of men to inordinate passion," of "obstructing their generative force," and of "removing the member accommodated to that act" *(Malleus,* p. 47).

"These women satisfy their filthy lusts not only in themselves, but even in the mighty ones of the age, of whatever state and condition; causing by all sorts of witchcraft the death of their souls through the excessive infatuation of carnal love, in such a way that for no shame or persuasion can they desist from such acts.... But indeed such hatred is aroused by witchcraft between those joined in the sacrament of matrimony and such freezing up of the generative forces, that men are unable to perform the necessary action for begetting offspring [*Malleus,* p. 48]."

15. Thomas, p. 569; King James I, *Daemonologie* (reprint ed., New York: Barnes & Noble, 1966; first published 1597), p. 43; Jean Bodin, *De la Demonomanie des Sorciers* (Paris, 1580); Alexander Roberts, *A Treatise of Witchcraft* (1616); Barbara Ehrenreich and Deirdre English, *Witches, Midwives, and Nurses* (Old Westbury, N.Y.: Feminist Press), p. 5; E. William Monter, "The Pedestal and the Stake," in Renate Bridenthal and Claudia Koonz, eds., *Becoming Visible* (Boston: Houghton Mifflin, 1977), pp. 130–32. Monter (pp. 132–33) has reported statistics on the percentages of women and widows tried for witchcraft between 1480 and 1700:

Women Tried for Witchcraft: 1480–1700

Region	Women Tried	Men Tried	% of Total Who Were Women
SW. Germany	1050	238	82%
W. Switzerland	893	237	80%
Venetian Republic	430	119	78%
Castile	324	132	71%
Belgium (Namur)	337	29	92%
England (Essex)	267	23	92%

Widows Tried for Witchcraft: 1480–1700

Region	Widows	Other Women	% of Total Who Were Widows
Neuchâtel	79	100	44%
Basel Bishropic	55	108	34%
Montbéliard	24	34	41%
Toul (Lorraine)	29	24	55%
Essex (England)	49	68	42%

16. Thomas, pp. 520–22. For attempts to determine the reality of witchcraft practices and the world view behind them see Julio Caro Baroja, *The World of Witches*, trans. O. N. V. Glendinning (Chicago: University of Chicago Press, 1965) and Margaret Murray, *The God of the Witches* (Garden City, N.Y.: Doubleday Anchor, 1960).

17. Johann Weyer, *De Praestigiis Daemonum* (first published 1563), excerpts in E. W. Monter, ed., *European Witchcraft* (New York: Wiley, 1969), p. 61. On Weyer, see D. P. Walker, *Spiritual and Demonic Magic from Ficino to Campanella* (London: Kraus Reprints, 1958), p. 152, and Ilza Veith, *Hysteria: The History of a Disease* (Chicago: University of Chicago Press, 1965), pp. 109–11; Lynn Thorndike, *History of Magic and Experimental Science* (New York: Columbia University Press, 1941), vol. 4, p. 515.

18. Johann Weyer, *Histoire Disputes et Discours* (Paris, 1885; first published 1579), vol. 1, pp. 300–303.

19. J. Bodin, "Demonomanie," excerpts in Monter, *European Witchcraft*, p. 55, see also pp. 47–55. See also Thorndike, pp. 525–26. On Renaissance views of the elements, humors, and their relationship to women's temperament, see Charles Carroll Camden, *The Elizabethan Woman* (New York: Elsevier, 1952), pp. 17–35, esp. p. 18, and Ruth Kelso, *Doctrine for the Lady of the Renaissance* (Urbana: University of Illinois Press), Chap. 2, "Women in the Scheme of Things," esp. pp. 16, 22. On Erastus, see Monter, *European Witchcraft*, p. 64.

20. Reginald Scot. *The Discoverie of Witchcraft* (Arundel, England: Centaur Press, 1964; first published 1584), pp. 55, 56, 64, 68, 75, 160, 89, 236.

21. Burton, *Melancholy*, vol. 1, p. 172.

22. Kelso, pp. 1–6, 22.

23. Sherry Ortner, "Is Female to Male as Nature Is to Culture?" in *Women, Culture, and Society*, ed. Michele Z. Rosaldo and Louise Lamphere (Stanford, Cal.: Stanford University Press, 1974), pp. 67–87, see pp. 69, 72, 73. For a critique of the nature-culture dichotomy, see Donna Haraway, "Animal Sociology as a Natural Economy of the Body Politic, Part 1: A Physiology of Dominance," *Signs 4* (Autumn, 1978): 21–36, esp. p. 23.

24. John Knox, *The First Blast of the Trumpet Against the Monstrous Regiment of Women*, in Edward Arber, ed. *The English Scholar's Library* (London, 1878; first published 1558), vol. 2, quotations on pp. 21, 14, 15, 20, 30, 27, 11, 32, 35. On Knox see Jasper Ridley, *John Knox* (Oxford, England: Clarendon Press, 1968), pp. 265–85.

25. On the controversy over female rule, see James E. Phillips, Jr., "The Background of Spenser's Attitude Toward Women Rulers," *Huntington Library Quarterly* 5 (1941–1942), see pp. 5–6; refutations of Knox as cited by Phillips were John Aylmer, *An Harborowe for Faithfull and Trewe Subiectes, Agaynst the Late Blowne Blaste, Concerninge the Government of Wemen* (London, 1559) reprinted in *The Mirror for Magistrates* (London, 1563); John Leslie, *A Defense of the Honour of the Right, Highe, Mightye and Noble Princesse Marie Quene of Scotlande . . .* (London, 1569); David Chambers, *Discours de la Le-*

*gitime Succession des Femmes aux Posses-
sions de Leurs Parens: Et du Gouvernement
des Princesses aux Empires et Royaumes*
(Paris, 1579).
26. John Calvin, "Commentary on I. Corinthians XI, 11," *Commentaries* (Edinburgh: Calvin Translation Society, 1844),
vol. 39, pp. 359, 360, and J. Calvin "Commentary on the Book of Genesis," *Commentaries,* vol. 1, pp. 129, 130, 172. On Calvin's
attitude toward "Mater Ecclesia," see Alexander Ganoczy, *Calvin, Theologian de
l'Eglise et du Ministère* (Paris: Cerf,
1964), pp. 421–23. On Protestant attitudes
toward women, see excerpts from Calvin,
Luther, and Bucer, in Julia O'Faolain and
Laura Martines, eds., *Not in God's Image*
(New York: Harper Torchbooks, 1973), pp.
195–203; Rosemary Radford Ruether, ed.,
*Religion and Sexism: Images of Women in
the Jewish and Christian Traditions,* (New
York: Simon and Schuster, 1974).
27. Natalie Z. Davis, "City Women and
Religious Change in Sixteenth Century
France," in D. McGuigan, ed., *A Sampler
of Women's Studies* (Ann Arbor: University of Michigan Press, 1973), pp. 29, 31;
Holy Bible, 1 Timothy 2:8–15; 1 Corinthians 11:8–9; Ephesians 4.22–28.
28. On French noblewomen, the Huguenot
movement, and Calvin's views on women's
rights, see Nancy L. Roelker, "The Appeal
of Calvinism to French Noblewomen in the
Sixteenth Century," *Journal of Interdisciplinary History* 2 (Spring 1972): 391–413.
On the reformers views on divorce and
women's roles in the Lollard and Anabap-
tist Reformation movements see Sherrin
Marshall Wyntjes, "Women in the Reformation Era," in Bridenthal and Koonz, eds.,
Becoming Visible, pp. 165–91. On women
and the separatists, see Keith Thomas,
"Women and the Civil War Sects," *Past
and Present,* no. 13 (April 1958): 13–46.
Also relevant is N. Z. Davis, "City Women
and Religious Change," p. 21.

CHAPTER 6:
PRODUCTION, REPRODUCTION
AND THE FEMALE

1. On women in Renaissance Italy, see
Joan Kelly-Gadol, "Did Women Have a
Renaissance?" in Renate Bridenthal and
Claudia Koonz, eds., *Becoming Visible:
Women in European History* (Boston:
Houghton Mifflin, 1977), pp. 137–64. The
discussion of England is based on Alice
Clark, *Working Life of Women in the Seventeenth Century,* (New York: Augustus
M. Kelley, 1968; first published 1919), pp.
6–13, 38–39, 42–46, 64–68, 86–92, 103–4,
146–49, 209–11, 216–29; Richard T. Vann,
"Toward a New Lifestyle: Women in Preindustrial Capitalism," in *Becoming Visible,*
pp. 192–216, esp. 200–205; G. E. Fussell
and K. R. Fussell, *The English Countrywoman: A Farmhouse Social History, A.D.
1500–1900* (London: Melrose, 1953);
Christina Hole, *The English Housewife in
the Seventeenth Century* (London: Chatto
& Windus, 1953).
2. Clark, pp. 265–69; Barbara Ehrenreich
and Deirdre English, *Witches, Midwives,
and Nurses* (Old Westbury, N.Y.: Feminist
Press; n.d.); Jean Donnison, *Midwives and
Medical Men* (New York: Shocken, 1977),
pp. 1–41; Irving S. Cutter and Henry R.
Viets, *A Short History of Midwifery*
(Philadelphia: Saunders, 1964), pp. 46–50;
Hilda Smith, "Gynecology and Ideology,"
in Berenice A. Carroll, ed., *Liberating Wom-
en's History* (Urbana: University of Illinois
Press, 1976), p. 110, and H. Smith, "Reason's Disciples: Seventeenth–Century English Feminists," unpublished doctoral dissertation, University of Chicago, 1975, pp.
179–92. On the controversy over the licensing of midwives, see James H. Aveling, *English Midwives, Their History and Prospects*
(reprint ed., London: Elliott, 1967; first
published, 1872), pp. 22–46, and J. Aveling,
*The Chamberlens and the Midwifery For-
ceps* (reprint ed., New York: AMS Press,
1977; first published 1882), pp. 34–48;
Thomas R. Forbes, *The Midwife and the
Witch* (New Haven, Conn.: Yale University
Press, 1966), pp. 112–55; Kate Campbell
Hurd Mead, *A History of Women in Medicine* (Haddam, Conn.: Haddam Press,
1938); Herbert R. Spencer, *The History of
British Midwifery from 1650 to 1800* (London: Bale, 1927), Introduction and Chap. 1,
especially pp. 3–6.
3. Quoted in Cutter and Viets, p. 49.

4. William Harvey, *Exercitationes de Generatione Animalium* (London, 1651), English trans., Harvey, *Works*, trans. Robert Willis (London: Sydenham Society, 1847), pp. 533–34.

5. Aveling, p. 36.

6. For a discussion of the state of gynecological knowledge, see Smith, "Gynecology and Ideology," p. 105; Arthur W. Meyer, *An Analysis of the "De Generatione Animalium" of William Harvey* (Stanford, Cal.: Stanford University Press, 1963), pp. 131, 153.

7. Clark, *Working Life*, pp. 271–73, 284; Jane Sharp, *The Midwives Book, or the Whole Art of Midwifery* (London, 1671); [Elizabeth Cellier], *A Scheme for the Foundation of a Royal Hospital . . . for the Maintenance of a Corporation of Skilfull Midwives* (n.p., 1687); [Elizabeth Cellier], *To Dr. ——— an Answer to His Queries, Concerning the College of Midwives* (London, 1688); Nicholas Culpeper, *A Directory for Midwives* (London, 1651); Donnison, *Midwives and Medical Men*, see pp. 17–19 on Sharp and Cellier; pp. 21–41 on the decline of the midwife; pp. 42–61 on the rise of male midwives.

8. On Harvey's vitalistic conceptual framework, see Walter Pagel, *William Harvey's Biological Ideas* (Basel: Karger, 1967); W. Pagel, *New Light on William Harvey* (Basel: Karger, 1978). On the mechanical metaphor in physiology and medicine, see René Descartes, *Treatise of Man* (written 1632–33), trans. Thomas Steele Hall (Cambridge, Mass.: Harvard University Press, 1972); H. Boerhaave, "Commentariolus," in G. A. Lindeboom, *Herman Boerhaave* (London: Methuen, 1968), and H. Boerhaave, *Aphorisms: Concerning the Knowledge and Cure of Diseases*, trans. J. Delacoste (London: Cowse and Innys, 1715; first published 1709); Julien LaMettrie, *L'Homme Machine* (Paris: Editions Bossard, 1921; first published 1748). On the healing power of nature, see Max Neuburger, *Wesen der Naturheilkraft* (Stuttgart, 1927).

9. Harvey, Willis trans., p. 217.

10. For discussions of Harvey's scientific ideas, see Elizabeth B. Gasking, *Investigations into Generation, 1651–1828* (Baltimore, Md.: Johns Hopkins University Press, 1966), Chap. 2; Gweneth Whitteridge, *William Harvey and the Circulation of the Blood* (New York: Elsevier, 1971); John Aubrey, *Brief Lives,* ed. Oliver Lawson Dick (Ann Arbor: University of Michigan Press, 1957), pp. 128, 130. On Harvey's political sympathies, see Christopher Hill, "William Harvey and the Idea of Monarchy," in Charles Webster, ed., *The Intellectual Revolution of the Seventeenth Century: Past and Present Series* (London: Routledge & Kegan Paul, 1974), pp. 160–81 especially p. 177; Gweneth Whitteridge, "William Harvey: A Royalist and No Parliamentarian," in Webster, ed., *Intellectual Revolution,* pp. 182–88; and Christopher Hill, "William Harvey (No Parliamentarian, No Heretic) and the Idea of Monarchy," in Webster, ed., *Intellectual Revolution,* pp. 189–96. Both Hill and Whitteridge agree that Harvey's political sympathies were royalist, at least until 1649 when Charles was beheaded. Harvey's *Exercitatio Anatomica de Motu Cordis et Sanguinis in Animalimus,* written 1628 (Venice, 1635) was dedicated to Charles I, who was "the foundation of his kingdoms, and the sun of his microcosm, the heart of his commonwealth, from whence all power and mercy proceeds."

11. Louis Wright, *Middle-Class Culture in Elizabethan England* (Chapel Hill: University of North Carolina Press, 1935), pp. 503–7; examples cited include, Thomas Heywood (d. 1641), *A Curtaine Lecture* (London, 1637); John Taylor (1580–1653) *Divers Crab-tree Lectures* (London?, 1639) and *A Juniper Lecture* (London, 1639); Richard Brathwaite, (1588?–1673) *Ar't Asleepe Husband?* (London, 1640); Mary Tattlewell and Ioane Hit-him-home, Spinsters, *The Woman's Sharp Revenge* (1640); and Henry Neville's (1620–1694) *An Exact Diurnal of the Ladies Parliament* ([London], 1647), and *The City-Dames Petition* ([London], 1647). On the role of women in the religious movements during the period of the English Civil War, see Keith Thomas, "Women in the Civil War Sects," *Past and Present,* no. 13 (April 1958): 45–46; on

women and science in the mid-seventeenth century, see Gerald Dennis Meyer, *The Scientific Lady in England, 1650–1760* (Berkeley and Los Angeles: University of California Press, 1955), p. 16; H. Smith, "Feminism in Seventeenth-Century England."

12. Needham, *History of Embryology* (Cambridge, England: Cambridge University Press, 1934), pp. 69, 95, 92; Galen, *Oeuvres,* ed. Charles Daremberg (Paris, 1854–1856), vol. 2, pp. 99, 103; Howard B. Adelmann, *Marcello Malpighi and the Evolution of Embryology* (Ithaca, N.Y.: Cornell University Press, 1966), vol. 2, p. 755. Followers of Galen who published in the sixteenth and seventeenth centuries, as discussed by Adelmann, included Alessandro Benedetti (ca. 1450–1512), *Anatomice: sive, Historia Corporis Humani* (Paris, 1514), Jacopo Berengario (d. 1550), *Carpi Commentaria cũ amplissimis additionibus super Anatomis Mũndini* (Bononiae, 1521), Andre Du Laurens (1558–1609), *Opera Anatomica in Quinque Libros Divisa* (Lugduni, 1593), Girolamo Capivaccio (1523–1589), *Opera Omnia* (Frankfurt, 1603), Jean Fernel (1497–1558), *Pathologie, Libri Septem* (Paris, 1638). On Aristotelian theories of the semen and the superiority of the male in generation, see F. J. Cole, *Early Theories of Sexual Generation* (Oxford, England: Clarendon Press, 1930), p. 38; Jacques Roger, *Les Sciences de la Vie dans la Pensêe francaise du XVIII Siecle* (Paris: Colin, 1963), p. 27, 28, 53–68; Aristotelians included Jean Riolan the Elder, *Ad Librum Fernelii de Procreatione Hominis Commentarius* (Paris, 1578); Severin Pineau, *De Integritatis et Corruptionis Virginum Notis* (Leyden, 1642); Ambrois Paré, *Les Oeuvres* (Paris, 1628), Luigi Bonacioli, *De Foetus Formatione* (1641); Adrian Spiegel, *De Formatio Foeto* (Frankfurt, 1631). Those who supported the more rapid development of the male embryo were Giovanni Costeo, *De Humani Conceptus Formatione* (Papiae, 1604); Fortunio Liceti (1577–1657), *De Monstrorum Caussis Natura et Differentiis Libri Duo* (Padua, 1616); Victor Cardelinus, *De Origine Foetus Libri Duo* (Vincentiae, 1628); Jacob Reuff (1500–1558),

De Conceptu et Generatione Hominis (Frankfurt, 1580), illustrated the semen and menstrual blood in the womb.

13. Girolamo Fabrizio d'Acquapendente, *De Formatione Ovi et Pulli Tractatus Accuratissimus* (Padua, 1621); English translation: Hieronymus Fabricius, *Embryological Treatises, The Formation of the Egg and the Chick,* trans. Howard B. Adelmann, 2 vols. (rev. ed., Ithaca, N.Y.: Cornell University Press, 1967; first published 1942):

"Hence the semen is not present in the egg, but what is still more to the point it is not even possible for it to be present, because the semen of the cock is incapable of reaching the egg ... the place it [the semen] reaches when it is introduced lies near the podex where the egg, already completed, has a shell or at least is covered by a dense tunic which the cock's semen cannot penetrate [vol. 1, p. 180]."

"From this discussion but especially too from what Aristotle writes, we gather, in the first place, that by its power, the semen of the male gives a certain quality to the material and the nutriment contained in the female; that the egg is vivified only by the force resident in the semen; and that Aristotle thinks that eggs are fertilized by the semen of the cock, which possesses an extraordinary fecundative power [vol. 1, p. 192]."

14. In his recent book, *New Light on William Harvey,* Walter Pagel postulates a cooperative interaction between male and female and a monistic ontology in Harvey's theory of generation (p. 27):

He granted the ovum a position of much higher dignity than it had enjoyed in Aristotle's view. He thereby shows himself more consequently faithful to the monist principle than its inaugurator Aristotle. For in Harvey's view a dualist action of something active (male) on something passive (female) was here replaced by the cooperation of two individual units of working matter. Somewhat related to this is Harvey's contention that the semen cooperated with the ovum at a distance.

In his earlier book, *William Harvey's Biological Ideas,* Pagel wrote (p. 44): "The male cannot therefore claim superiority as the source of semen. Instead, each one, male as well as female, is operative as a parent. The female part is just as much *causa generationis* as the male since it stimulates the latter in various ways." See also Hans Fischer, cited by Pagel, *ibid.*, "There was no amphimixis, no coming together of the male and female components, but an almost parthenogenetic generation in which the male has a stimulating effect, but no more." For an alternative to my interpretation see also Gasking, Chap. 2, pp. 16–36.

15. Harvey, *Works.* Quotations on pp. 278, 290; see also pp. 281, 288, 368.
16. *Ibid.,* p. 362.
17. *Ibid.,* pp. 290, 321, 368.
18. *Ibid.,* pp. 298, 299, see also p. 294; Adelmann, *Malpighi,* vol. 1, p. 40.
19. *Ibid.,* p. 478.
20. *Ibid.,* p. 277. See also pp. 295, 317, 357.
21. *Ibid.,* p. 322. See also pp. 315, 321, 372. For a discussion of contagion, see Pagel, *Harvey's Biological Ideas,* p. 25, and *New Light,* pp. 43–44. See also Needham, pp. 125–26.
22. Harvey, *Works,* pp. 575, 315.
23. *Ibid.,* p. 576.
24. Paracelsus, *The Hermetic and Alchemical Writings of Paracelsus,* ed. Arthur Edward Waite (London: Elliott, 1894), vol. 1, pp. 121–22. Paracelsus also wrote (p. 123) that "a menstrous woman, ... also carries poison in her eyes in such a way that from her very glance the mirror becomes spotted and stained.... By her breath as well as by her look, she affects many objects, rendering them corrupted and weak and also by her touch"; see also Reginald Scot, *The Discoverie of Witchcraft,* (Arundel, England: Centaur Press, 1964; first published 1584), p. 227.
25. Harvey, *Works,* pp. 577–78.
26. *Ibid.,* p. 578.
27. Adelmann, *Malpighi,* vol. 2, pp. 793, 759, 761, 778, 796, 766–77. Quotes on pp. 778, 796. Descartes, *Oeuvres,* ed. Charles Adam and Paul Tannery (Paris: Cerf, 1909), vol. 2, pp. 274–76. Emilio Parisano (1567–1643), *Nobilium Exercitationum Libri Duodecim de Subtilitate* (Venetiis,

1621); Nathaniel Highmore, *History of Generation* (London, 1651), p. 108; Anthony Everard, *Novus et Genuinus Hominus Brutique Animalis Exortus* (1661). On the microscope and theories of preformation see Peter J. Bowler, "Preformation and Pre-Existence in the Seventeenth Century," *Journal of the History of Biology* 4 (Fall 1971), p. 224. For a discussion of the differences between the Aristotelian and atomistic approaches to generation, see Pagel, *New Light,* pp. 23–27. On spermists and ovists see Needham, pp. 205–11.
28. Stephanie A. Shields, "Functionalism, Darwinism, and the Psychology of Women: A Study in Social Myth," *American Psychologist* 30 (July 1975): 739–54; Rosalind Rosenberg, "In Search of Women's Nature, 1850–1920," *Feminist Studies* 3 (1975): 14–54; Diana Long Hall, "Biology, Sex Hormones, and Sexism," *Philosophical Forum* 5 (1973–4): 81–86; Elizabeth Fee, "The Sexual Politics of Victorian Social Anthropology," in Mary Hartman and Lois Banner, eds., *Clio's Consciousness Raised* (New York: Harper Colophon, 1974), pp. 86–102; Estelle R. Ramey, "Sex Hormones and Executive Ability," *Annals of the New York Academy of Sciences* 208 (March 1973): 237–45.

CHAPTER 7:
DOMINION OVER NATURE

1. Treatments of Francis Bacon's contributions to science include Paolo Rossi, *Francis Bacon: From Magic To Science* (London: Routledge & Kegan Paul, 1968); Lisa Jardine, *Francis Bacon: Discovery and the Art of Discourse* (Cambridge, England: Cambridge University Press, 1974); Benjamin Farrington, *Francis Bacon: Philosopher of Industrial Science* (New York: Schumann, 1949); Margery Purver, *The Royal Society: Concept and Creation* (London: Routledge & Kegan Paul, 1967).
2. Farrington, *Francis Bacon,* p. 82. James Spedding, *The Letters and the Life of Francis Bacon,* 7 vols. (London: Longmans, Green, Reader, and Dyer, 1869), vol. 3, pp. 56–66.
3. Louis Wright, "The Popular Controversy Over Women," in *Middle-Class Culture*

in *Elizabethan England* (Chapel Hill: University of North Carolina Press, 1935), Chap. 13, pp. 493, 494; Anon., *Hic Mulier, or The Man-Woman, Being a Medicine to Cure the Coltish Disease of the Staggers in the Masculine-Feminines of Our Times* (London, 1620); Lucy Ingram Morgan, "The Renaissance Lady in England," unpublished doctoral dissertation, University of California at Berkeley, 1932.

4. "Letter of John Chamberlain," Jan. 25, 1620. Quoted in Wright, p. 493.

5. Thomas Overbury, *Miscellaneous Works*, ed. E. F. Rimbault (London: Smith, 1856), quotation on p. xxxvii; see also Spedding, *Letters and Life of Francis Bacon*, vol. 5, pp. 296–305, esp. 297, 298 n.; Violet A. Wilson, *Society Women of Shakespeare's Time* (London: Lane, Bodley Head, 1924), p. 205; Wright, p. 491.

6. James I, *Daemonologie* (New York: Barnes & Noble, 1966; first published 1597); Keith Thomas, *Religion and the Decline of Magic* (New York: Scribner's, 1971), p. 520; Wallace Notestein, *A History of Witchcraft in England from 1558 to 1718* (New York: Apollo Books, 1968), p. 101; Ronald Seth, *Stories of Great Witch Trials* (London: Baker, 1967), p. 83.

7. Bacon, "De Dignitate et Augmentis Scientiarum," (written 1623), *Works,* ed. James Spedding, Robert Leslie Ellis, Douglas Devon Heath, 14 vols. (London: Longmans Green, 1870), vol. 4, p. 296. The ensuing discussion was stimulated by William Leiss's, *The Domination of Nature* (New York: Braziller, 1972), Chap. 3, pp. 45–71.

8. Bacon, "Preparative Towards a Natural and Experimental History," *Works,* vol. 4, p. 263. Italics added.

9. Bacon, "De Dignatate," *Works,* vol. 4, p. 298. Italics added.

10. Bacon, "The Great Instauration" (written 1620), *Works,* vol. 4, p. 20; "The Masculine Birth of Time," ed. and trans. Benjamin Farrington, in *The Philosophy of Francis Bacon* (Liverpool, England: Liverpool University Press, 1964), p. 62; "De Dignitate," *Works,* vol. 4, pp. 287, 294.

11. Quoted in Moody E. Prior, "Bacon's Man of Science," in Leonard M. Marsak, ed., *The Rise Of Modern Science in Rela-tion to Society* (London: Collier-Macmillan, 1964), p. 45.

12. Rossi, p. 21; Leiss, p. 56; Bacon, *Works,* vol. 4, p. 294; Henry Cornelius Agrippa, *De Occulta Philosophia Libri Tres* (Antwerp, 1531): "No one has such powers but he who has cohabited with the elements, vanquished nature, mounted higher than the heavens, elevating himself above the angels to the archetype itself, with whom he then becomes cooperator and can do all things," as quoted in Frances A. Yates, *Giordano Bruno and the Hermetic Tradition* (New York: Vintage Books, 1964), p. 136.

13. Bacon, "Novum Organum," Part 2, in *Works,* vol. 4, p. 247; "Valerius Terminus," *Works,* vol. 3, pp. 217, 219; "The Masculine Birth of Time," trans. Farrington, p. 62.

14. Bacon, "The Masculine Birth of Time," and "The Refutation of Philosophies," trans. Farrington, pp. 62, 129, 130.

15. Bacon, "De Augmentis," *Works,* vol. 4, p. 294; see also Bacon, Aphorisms," *Works,* vol. 4.

16. "De Augmentis," *Works,* vol. 4, pp. 320, 325; Plato, "The Timaeus," in *The Dialogues of Plato,* trans. B. Jowett (New York: Random House, 1937), vol. 2, p. 17; Bacon, "Parasceve," *Works,* vol. 4, p. 257.

17. Bacon, "De Augmentis," *Works,* vol. 4, pp. 343, 287, 343, 393.

18. Bacon, "Novum Organum," *Works,* vol. 4, p. 246; "The Great Instauration," *Works,* vol. 4, p. 29; "Novum Organum," Part 2, *Works,* vol. 4, p. 247.

19. Alain de Lille, *De Planctu Naturae,* in T. Wright, ed., *The Anglo-Latin Satirical Poets and Epigrammatists* (Wiesbaden: Kraus Reprint, 1964), vol. 2, pp. 441, 467; Thomas Kuhn, "Mathematical vs. Experimental Traditions in the Development of Physical Science," *Journal of Interdisciplinary History* 7, no. 1 (Summer 1976): 1–31, see p. 13. On the Accademia del Cimentio's experiments see Martha Ornstein [Bronfenbrenner], *The Role of Scientific Societies in the Seventeenth Century* (reprint ed., New York: Arno Press, 1975), p. 86.

20. Bacon, "Thoughts and Conclusions on

the Interpretation of Nature or A Science of Productive Works," trans. Farrington, *The Philosophy of Francis Bacon,* pp. 96, 93, 99.

21. Bacon, "De Augmentis," *Works,* vol. 4, pp. 294; "Parasceve," *Works,* vol. 4, pp. 257; "Plan of the Work," vol. 4, pp. 32; "Novum Organum," *Works,* vol. 4, pp. 114, 115.

22. Peter Laslett, *The World We Have Lost: England Before the Industrial Age,* 2nd ed. (New York: Scribner's, 1965), Chap. 1, pp. 2, 4, 5, 7–9. Lawrence Stone, *The Family, Sex, and Marriage in England: 1500–1800* (New York: Harper & Row, 1977); L. Stone, "The Rise of the Nuclear Family in Early Modern England: The Patriarchal Stage," in Charles E. Rosenberg, ed., *The Family in History* (Philadelphia: University of Pennsylvania Press, 1975), pp. 13–57. Also useful as a perspective on lower-class women, although dealing primarily with the nineteenth century, is Joan W. Scott and Louise Tilly, "Women's Work and the Family in Nineteenth-Century Europe," in C. E. Rosenberg, ed., *The Family in History,* pp. 145–178.

23. James I, *The Political Works of King James I,* ed. Charles H. McIlwain (Cambridge, Mass.: Harvard University Press, 1918; first published 1616), p. 307; quoted in Stone, "Rise of the Nuclear Family," p. 54.

24. Bacon, "The New Atlantis," *Works,* vol. 3, pp. 129–66. Quotes in order on pp. 147, 148–49, 150, 148.

25. Charles G. Sellers and others, *As It Happened: A History of the United States* (New York: McGraw–Hill, 1975), pp. 10, 17; Great Britain Laws, Statutes, etc., *The Statutes of the Realm, 1101–1713,* 10 vols. (London, 1810–28), vol. 1. pp. 380–82.

26. Quoted in Sellers, p. 7. See Phillip Stubbs, *Anatomy of the Abuses in England,* ed. F. J. Furnivall (London, 1879), pp. 27–34.

27. Bacon, "The New Atlantis," *Works,* vol. 3, quotes on pp. 131, 135, 155.

28. Tommaso Campanella, "The City of the Sun," in Henry Morley, ed. *Ideal Commonwealths* (New York: Colonial Press, 1901), pp. 154–55; Johann Valentin Andreä, *Christianopolis,* trans. Felix Emil Held (Urbana: University of Illinois Press, 1914), p. 171.

29. William Ashley points out that Bacon opposed the enclosures by English landowners, who turned common lands into sheep-grazing pastures and thereby created large groups of homeless cottagers and vagrant poor. But Bacon did not act out of sympathy for the poor. His rationale was to support the state's need for stable bands of free, nonservile foot soldiers to avoid the use of foreign mercenaries who had no allegiance to England. (See William Ashley, *The Economic Organization of England* [New York: Longmans Green, 1926], pp. 113–114.) The problem of poverty in England, which reached a crisis in the late sixteenth century, was discussed in a parliamentary debate in 1597, in which Bacon participated, arguing that poverty and idleness impoverished the realm. (W. K. Jordan, *Philanthropy in England,* 1480–1660 [New York: Russell Sage Foundation, 1959], p. 94.) The results were a series of Elizabethan statutes legislating punishment and relief: 39 Eliz. C. 2–5, continued by Eliz. C. 9 and I Jac. I, C. 25. Among other provisions, these statutes defined "rogues," "vagabonds," and "sturdy beggars" as "persons able in body, using loitering, refusing to work for common wages, not having otherwise to maintain themselves." On apprehension, they were to "be stripped naked from the middle upwards, and openly whipped until [the] body be bloody." Then they were to be sent back to the parish where they were born to "labor as a true subject ought to do." The poor were to be provided for by the erection of hospitals and working houses to care for the crippled, deaf and dumb, poor children, and so on, through the taxation of "every inhabitant, parson, vicar . . . occupier of lands, houses . . . [and] propriations of tithes, coal mines, or saleable underwoods." (See Richard Burn, *The History of the Poor Laws* [London, 1764], pp. 40–43, 91–92.) These laws were based on the assumption that anyone

who was able to work could find employment, a presupposition that was false at the time.

30. The following discussion of the effects of early capitalist organization in the textile industry draws on Laslett, pp. 15–17; Eugene F. Rice, Jr., *The Foundations of Early Modern Europe, 1460–1559* (New York: Norton, 1970), pp. 52–53; E. Lipson, *The Economic History of England* (London: Black, 1943), vol. 2, pp. 9, 11–15, 17, 31; E. Lipson, *A Short History of Wool and Its Manufacture (Mainly in England)* (Cambridge, Mass.: Harvard University Press, 1953); Henry Kamen, *The Iron Century: Social Change in Europe, 1550–1660* (London: Weidenfeld and Nicolson, 1971), p. 114. For late sixteenth-century laws regulating the textile industry, see R. H. Tawney and Eileen Power, *Tudor Economic Documents* (London: Longmans Green, 1924), vol. 1, pp. 169–228; see also Thomas Deloney, "The Pleasant History of John Winchcomb, in his Younger Years Called Jack of Newburie," in F. O. Mann, ed., *The Works of Thomas Deloney* (Oxford, 1912).

31. Rice, pp. 53–54.

32. The discussion of capitalist organization in the mining industry draws on Lipson, *Economic History*, vol. 2, pp. 114, 162; John U. Nef, "Coal Mining and Utilization," in Charles Singer and others, eds., *A History of Technology* (New York: Oxford University Press, 1957), vol. 3, p. 77; J. U. Nef, *The Rise of the British Coal Industry* (London: Routledge & Kegan Paul, 1932), vol. 1. On laws regulating the mining industry see Tawney and Power, pp. 229–92.

33. Henry Hamilton, *The English Brass and Copper Industries to 1880* (London: Cass, 1967; first published 1926), pp. 70, 75, 76.

34. For example, see Deloney. On declining upward mobility for journeymen see J. U. Nef, *Industry and Government in France and England, 1540–1640* (Philadelphia: American Philosophical Society, 1940) pp. 17–19; Christopher R. Friedrichs, "Capitalism, Mobility, and Class Formation in the Early Modern German City," *Past and Present*, no. 69 (Nov. 1975): 24–49; E. F.

Rice, *Foundations of Early Modern Europe*, pp. 48–49; Natalie Z. Davis, *Society and Culture in Early Modern France* (Stanford, Cal.: Stanford University Press, 1975), pp. 4–15. See also Lipson, *Economic History*, vol. 2, pp. 35; Hamilton, p. 71.

35. Walter E. Houghton, Jr., "The History of Trades: Its Relation to Seventeenth Century Thought," in Philip P. Wiener and Aaron Noland, eds., *Roots of Scientific Thought* (New York: Basic Books, 1953), pp. 355–60. Bacon, *Works*, vol. 4, pp. 74–75.

36. Edgar Zilsel, "The Genesis of the Concept of Scientific Progress," in *Roots of Scientific Thought*, pp. 251–55, quotation on p. 275. A. C. Keller, "Zilsel, the Artisans, and the Idea of Progress in the Renaissance," in *Roots of Scientific Thought*, pp. 281–86. Paolo Rossi, *Philosophy, Technology, and the Arts in the Early Modern Era* (New York: Harper & Row, 1970), Chap. 1, 2, pp. 1–99. J. B. Bury, *The Idea of Progress* (New York: Dover, 1955). For a criticism of the scholar-craftsman theory and its relation to technological progress, see A. Rupert Hall, "The Scholar and the Craftsman in the Scientific Revolution," in Marshall Clagett, ed., *Critical Problems in the History of Science* (Madison: The University of Wisconsin Press, 1959), pp. 3–23. As cited by Zilsel, Keller, and Rossi, sixteenth-century treatises advocating cooperative sharing of knowledge for human progress, included: Master craftsmen: Kaspar Brunner, "Gründlicher Bericht des Buchsengiessens" (1547); *Archive für die Geschichte der Naturwissenschaften und Technik* 7 (1916), p. 171; Robert Norman, *The Newe Attractive* (London, 1581), preface and dedication; Peter Apianus, *Quadraus Astronomicus* (Ingolstadt, 1932), dedication; Ambroise Paré (1509–1590), *Oeuvres*, ed. J. F. Malgaigne (Paris, 1840; first published 1575), introduction; Gerard Mercator (1512–1594), *Atlas* (1595), 4th ed. (Antwerp, 1630), introduction to maps of France. Military engineers: William Bourne (d. 1583), *Inventions or Devices* (London, 1587), Preface to the Reader; Niccolo Tartaglia (1499–1557), *Questi et*

Inventioni (Venice, 1546), dedication; S. Stevinus (1548–1620), *Hypomnemata Mathematica* (written 1605–1608) in *Oeuvres Mathematica,* ed. Girard (Leyden, 1634), vol. 2, pp. 111 ff.; Bonaiuto Lorini, *Delle Fortificazioni* (Venice, 1597). Humanists and academics: Abraham Ortelius (1527–1598), *Theatrum Orbis Terrarum* (Antwerp, 1570); François Rabelais (1490–1553), *Gargantua* and *Pantagruel* (Paris 1533), last chapter; Jean Bodin, *Methodus and Facilem Historiarum Cognitionem* (Paris, 1566), Chap. 7; Loys Leroy, *Les Politiques d' Aristotle* (Paris, 1568), argument to Book II. Political theorist J. Schaar has pointed out that the full humanization of life logically implies a human environment filled with humans at the expense of nature.

37. Bacon, "The New Atlantis," *Works* vol. 3, subsequent quotations on pp. 165, 154, 155. On politics and science in "The New Atlantis," see Joseph Haberer, *Politics and the Community of Science* (New York: Van Nostrand Reinhold, 1969), pp. 46, 47; see M. E. Prior, "Bacon's Man of Science," in L. M. Marsak, ed., pp. 41–53; P. Rossi, *Francis Bacon,* Chap. 1. On critiques of technology, see John McDermott, "Technology: The Opiate of the Intellectuals," *New York Review of Books,* July 31, 1969; Theodore Roszak, *Where the Wasteland Ends* (Garden City, N.Y.: Doubleday, 1963), Chap. 2.

38. Bacon, "The New Atlantis," *Works,* vol. 3, quotations on pp. 157, 158, 159.

39. G. della Porta, *Natural Magic,* ed. D. J. Price (facsimile of 1658 ed., New York: Basic Books, 1957; first published 1558), pp. 27, 29, 31–40.

40. Bacon, *Works,* vol. 3, quotations on pp. 159, 158. Cf. Della Porta, pp. 59, 61, 62.

41. Bacon, *Works,* vol. 3, p. 158. Cf. Della Porta, pp. 61–62, 73, 74–75, 81, 95–99.

42. Bacon, *Works,* vol. 3, p. 159.

43. Henry Cornelius Agrippa, *The Vanity of Arts and Sciences* (London, 1694; first published 1530), pp. 252–53.

44. Bacon, *Works,* vol. 3, pp. 157, 158.

45. *Ibid.,* p. 156.

46. Lewis Roberts, *The Treasure of Traffike, Or A Discourse of Foreign Trade* (London, 1641). Quoted in Charles Webster, *The Great Instauration: Science, Medicine and Reform, 1626–1660* (London: Duckworth, 1975), p. 356; for more details on the Baconian program in mid-seventeenth-century England, the reader is referred to Webster's thorough, scholarly study.

47. Quoted in Walter E. Houghton, "The History of Trades: Its Relation to Seventeenth Century Thought," in *Roots of Scientific Thought* p. 361.

48. [Attributed to Samuel Hartlib], *A Description of the Famous Kingdome of Macaria,* intro. by Richard H. Dillon (facsimile ed., Sausalito, Cal.: Elan, 1961; first published 1641), quotations on pp. 4, 5, 8, 5, 2. On Gabriel Plattes as the probable author, see Charles Webster, "The Authorship and Significance of Macaria," in C. Webster, ed., *The Intellectual Revolution of the Seventeenth Century* (London: Routledge & Kegan Paul), pp. 369–85. See also Houghton, p. 361, and Webster, *Great Instauration,* pp. 87, 368–69.

49. Thomas Sprat, *History of the Royal Society,* 4th ed. (London, 1734; first published 1667), pp. 129–30, 190; Houghton, pp. 370, 377. On the interest of the Royal Society in practical application and technology, see Robert K. Merton, *Science, Technology, and Society in Seventeenth Century England* (New York: Fertig, 1970; first published 1938).

50. René Descartes, "Discourse on Method," Part 4, in E. S. Haldane ånd G. R. T. Ross, eds., *Philosophical Works of Descartes* (New York: Dover, 1955), vol. 1, p. 119.

51. Joseph Glanvill, *Plus Ultra* (Gainesville, Fla.: Scholar's Facsimile Reprints, 1958; first published 1668), quotations on pp. 9, 87, 13, 56, 104, 10.

52. Robert Boyle, *Works,* ed. Thomas Birch (Hildesheim, W. Germany: Olms, 1965; first published 1772), vol. 1, p. 310. On Boyle's mechanical philosophy, see Marie Boas, "The Establishment of the Mechanical Philosophy," *Osiris* 10 (1952): 412–541; Frederick O'Toole, "Qualities and Powers in the Corpuscular Philosophy of Robert Boyle," *Journal of the History of*

Philosophy 12 (July 1974): 295–316; Margaret J. Osler, "John Locke and Some Philosophical Problems in the Science of Boyle and Newton," unpublished doctoral dissertation, Indiana University, 1968; Robert Kargon, "Walter Charleton, Robert Boyle, and the Acceptance of Epicurean Atomism in England," *Isis* 55 (1964): 184–92.

CHAPTER 8:
THE MECHANICAL ORDER

1. Richard Popkin, *The History of Scepticism from Erasmus to Descartes* (New York: Humanities Press, 1964), pp. 1–16, 18; Charles B. Schmitt, "The Recovery and Assimilation of Ancient Scepticism in the Renaissance," *Revista Critica di Storia della Filosofia* 4 (1972): 365–384. On the Reformation and the rise of capitalism, the classic treatments are Max Weber, *The Protestant Ethic and the Spirit of Capitalism* (New York: Scribner's, 1958) and R. H. Tawney, *Religion and the Rise of Capitalism* (New York: Harcourt Brace Jovanovich, 1926). A large scholarship supporting, refining, and refuting this thesis has grown out of Weber's initial essay. On the challenge of Calvinism to the hierarchical view of the cosmos, see S. F. Mason, "The Scientific Revolution and the Protestant Reformation," *Annals of Science* 9 (1953): 64–81. On the crisis in community in the Protestant sects, see Sheldon Wolin, *Politics and Vision* (Boston: Little, Brown, 1960). On the problem of social upheaval, see Theodore K. Rabb, *The Struggle for Stability in Early Modern Europe* (New York: Oxford University Press, 1975).

2. Robert Lenoble, *Mersenne ou la Naissance du Mécanisme* (Paris: Vrin, 1943), pp. 43, 170; Margaret Osler, "Descartes and Charleton on Nature and God," *Journal of the History of Ideas* 40 (Sept. 1979): 445–56; M. Osler, "Gassendi, Descartes, and the Foundations of the Mechanical Philosophy," forthcoming. On the history of atomism the pioneering study is Kurd Lasswitz, *Geschichte der Atomistik vom Mittelalter bis Newton,* 2 vols. (Hamburg: Voss, 1890). Also important are Marie Boas, "The Establishment of the Mechanical Phi-

losophy," *Osiris* 10 (1952): 412–541; Robert Hugh Kargon, *Atomism in England from Hariot to Newton* (Oxford, England: Clarendon Press, 1966); Charles Harrison, "Bacon, Hobbes, Boyle and the Ancient Atomists," *Harvard Studies and Notes in Philology and Literature* 15 (1933): 191–218; C. Harrison, "The Ancient Atomists and English Literature of the Seventeenth Century," *Harvard Studies in Classical Philology* 45 (1934): 1–79; 13–27; Grant McColley, "Nicholas Hill and the Philosophia Epicurea," *Annals of Science* 4 (1939): 390–405; J. C. Gregory, *A Short History of Atomism: Democritus to Bohr* (London: Black, 1931); J. R. Partington, "The Origins of the Atomic Theory," *Annals of Science* 4 (1939): 245–282.

3. Richard Popkin, "Father Mersenne's War Against Pyrrhonism," *The Modern Schoolman* 34 (Jan. 1957): 61–77; R. Popkin, *History of Scepticism,* pp. 139–149; M. Mersenne, *Correspondance,* vol. 1: *1617–1627,* (Paris: Presses Universitaires de France, 1945), p. 51.

4. Lenoble, pp. 38, 153, 154, 155–161; See M. Mersenne, *Questiones in Genesim* (Paris, 1623); M. Mersenne, *L'Impieté des Deistes. Athées et Libertins de ce Temps* (Paris, 1624).

5. Frances Yates, *The Rosicrucian Enlightenment* (London and Boston: Routledge & Kegan Paul, 1972), pp. 112–13, 116; Adrien Baillet, *Vie de Monsieur Descartes,* abridged edition (Paris: Le Table Ronde, 1945; first published 1692), p. 51; Anon., *Fama Fraternalis,* (1614), reprinted in Yates, appendix, pp. 243, 251.

6. Quoted in Yates, p. 41; Anon., *Confessio* (1615), reprinted in Yates, pp. 251, 256.

7. François Garasse, *La Doctrine Curieuse des Beaux Espirits de ce Temps, ou Prétendus Tels, Contenant Plusieurs Maximes Pernicieuses à la Religion, à l'Estat et aux Bonnes Moeurs, Combattue et Renversée* (Paris, 1623). On Mersenne's attack on Fludd, see Mersenne, *Correspondance,* vol. 1, p. 154; Yates, pp. 40, 53; Joseph Bougerel, *Vie de Pierre Gassendi* (Paris, 1737), pp. 35, 36; Baillet, p. 51.

8. On Gassendi, see Popkin, *History of Scepticism,* pp. 88, 89, 90, 98; Gaston Sor-

tais, *La Philosophie Moderne Dêpuis Bacon Jusqu'à Leibniz*, 2 vols. (Paris: Lethielleux, 1922), vol. 2; p. 10; Bernard Rochot, *Les Travaux de Gassendi sur Epicure et sur l'Atomisme, 1619–1658* (Paris: Vrin, 1944), pp. 5, 29.

9. On Gassendi's response to Fludd, see Rochot, pp. 10, 20, 21, 25; Bougerel, pp. 35–38; Lenoble, pp. 33, 35, 36.

10. Pierre Gassendi, "Examen Philosophiae Roberti Fluddi," in *Opera Omnia*, 6 vols. (facsimile ed., Stuttgart-Bad Cannstatt: Fromann, 1964; first published 1658), vol. 3, pp. 211–68, esp. 214–15; Mersenne, *Correspondance*, vol. 2, pp. 444–45; 217–18, 221, 230. For a discussion, see Bougerel, pp. 72, 78; Sortais, *La Philosophie Moderne*, vol. 2; pp. 41–48; J. G. Brule, *Histoire de la Philosophie Moderne*, vol. 3, pp. 160–61.

11. Rochot, pp. 7, 24, 29; Reijer Hooykaas, *Religion and the Rise of Modern Science* (Grand Rapids, Mich.: Eerdmans, 1972), pp. 16–17; Marie Boas, "The Establishment of the Mechanical Philosophy," pp. 422–33. Nicholas Hill, *Philosophia Epicurea* (Paris, 1601); Sebastian Basso, *Philosophia naturalis adversus Aristotelem* (Geneva, 1621); Daniel Sennert, *Hypomnemata physica* (Lugduni, 1631); D. Sennert, *De Chymicormum cum Aristotelicis et Galenicis consensu ac dissensu* (Wittenberg, 1619); Claude Berigard, *Circulus Pisanus* (Utini, 1643).

12. Pierre Gassendi, "Philosophiae Epicuri Syntagma" (first published 1649), in *Opera*, vol. 3, pp. 1–94, see pp. 19–20. Sortais, p. 108.

13. Pierre Gassendi, "Syntagma Philosophicum" (first published 1658), *Opera*, vol. 1. Quotations from P. Gassendi, *The Selected Works of Pierre Gassendi*, trans. Craig B. Bush (London: Johnson Reprint, 1972), pp. 380–434, quotations on pp. 411–22, 417.

14. Gassendi, "Syntagma Philosophicum" in *Opera*, vol. 1, pp. 158, 450; vol. 2, pp. 3, 822–23; Sortais, pp. 121–22, 154.

15. Margaret Osler, "Certainty, Scepticism and Scientific Optimism: The Roots of 18th-Century Attitudes Towards Scientific Knowledge," in Paula Backscheider, ed., *Probability, Time, and Space in Eighteenth Century Literature* (New York: AMS Press, 1979), and M. Osler, "Gassendi, Descartes, and the Foundations of the Mechanical Philosophy," forthcoming. I am grateful to Dr. Osler for preprints of these articles. Walter Charleton, *Physiologia Epicuro-Gassendo-Charltonia: or a Fabrick of Science Natural upon the Hypothesis of Atoms* (New York: Johnson Reprint, 1966; first published 1654). For a discussion, see R. H. Kargon, intro. to 1966 ed., *ibid.*, and Sortais, p. 120. On Charleton, see Nina Rattner Gelbart, "The Intellectual Development of Walter Charleton," *Ambix* 18, no. 3 (November 1971): 149–68.

16. René Descartes, "Discours de la Methode" (first published 1637) in *Oeuvres*, ed. Charles Adam and Paul Tannery, 12 vols. (Paris: Cerf, 1897–1913), vol. 6, pp. 4, 8; quoted and discussed in Theodore K. Rabb, *The Struggle for Stability in Early Modern Europe* (New York: Oxford University Press, 1975), pp. 38–39; Baillet, pp. 39–41, 50–53; Yates, *Rosicrucian Enlightenment*, pp. 114–17.

17. R. Descartes, "Lettre à Mersenne," April 15, 1630, in *Oeuvres*, vol. 1, p. 145. For a recent discussion of this issue, see Harry Frankfurt, "Descartes on the Creation of the Eternal Truths," *Philosophical Review* 84 (January, 1977): 36–57.

18. R. Descartes, "Meditationes de Prima Philosophia" (1641), *Oeuvres*, vol. 7, pp. 62, 65–67, 71–90. For a recent discussion, see Osler, "Certainty, Scepticism and Scientific Optimism."

19. Descartes, "Le Monde," *Oeuvres*, vol. 11, pp. 1–118. R. Descartes, *De Homine Figuris et Latinitale Donatus* (Leyden, 1662); First French edition: *L'Homme de René Descartes* (Paris, 1664); *Oeuvres*, vol. 11, pp. 119–215; English translation: R. Descartes, *Treatise of Man*, trans. Thomas Steele Hall (Cambridge, Mass.: Harvard University Press, 1972). For a discussion, see Phillip R. Sloan, "Descartes, the Sceptics, and the Rejection of Vitalism in Seventeenth-Century Physiology," *Studies in History and Philosophy of Science* 8 (1977): 1–28, especially pp. 4, 14.

20. R. Descartes, "Principia Philosophiae," in *Oeuvres*, vol. 9, Part 2, prin. 37–38, pp.

84–85. On the *spiritus mundi* and the ether, see Richard S. Westfall, "Newton and the Hermetic Tradition" in Allen G. Debus, ed., *Science, Medicine, and Society: Essays to Honor Walter Pagel,* 2 vols. (New York: Science History Publications, 1972), vol. 2, pp. 186–87.

21. Descartes, "Principia Philosophiae," *Oeuvres,* vol. 9, part 4, prin. 187, p. 309, see also prin. 198, pp. 316–17.

22. Descartes, "Les Passions de l'Ame," *Oeuvres,* vol. 11, articles 19, 20, pp. 343–44. English translation: "The Passions of the Soul," in *Philosophical Works,* trans. E. S. Haldane and G. R. T. Ross (New York: Dover, 1955), vol. 1, pp. 340–41.

23. Descartes, "Lettre à Mersenne," April 15, 1630, *Oeuvres,* vol. 1, p. 145.

24. John U. Nef, *Industry and Government in France and England, 1540–1640* (Philadelphia: American Philosophical Society, 1940), pp. 10, 24, 72.

25. George Croom Robertson, *Hobbes* (Edinburgh: Blackwood, 1905), pp. 26–53. Useful commentaries on Hobbes' scientific conception of nature include Marjorie Grene, "Hobbes and the Modern Mind," in M. Grene, ed., *The Anatomy of Knowledge* (Amherst: University of Massachusetts Press, 1969), pp. 1–28; Wilbur Applebaum, "Boyle and Hobbes," in the *Journal of the History of Ideas* 25 (January–March, 1964): 117–19; A. E. Bell, "Modern Science and Thomas Hobbes," *Nature* 149 (June 20, 1949): 688–90; Robert Woodfield, "Hobbes on the Laws of Nature and the Atheist," *Renaissance and Modern Studies* 15 (1971): 34–43; Quentin Skinner, "Thomas Hobbes and the Nature of the Early Royal Society," *The Historical Journal* 12 (1969): 217–39. On the Cavendish circle, see Jean Jacquot, "Sir Charles Cavendish and His Learned Friends," Part 1, *Annals of Science* 8 (1952): 13–27; Part 2, *Ibid.,* pp. 175–91.

26. R. Kargon, *Atomism in England,* p. 60, 63, 68, 65–66.

27. Thomas Hobbes, "A Short Tract on First Principles" (known as the "Little Treatise"), in Ferdinand Tönnies, ed., *The Elements of Law* (New York: Barnes & Noble, 1969), appendix 1, pp. 193–201;

quotations on pp. 193, 199. For an analysis of the "Little Treatise," see Frithiof Brandt, *Thomas Hobbes Mechanical Conception of Nature* (Copenhagen: Levin and Munksgaard, 1923), pp. 9–85.

28. "Hobbes to Mersenne," 17 February 1648, quoted in Kargon, p. 58. Hobbes was widely viewed in England as holding Epicurean ideas concerning a materialist formation of the world from a chaos of atoms. See Thomas Franklin Mayo, *Epicurus in England, 1650–1725* (Dallas, Tex.: Southwest Press, 1934), Chap. 8. On the differences between Hobbes and Epicurus, see Charles T. Harrison, "Bacon, Hobbes, Boyle and the Ancient Atomists," *Harvard Studies and Notes in Philology and Literature* 15 (1933): 191–218, and "The Ancient Atomists and English Literature of the Seventeenth Century," *Harvard Studies in Classical Philology* 45 (1934): 1–79.

29. Thomas Hobbes, "Concerning Body" ["De Corpore"], in William Molesworth, ed., *English Works,* 11 vols. (reprint edition, Aalen, W. Germany: Scientia, 1966), vol. 1, p. 426: "First therefore, I suppose that the immense space, which we call the world, is the aggregate of all bodies which are either consistent and visible, as the earth and the stars; or invisible, as the small atoms which are disseminated through the whole space between the earth and the stars; and lastly, that most fluid ether, which so fills all the rest of the universe, as that it leaves in it no empty place at all." For a discussion, see Kargon, pp. 58–59.

30. "Charles Cavendish to John Pell, Aug. 2, 1948," quoted in Helen Hervey, "Hobbes and Descartes in the Light of Some Unpublished Letters of the Correspondence between Sir Charles Cavendish and Dr. John Pell," *Osiris* 10 (1952): 84. On Descartes' animosity stemming from Hobbes' response to the former's *Meditations,* see Hervey, p. 73.

31. Hobbes, "De Corpore," *English Works* vol. 1, pp. 115, 124, 205, 390, 510, quotations on pp. 120, 127–29. For discussion, see Brandt, pp. 265, 289.

32. Hobbes, *English Works,* vol. 1, pp. 206–7, 216, 389–90, 477, 407, 409; Brandt, pp. 300–301, Howard Bernstein, "Conatus,

Hobbes, and the Young Leibniz," *Studies in History and Philosophy of Science* 9 (1979); J. N. Watkins, *Hobbes' System of Ideas* (London: Hutchinson University Library, 1965), pp. 120–37.

33. Hobbes, "Leviathan," in *English Works,* vol. 3, pp. 140–41. See also p. 156. For an analysis of mechanical ideas in Hobbes' political theory, see Wolin, *Politics and Vision,* Chap. 8.

34. Hobbes, "The Philosophical Rudiments Concerning Government and Society" ["De Cive," 1647] in *English Works,* vol. 2, p. 11.

35. *English Works,* vol. 2, p. 11.

36. *English Works,* vol. 3, p. 145. Garrett Hardin, "The Tragedy of the Commons," *Science* 162 (1968): 1243–1248. On the competitive economic base of the English economy and its implications for Hobbes' political theory see C. B. Macpherson, *The Political Theory of Possessive Individualism: Hobbes to Locke* (New York: Oxford University Press, 1962). For a critique, see Quentin Skinner, "Some Problems in the Analysis of Political Thought and Action," *Political Theory* 2 (Aug. 1974): 277–303; Q. Skinner, "Meaning and Understanding in the History of Ideas," *History and Theory* 8 (1969): 1–53, and Keith Thomas, "The Social Origins of Hobbes' Political Thought," in K. C. Brown, ed., *Hobbes' Studies* (Oxford, England: Blackwell, 1965), pp. 185–236.

37. *English Works,* vol. 3, p. 158. See also Wolin, Chap. 8.

38. *English Works,* vol. 3, p. *ix.*

39. *English Works,* vol. 3, pp. *ix, x.* On Hobbes' use of the organic metaphor as mechanical, see T. D. Weldon, *States and Morals* (New York: McGraw–Hill, 1947), pp. 26, 45.

40. J. U. Nef, *Government and Industry,* pp. 8–9, 100–101, 113, 130–33; On the wider social and intellectual causes of the English Civil War, see Christopher Hill, *Puritanism and Revolution* (New York: Schocken Books, 1958); C. Hill, *The Intellectual Origins of the English Revolution* (Oxford, England: Oxford University Press, 1965); and Lawrence Stone, *The Causes of the English Revolution, 1529–1642* (New York: Harper & Row, 1972).

41. On the reception of Hobbes' ideas in England, see Samuel I. Mintz, *The Hunting of Leviathan: Seventeenth-Century Reactions to the Materialism and Moral Philosophy of Thomas Hobbes* (Cambridge, England: Cambridge University Press, 1962). On the more favorable French reactions to his politics, see Nannerl Keohane, *Power and Participation in French Political Philosophy* (Princeton, N. J.: Princeton University Press, 1980). On Hobbes' laws of nature, see *English Works,* vol. 3, pp. 116–47. On Hobbes' nominalism and theory of language, see J. W. N. Watkins, *Hobbes' System of Ideas* (London: Hutchinson University Library, 1965), pp. 138–62.

42. Teresa Brennan and Carole Pateman, "Mere Auxiliaries to the Commonwealth: Women and the Origins of Liberalism," *Political Studies* 27 (June 1979): 183–200. Hobbes, "Leviathan," *English Works,* vol. 3, Chap. 20; Hobbes, "De Cive," *English Works,* vol. 2, Chap. 9, p. 10.

CHAPTER 9:
MECHANISM AS POWER

1. On the internal history of mechanics and mathematics, see E. J. Dijksterhuis, *The Mechanization of the World Picture,* trans. C. Dikshoorn (Oxford, England: Clarendon Press, 1961). For a discussion of the foundations of knowledge in everyday life and the role of symbolism, see Peter L. Berger and Thomas Luckmann, *The Social Construction of Reality* (Garden City, N.Y.: Doubleday, 1966). On autonomous and nonautonomous machines, see Joseph Weizenbaum, "On the Impact of the Computer on Society," *Science* 176 (1972): 609–14, see p. 610.

2. The discussion of machine technology is drawn from Albert Payton Usher, *History of Mechanical Inventions* (Cambridge, Mass.: Harvard University Press, 1954); pp. 198–205, 232–35, 268–69; A. P. Usher, "Machines and Mechanisms," in Charles Singer, ed., *History of Technology* (New York and London: Oxford University Press, 1956), vol. 3, pp. 324–46, see p. 328; Fre-

derich Klemm, *A History of Western Technology* (Cambridge, Mass., M.I.T. Press, 1954), p. 136; Rex Wailes, "Windmills," in Charles Singer, ed., *History of Technology*, vol. 3, pp. 89–109; John Reynolds, *Windmills and Watermills* (London: Evelyn, 1974); Agostino Ramelli, *Le Diverse et Artificiose Macchine* (Paris, 1588); English translation: A. Ramelli, *Various and Ingenious Machines*, trans. Martha Teach Gnudi (Baltimore, Md.: Johns Hopkins University Press, 1976), Chap. 73, p. 206; William Coles Finch, *Watermills and Windmills: An Historical Survey of Their Rise, Decline and Fall as Portrayed by Those of Kent* (London: Daniel, 1933); John U. Nef, "The Progress of Technology and the Growth of Large-Scale Industry in Great Britain, 1540–1640," *Economic History Review* 5 (1934): 3–24, see pp. 12–13.
3. Georg Agricola, *De Re Metallica*, trans. Herbert C. Hoover and Lou H. Hoover (New York: Dover, 1950; first published 1556); Vannuccio Biringuccio, *Pirotechnica* (Venice, 1540); Jacques Besson, *Theatrum Instrumentorum et Machinarum* (Lugduni, 1582; first published 1569); A. Ramelli, *Various and Ingenious Machines;* H. Arthur Klein, "Pieter Bruegel the Elder as a Guide to 16th Century Technology," *Scientific American* (March 1978): 134–40.
4. A. P. Usher, *A History of Mechanical Inventions*, pp. 199–209; Carlo Cipolla, *Clocks and Culture, 1300–1700* (London: Collins, 1967), pp. 37–69; E. P. Thompson, "Time, Work-Discipline, and Industrial Capitalism," *Past and Present*, no. 38 (Dec. 1967): 56–97, see pp. 63–65.
5. A. Deverny and A. Menut, "Maistre Nicole Oresme; Le Livre du Ciel et du Monde," *Medieval Studies* 4 (1942): 170. Translation of passage, Richard C. Olson, in R. C. Olson, ed., *Science as Metaphor* (Belmont, Cal.: Wadsworth, 1971), p. 59; Derek J. de Solla Price, "Automata and the Origins of Mechanism and Mechanistic Philosophy," *Technology and Culture* 5, no. 1 (Winter 1964): 9–23; Lynn White, Jr., *Medieval Technology*, pp. 125, 174; Lewis Mumford, *Technics and Civilization* (New York: Harcourt Brace, 1934), p. 174.

6. Lynn White, Jr., "The Iconography of Temperantia and the Virtuousness of Technology," in Theodore K. Rabb and Jerrold E. Seigel, eds., *Action and Conviction in Early Modern Europe* (Princeton, N.J.: Princeton University Press, 1969), pp. 197–219, pp. 207–13; Klein, pp. 137–39. On the clock in the Renaissance, see Otto Mayr, "Automatenlegenden in der Spätrenaissance," *Technikgeschichte* 41, no. 1 (1974): 20–32; On the clock metaphor in seventeenth-century science and philosophy, see Laurens Laudan, "The Clock Metaphor and Probabilism: The Impact of Descartes on English Methodological Thought, 1650–1665," *Annals of Science* 22, no. 2 (June 1966): 73–104.
7. Rebert Lenoble, *Mersenne ou la Naissance du Mécanisme* (Paris: Vrin, 1943), p. 537.
8. R. Descartes, "The Meditations," in *Meditations and Selections from the Principles of Philosophy* (La Salle, Ill.: Open Court, 1952), p. 98.
9. Quoted in Edwin Burtt, *The Metaphysical Foundations of Modern Physical Science* (Garden City, N.Y.: Doubleday, 1954), p. 202.
10. O. Mayr, "From the Clock-Work Universe to Checks and Balances," forthcoming.
11. Boris Hessen, *The Social and Economic Roots of Newton's Principia* (New York: Fertig, 1971); Edgar Zilsel, "The Genesis of the Concept of Scientific Progress," in Philip P. Wiener and Aaron Noland, eds., *Roots of Scientific Thought* (New York: Basic Books, 1953), pp. 251–75; Paolo Rossi, *Philosophy, Technology and the Arts in the Early Modern Era*, trans. Salvator Attanasio (New York: Harper & Row, 1970); J. U. Nef, *Industry and Government in France and England, 1540–1640* (Ithaca, N.Y.: Cornell University Press, 1957).
12. On the social origins of symbolic and conceptual universes from everyday experience and their function as legitimations of social values, see P. Berger and T. Luckmann, pp. 92–198. On the sociology of scientific knowledge, see Everett Mendelsohn,

"The Social Construction of Scientific Knowledge," in E. Mendelsohn and P. Weingart, eds., *The Social Production of Scientific Knowledge* (Boston: Reidel, 1977), pp. 3–26; Wolfgang van den Daele, "The Social Construction of Science," in Mendelsohn and Weingart, eds., *Social Production,* pp. 27–54; Phyllis Colvin, "Ontological and Epistemological Commitments and Social Relations in the Sciences," in Mendelsohn and Weingart, eds., *Social Production,* pp. 103–28.

13. Martin Heidegger, *The Question Concerning Technology* (New York: Harper & Row, 1977), pp. 21, 23. Also important are M. Heidegger, "The Age of the World Picture," in *The Question Concerning Technology,* esp. pp. 127–36 on Descartes and the modern scientific world-picture; and M. Heidegger, "The Principle of Identity," in *Identity and Difference,* trans. J. Stambaugh (New York: Harper & Row, 1969), pp. 23–41. Important commentaries include Hubert Dreyfus, *What Computers Can't Do* (New York: Harper & Row, 1972); Harold Alderman, "Heidegger: Technology as Phenomenon," *The Personalist* 51 (Autumn 1970): 535–45, and H. Alderman, "Heidegger's Critique of Science," *The Personalist* 50 (Autumn 1969): 449–548.

14. Emile Meyerson, *Identity and Reality* (New York: Humanities Press, 1964), p. 28.

15. Descartes, "Discours de la Methode," in *Oeuvres,* ed. Charles Adam and Paul Tannery (Paris: Cerf, 1897–1913), vol. 6, pp. 7–8; English translation: Descartes, "Discourse on Method," in E. S. Haldane and G. R. T. Ross, eds., *The Philosophical Works of Descartes* (New York: Dover, 1955), vol. 1, p. 85.

16. Descartes, "Principia Philosophiae," *Oeuvres,* vol. 9, part 2, prin. 53, p. 93.

17. Descartes, "Discourse," in *Philosophical Works,* quotations on pp. 92, 93, 87, 89.

18. Thomas Hobbes, "De Cive," *English Works* (reprint ed., Aalen, W. Germany: Scientia, 1966; first published 1834), vol. 2, p. *xiv.*

19. Hobbes, "Leviathan," *English Works,* vol. 3, quotations from Chaps. 4, pp. 18, 20; 3, p. 17.

20. *Ibid.,* Chap. 5, pp. 29, 30. Napier's bones (invented 1617) were a set of rods comprising a form of the multiplication table. In 1642, after some fifty attempts, Blaise Pascal constructed an adding machine that would carry forward the tens when operated by a stylus that turned six decade wheels. A smaller model was constructed by Morland in 1661, and a more sophisticated rapid machine was designed by Leibniz in 1683, although it was impossible to construct at the time. See Henri Michel, *Scientific Instruments in Art and History,* trans. R. E. W. Maddison and F. R. Maddison (New York: Viking Press, 1966), pp. 20–22, 46–47, and illustrations 15–19. It is probable that Hobbes, who was in Paris during the 1640s, knew of Pascal's machine.

21. Heidegger, *Der Satz vom Grund,* quoted in Dreyfus, *What Computers Can't Do,* p. 242, n. 16. Dreyfus gives an important discussion of the ontological, epistemological, and methodological assumptions underlying computer technology and computer models of the brain. On the brain as a machine, see Colin Blackmore, *Mechanics of Mind* (Cambridge, England: Cambridge University Press, 1977), esp. Chap. 1.

22. Weizenbaum, "On the Impact of the Computer," p. 610.

23. Robert Kargon, "Walter Charleton, Robert Boyle, and the Acceptance of Epicurean Atomism in England," *Isis* 55 (1964): 184–92; Marie Boas, "The Establishment of the Mechanical Philosophy," *Osiris* 10 (1962): esp. 442–520; Marjorie Nicolson, "The Early Stage of Cartesianism in England," *Studies in Philology* 26 (1929): 356–74; Sterling Lamprecht, "The Role of Descartes in Seventeenth Century England," in Columbia Department of Philosophy, ed., *Studies in the History of Ideas* (New York: Columbia University Press, 1935), vol. 3, pp. 181–240; Paul Mouy, *Le Developpement de la Physique Cartesienne, 1646–1712* (Paris: Vrin, 1934); P. Mouy, "L'influence de Descartes sur le Developpement de la Physique," *Scientia* 48 (1930): 227–36; A. R. Hall, "Cartesian Dynamics," *Archive for History of Exact Sciences* 1 (1961): 172–78; L. M. Marsak, "Cartesianism in Fontenelle and French Science, 1686–1752," *Isis* 50 (1969): 51–60; E. J. Aiton, *The Vortex Theory of Planetary Motions* (New York: Elsevier, 1972); A. Koyré, "Newton and Descartes," in *Newtonian Studies* (Chicago: University of Chicago Press, 1965); Carolyn [Merchant] Iltis, "The Decline of Cartesianism in Mechanics: The Leibnizian–Cartesian Debates," *Isis* 64 (1973): 356–73. On Hobbes' political theory in France, see Nannerl Keohane, *Power and Participation in French Political Philosophy* (Princeton, N.J.: Princeton University Press, 1980).

CHAPTER 10:
THE MANAGEMENT OF NATURE
1. John Evelyn, *Silva, or A Discourse of Forest Trees* (York, England, 1776), p. 1. See also C. J. Glacken, *Traces on the Rhodian Shore: Nature and Culture in Western Thought from Ancient Times to the End of the Eighteenth Century* (Berkeley and Los Angeles: University of California Press, 1967), pp. 484–91; Thomas Birch, *History of the Royal Society* (London, 1756), vol. 1, pp. 110–20.
2. Quoted in H. C. Darby, "The Clearing of the Woodland in Europe," in *Man's Role in Changing the Face of the Earth,* 2 vols. (Chicago: University of Chicago Press, 1956), vol. 1, p. 201.
3. Evelyn, *Silva,* p. 1, footnote by A. Hunter, and p. 571.
4. Margaret C. Jacob, *The Newtonians and the English Revolution, 1689–1720* (Ithaca, N.Y.: Cornell University Press, 1976), Chap. 1 and pp. 47–52, 81–87. On the Latitudinarians, see Simon Patrick, "A Brief Account of the New Sect of Latitude-Men: Together with Some Reflections upon the New Philosophy," in *The Phoenix* 2 (1708): 499–518; Edward Fowler, *The Principles and Practices of Certain Moderate Divines of the Church of England Abusively Called Latitudinarians,* 2nd ed. (London, 1671), especially pp. 332–33; Barbara Shapiro, "Latitudinarianism and Science in Seventeenth Century England," *Past and Present,* no. 40 (1968): 16–41; Sterling P. Lamprecht, "Innate Ideas in the Cambridge Platonists," *Philosophical Review* 35 (1926): 553–573; Marjorie Nicolson, "Christ's College and the Latitude-Men," *Modern Philosophy* 18 (1929): 50–51. The Latitudinarians were a loose grouping of moderate Low Churchmen noted for their interest in peacefully coexisting with both the Puritans during the Commonwealth era and with the Anglican establishment that resumed control of the church after the Restoration. They believed it was the duty of all churchgoers to submit to the right of civil government to establish forms of religious worship. But because they did not wish to lay the blame for failing on established civil, secular, and sacred institutions, as had the Civil War Puritan dissenters and radicals, they held to a doctrine of individual responsibility and accountability for one's own actions. Because they feared the weakness that could be brought about by divisiveness and discord among sects, they held that the "church cannot be without unity and uniformity" (Patrick, p. 506). Therefore compromise in doctrinal and liturgical matters was stressed, so that greater emphasis could be placed on more significant moral issues. Plainness in the language of both pulpit and ordinary discourse was emphasized in order to make religious ideas clear to all churchgoers. The rational impartial weighing of extremes in the search for truth was a distinguishing characteristic of the Latitudinarian.
5. "Life of Evelyn," in *Silva,* introduction, p. 7; John Evelyn, *Navigation, Commerce, Their Origin and Progress* (London, 1674).
6. Samuel P. Hayes, *Conservation and the Gospel of Efficiency: The Progressive Conservation Movement, 1890–1920* (Cambridge, Mass.: Harvard University Press, 1959); Donald Worster, *Nature's Economy* (San Francisco: Sierra Club Books, 1977).
7. Evelyn, *Silva,* pp. 567, 577, 567. On the timber crisis in the shipbuilding and iron industries as an impetus to New England settlement, see Charles F. Carroll, *The Timber Economy of New England* (Providence, R. I.: Brown University Press, 1973). On England's regulation of the colonial American iron industry, see Joseph M. Petulla, *American Environmental History,* (San Francisco: Boyd and Fraser, 1977), pp. 61–65, and

Douglas A. Fisher, *The Epic of Steel* (New York: Harper & Row, 1963).

8. Evelyn, *Silva*, quotations in order on pp. 567–68, 587, 14. On the treatment of soils, see John Evelyn, *Terra: A Philosophical Discourse of Earth, Relating to the Culture and Improvement of It for Vegetation and the Propagation of Plants* (York, England, 1778).

9. R. G. Albion, *Forests and Sea-Power* (Cambridge, Mass.: Harvard University Press, 1926), pp. 131–35.

10. John Crombie Brown, ed. and trans., *The French Forest Ordinance of 1669,* (Edinburgh: Oliver and Boyd, 1883), pp. 13–20, 40–47.

11. John Evelyn, *Fumifugium* (reprint of 1772 edition, Oxford, England: Old Ashmolean Reprints, 1930; first published 1661), quotations in order on pp. 44, 18, 19, 24, 29, 30.

12. John Graunt, *Natural and Political Observations Mentioned in a Following Index and Made Upon the Bills of Mortality ... with Reference to the Government, Religion, Trade, Growth, Air, Diseases, and Several Changes of the Said City,* 5th ed., rev., 1676, reprinted in William Petty, *The Economic Writings of Sir William Petty,* ed. Charles Hull (Cambridge, England, 1899), vol. 2, pp. 393–94. See also William Te Brake, "Air Pollution and Fuel Crises in Pre–Industrial London," *Technology and Culture* 16 (1975): 337–59, see p. 338.

13. Evelyn, *Fumifigium,* p. 25; Graunt, *Observations,* p. 6; Corbyn Morris, "Observations Political and Natural to Thomas Potter," in *A Collection of the Yearly Bills of Mortality from 1657–1758 Inclusive* (London, 1759), p. 88.

14. Evelyn, *Fumifugium,* pp. 34–35, 47, 48.

15. Henry More, *Enthusiasmus Triumphatus* (first published, 1656), in *A Collection of Several Philosophical Writings,* 4th ed. (London, 1712), separate pagination, pp. 34, 30. For a discussion of More's reaction to enthusiasm, see Frederic B. Burnham, "The More–Vaughan Controversy: The Revolt Against Enthusiasm," *Journal of the History of Ideas* 35 (1974): 33–49. On the response of the Cambridge Platonists to the Civil War and to the Commonwealth, see John Tullock, *Rational Theology and Christian Philosophy in the Seventeenth Century* (Edinburgh and London, 1872), vol. 2, pp. 219–22, 203, 207–10, and Lamprecht, "Innate Ideas in the Cambridge Platonists." Relevant works on More include Alan Gabbey, "Philosophia Cartesiana Triumphata: Henry More 1646–1671," forthcoming; Ernst Cassirer, *The Platonic Renaissance in England,* trans. James P. Pettigrove, (London: Nelson, 1953); Rosemary Colie, *Light and Enlightenment: A Study of the Cambridge Platonists and the Dutch Arminians* (Cambridge, England: Cambridge University Press, 1957); John Hoyles, *The Waning of the Renaissance, 1640–1740: Studies in the Thought and Poetry of Henry More, John Norris and Isaac Watts* (The Hague: Nijhoff, 1971); Serge Hutin, *Henry More, Essai sur les Doctrines Theosophiques chez les Platoniciens de Cambridge* (Hildesheim, W. Germany: Ohms, 1966); Geoffrey P. H. Pawson, *The Cambridge Platonists* (London: Society for Promoting Christian Knowledge, 1930); F. J. Powicke, *The Cambridge Platonists,* (London: Dent, 1926); J. A. Stewart, "The Cambridge Platonists," in *The Encyclopedia of Religion and Ethics,* ed. James W. Hastings (New York: Scribner's, 1912), vol. 3, pp. 167–73; Robert Greene, "Henry More and Robert Boyle on the Spirit of Nature," *Journal of the History of Ideas* 23 (1962): 451–74; Aharon Lichtenstein, *Henry More: The Rational Philosophy of a Cambridge Platonist* (Cambridge, Mass.: Harvard University Press, 1962).

16. More, "The Immortality of the Soul" (first published 1659), in *A Collection,* separate pagination, pp. 29–30, 160.

17. More, "An Antidote Against Atheism" (first published, 1653), in *A Collection,* separate pagination, p. 38; on plastic natures, see William B. Hunter, "The Seventeenth Century Doctrine of Plastic Natures," *Harvard Theological Review* 43 (1950): 197–213.

18. Ralph Cudworth, *The True Intellectual System of the Universe,* (1678) 2 vols. (New York: Gould and Newman, 1938), vol. 1, p. 242. Relevant works on Cudworth's philosophy include Tulloch, vol. 2; Gunnar Aspelin, "Ralph Cudworth's Interpretation of Greek Philosophy," *Acta Universitatis Gotoburgensis, Göteborgs, Högskolas Arsskrift* 49 (1943): 1–47; Lydia Gysi, *Platonism and Cartesianism in the Philosophy of Ralph Cudworth* (Bern: Lang, 1962); John Passmore, *Ralph Cudworth: An Interpretaton* (Cambridge, England: Cambridge University Press, 1951); Danton B. Sailor, "Cudworth and Descartes," *Journal of the History of Ideas* 23 (January–March 1962): 133–140; J. A. Stewart, "The Cambridge Platonists."

19. More, "Antidote," in *A Collection,* pp. 49, 53, 316, 237; "The Immortality of the Soul" in *A Collection,* pp. 12, 29, 30; Hunter, p. 200. More denied to the spirit of nature reason, free will, and sensation, for to endow it thus would give it too much freedom to prepare any fanciful forms or monsters it might please. By negating sensation in the spirit of nature, More rejected animism of the sort that had assigned sense to the earth, the sun, and other heavenly bodies. He further denied that the sun and stars were endowed with the higher functions of imagination, memory, reason, or providence, arising ultimately out of the activity of nimble atoms and particles of matter, a position epitomized in the atheistic Epicurean philosophy that "the sun and the stars are the most intellectual beings in the world, and in them is that knowledge, counsel, wisdom, by which all sublunary things are framed and governed," and also presupposed by the Renaissance philosopher Pomponasse, who attributed intelligence to the stars. In this philosophy, the imaginative faculty of the celestial bodies imprinted shapes on terrestrial matter, filling "the sea with fishes, the fields with beasts, and the air with fowls" as well as "raising innumerable sorts of flowers, herbs, and trees out of the ground," "filling the whole earth with vital motion" (More, "Antidote," p. 49). Although More frequently referred to the spirit of nature as the "inferior" soul of the world" and "universal soul of the world," he did not mean by this the *anima mundi* of the Averroists, which had the power of perception and memory (More, "Antidote," pp. 213–14, 219–20).

20. More, "Notes on Chapter VIII," "Antidote," p. 31.

21. Cudworth, *The True Intellectual System,* vol. 1, p. 225; see also pp. 243, 244, 208. The cosmoplastic form of Stoicism pursued by philosophers such as Boethius (480–524 A.D.) attributed the order observed in the whole to a single overriding "plastic or plantal nature," (*True Intellectual System,* vol. 1, p. 191). This philosophy, which rejected both the fate and chance attributed to nature by the Stoic, Zeno, and the animality of the world, supposed "one plastic or spermatic nature, one plantal or vegetative life in the whole world, as the highest principle." The cosmos was "a body governed by a plastic or vegetative nature, as trees, plants, and herbs" (p. 193). From it, Cudworth accepted the idea of a general plastic or vegetative principle guiding the development of nature from within, but asserted its incorporeal rather than corporeal character.

Strato, likewise, did not make the mistake of attributing animality to the world. Although an atheist and corporealist, his deity was nature without consciousness. It was living and active with an "inward plastic life" but without knowledge or sensation, forming existences by "certain inward natural forces and activities" (p. 161). Secondly, plastic life was in each particular existence, rather than "one common life . . . ruling over the whole mass of matter and corporeal universe" (p. 161). The world evolved therefore according to a combination of chance operating in the whole and plastic natures in each individual part. From the Stratonic form of hylozoism, Cudworth accepted the possibility of individual plastic natures without consciousness and sensation, but rejected its corporeal atheistic character and chance operation.

22. More, "Immortality," pp. 15, 25, 28.

23. More, "Antidote Against Atheism," examples and quotations on pp. 187–89, 190–92, 214, 216, 223.

24. John Ray, *The Wisdom of God Manifested in the Works of the Creation,* 10th ed. (London: Innys and Manby, 1935; first published 1691), pp. 37–38, 31–34. On Ray and his work as a biologist, see Charles E. Raven, *John Ray, Naturalist: His Life and Works* (Cambridge, England: Cambridge University Press, 1942), and an essay review of Raven's book by Agnes Arber, "A Seventeenth-Century Naturalist: John Ray," *Isis* 34 (1943): 319–24. Also relevant is Charles Raven, *English Naturalists* (Cambridge, England: Cambridge University Press, 1947); Joan M. Eyles, "John Ráy, F.R.S. (1627–1705)," *Nature* 175 (1955): 103–8.

25. Ray, *Wisdom*, pp. 42, 46, 75–76.
26. *Ibid.*, pp. 176–77, ,18, 161, 163–64, 171, 164.
27. *Ibid.*, pp. 96, 98. On the disruption of the Indian lifestyle, see Wilcomb E. Washburn, *The Indian in America* (New York: Harper & Row, 1975); Calvin Martin, "The European Impact on the Culture of a Northeastern Algonquian Tribe: An Ecological Interpretation," *William and Mary Quarterly* 31 (Jan. 1974): 3–26.
28. Ray, *Wisdom*, p. 206, 215.
29. William Derham, *Physico-Theology: or A Demonstration of the Being and Attributes of God, from His Works of Creation*, 6th ed. (London: Innys, 1728; first published 1713). Jacob, pp. 143–44, 146, 176–80. In the discussion that follows, I have also drawn on William Coleman, "Providence, Capitalism, and Environmental Degradation: English Apologetics in an Era of Economic Revolution," *Journal of the History of Ideas* 37 (January-March, 1976): 27–44, see pp. 39–41; Frank Egerton, "Observations and Studies of Animal Populations Before 1860," unpublished doctoral dissertation, University of Wisconsin, 1967, pp. 127–44; John Passmore, *Man's Responsibility for Nature: Ecological Problems and Western Traditions* (New York: Scribner's,1974); Clarence Glacken, *Traces on the Rhodian Shore: Nature and Culture in Western Thought from Ancient Times to the End of the Eighteenth Century* (Berkeley: University of California Press, 1967); C. Glacken, "This Growing Second World Within the World of Nature," in F. R. Fosberg, ed., *Man's Place in the Island Ecosystem: A Symposium of the Tenth Pacific*

Scientific Congress, Honolulu, Hawaii, 1961 (Honolulu: Bishop Museum Press, 1963), pp. 75–100, see pp. 83–84.
30. Derham, *Physico-Theology*, subsequent quotations in order on pp. 257, 260, 280.
31. *Ibid.*, subsequent quotations in order on pp. 111, 68, 54.
32. *Ibid.*, subsequent quotations in order on pp. 37, 216, 59, 68.
33. *Ibid.*, subsequent quotations on pp. 166–67.
34. Donald Worster, *Nature's Economy* (San Francisco: Sierra Club Books, 1977), pp. 239–42, and Chap. 14; Donella H. Meadows, Dennis L. Meadows, Jorgen Randers, William W. Behrens III, *The Limits to Growth* (New York: Signet Books, 1972). For a critique of the systems approach to environmental problems, see Huey D. Johnson, "The Flaws of RARE II," *The Sierra Club Bulletin* 64 (May–June, 1979): 8–10.
35. Aldo Leopold, "The Land Ethic," *A Sand County Almanac* (New York: Ballantine, 1970; first published 1949), pp. 237–63; Ecology Action, "The Unanimous Declaration of Interdependence," in Theodore Roszak, ed., *Sources* (New York: Harper & Row, 1972), pp. 388–89; Holmes Rolston III, "Is There an Ecological Ethic?" *Ethics* 85, no. 2 (January 1975): 93–109.

CHAPTER 11:
WOMEN ON NATURE

The following discussion of Anne Conway is reprinted with modifications from Carolyn Merchant. "The Vitalism of Anne Conway: Its Impact on Leibniz's Concept of the Monad," *The Journal of the History of Philosophy* (July 1979) by permission of the editor. Portions of this chapter also appeared in C. Merchant, "The Vitalism of Francis Mercury Van Helmont," *Ambix* (November 1979).
1. The Latin translation of Anne Conway's book first appeared anonymously in a collection of three works with separate title pages: Francis Mercury Van Helmont, *Opuscula Philosophica Quibus Continentur Principia Philosophiae Antiquissimae et*

Recentissimae. Ac Philosophiae Vulgaris Refuta [Auctore J. Gironnet] *Quibus subjuncta sunt cc. problemata de revolutione Animarum humanarum.* (Amsterdam, 1690). The English retranslation was [Anne Conway, supposed author] *The Principles of the Most Ancient and Modern Philosophy, Concerning God, Christ, and the Creatures, viz. of spirit and matter in general, whereby may be resolved all those problems or difficulties which neither by the school nor common modern philosophy nor by the Cartesian, Hobbesian or Spinosian could be discussed. Being a little treatise published since the author's death translated out of the English into Latin with annotations taken from the ancient philosophy of the Hebrews, and now again made into English by J. C. Medicinae professor. Printed in Latin at Amsterdam by M. Brown, 1690 and reprinted at London 1692.* [Authorship attributed to Lady Anne Conway, trans. probably by J. Crull.] This English title is a direct translation of the Latin title of the 1690 edition published in Van Helmont's *Opuscula philosophica: Principia philosophiae antiquissimae et recentissimae de Deo Christ et creatura id est spiritu et materia in genere . . . Opusculum post humum e lingua, anglicana latinate donatum, cum annotationibus ex antiqua Hebraeorum philosophia desumtis* (Amsterdam, 1690). Preface to the English translation of 1692 states that the book is the work of "a certain English countess, a woman learned beyond her sex, being very well skilled in the Latin and Greek tongues, and exceedingly well versed in all kinds of philosophy." The translator's introduction states:

Being some time since in Holland, and in conference with the renowned F.M.B. Van Helmont, then resident of Amsterdam, it so happened that I demanded of the said Helmont, if he had published, or did intend to publish any new books of his own, or others works, who presently directed me where I might procure certain books, published by his order, which accordingly I did, two whereof were extant in Latin, the other in Nether-Dutch; this being the works of an English Countess (after brief perusal) I

have endeavoured to render into an English style, as familiar as the language would conveniently admit, without some abuse to the author. Heinrich Ritter incorrectly attributes Conway's *Principia Philosophia* in the *Opuscula Philosophica* to Francis Mercury Van Helmont. See H. Ritter, *Geschichte der Philosophie* (Hamburg, 1853), vol. 12, pp. 3–47, p. 7, note 1. Although Ritter used three other books by Van Helmont, the most substantial part of his account of Van Helmont's ideas is based on Conway's book and is therefore almost wholly unreliable. For Ludwig Stein's acceptance of Ritter's interpretation see L. Stein, *Leibniz und Spinoza* (Berlin: Reimer, 1890), p. 212, note 1, in which Stein cites Helmont on the monad from "*Princ. phil.* III, 9, p. 25 . . . angeführt von Ritter, *Gesch. d. Phil.* XII, 22." The reference is to [Conway], *Principles*, Chap. 3. Sec. 9 (p. 28 in the English translation). I shall discuss this more fully later.

2. Marjorie Nicolson, "The Real Scholar Gipsy," *Yale Review* (January 1929): 347–63, see p. 356. On Anne Conway's life and philosophy; see Gilbert Roy Owen, "The Famous Case of Lady Anne Conway," *Annals of Medical History* 9 (1937): 567–71; Alan Gabbey, "Anne Conway et Henri More, Lettres sur Descartes," *Archives de Philosophie* 40 (1977): 379–404; Alison Coudert, "A Quaker–Kabbalist Controversy," *Journal of the Warburg and Courtauld Institutes* 39 (1976): 171–89, and "A Cambridge Platonist's Kabbalist Nightmare," *Journal of the History of Ideas* 36 (1975): 633–52; Alison Gottesman [Coudert], "Francis Mercurius Van Helmont: His Life and Thought," unpublished doctoral dissertation, University of London, 1972; Joseph Politella, *Platonism, Aristotelianism, and Cabalism in the Philosophy of Leibniz* (Philadelphia: Politella, 1938), pp. 13–19, 55–57.

3. Marjorie Nicolson, *Conway Letters: The Correspondence of Anne, Viscountess Conway, Henry More and their Friends, 1642–1684* (New Haven, Conn.: Yale University Press, 1930), pp. 1–9, 39–51, 116–18, 244–61, 316–18, 381–83, 407–8. Coudert, "Kabbalist Nightmare."

4. Richard T. Vann, "Toward a New Lifestyle: Women in Preindustrial Capitalism," in Renate Bridenthal and Claudia Koonz, eds., *Becoming Visible* (Boston: Houghton Mifflin, 1977), pp. 210–11. R. T. Vann, *The Social Development of English Quakerism, 1655–1755* (Cambridge, Mass.: Harvard University Press, 1969), pp. 1, 10, 15, 32.

5. A. Gottesman [Coudert] "Francis Mercurius Van Helmont," pp. 463, 584–85, 597.

6. Knorr von Rosenroth, ed., *Kabbalah Denudata,* 3 vols. (Sulzbach, 1677–1678), vol. 1, Pt. 2, p. 308. Francis Mercury Van Helmont, *A Cabbalistical Dialogue in Answer to the Opinion of a Learned Doctor in Philosophy and Theology that the World Was Made of Nothing. As it is Contained in the Second Part of the Cabbala Denudata and Apparatus in Lib. Sohar, p. 308, etc., 1677. To Which Is Subjoyned a Rabbinical and Paraphrastical Exposition of Genesis I, Written in High Dutch by the Author of the Foregoing Dialogue, First Done into Latin but Now Made into English* (London: Clark, 1682); Nicolson, *Conway Letters,* p. 453.

7. Gottfried Wilhelm von Leibniz, *Correspondance de Leibniz avec l'Electrice Sophie de Brunswicke-Lunebourg,* ed. O. Klopp (Hanover, 1874), vol. 2, p. 8, letter of Sept. 1696; G. W. Leibniz, *Philosophischen Schriften,* ed. C. I. Gerhardt (Berlin, 1875–1890), vol. 3, pp. 176, 180; Politella, p. 16; Nicolson, *Conway Letters,* p. 455.

8. Leibniz, *Philosophischen Schriften,* vol. 3, p. 217.

9. G. W. Leibniz, *New Essays Concerning Human Understanding* (written 1697), trans. A. G. Langley (Lasalle, Ill.: Open Court, 1949; first published 1765), p. 67.

10. Conway, *Principles,* pp. 140, 147, 104, 126, 132.

11. *Ibid.,* pp. 127–28, 132.

12. *Ibid.,* p. 147.

13. *Ibid.,* p. 82, 112, 114–15, 106–7, 110.

14. *Ibid.,* pp. 118, 123, 107.

15. *Ibid.,* p. 113. She cited *Kabbalah Denudata,* vol. 2, *Tract. Ult.,* p. 6, sec. 13.

16. Henry More, *Conjectura Cabbalistica: or a Conjectural Essay of Interpreting the Mind of Moses, in the Three First Chapters of Genesis, According to a Threefold Cabbala: viz. Literal, Philosophical, Mystical, or, Divinely Moral* (first published 1653), in *A Collection of Several Philosophical Writings of Dr. Henry More* (London, 1712). On this work, see Marjorie Nicolson, "Milton and the *Conjectura Cabbalistica,*" *Philological Quarterly* 6 (1927): 1–18; Gottesman [Coudert], "Frances Mercurius Van Helmont," pp. 519, 526, 536; Nicolson, *Conway Letters,* p. 83, letter of 9 Aug. 1653. On the interest of Cudworth, More, and Conway in the "ancient philosophy," see Gunnar Aspelin, "Ralph Cudworth's Interpretation of Greek Philosophy," in *Acta Universitatis Gotoburgensis, Göteborgs Högskolas Arsskrift* 49, no. 1 (1943): 1–47; J. E. McGuire and P. M. Rattansi, "Newton and the Pipes of Pan," *Notes and Records of the Royal Society* 21 (1966): 108–43; Politella, *Platonism, Aristotelianism, and Cabalism.*

17. Conway, p. 148.

18. *Ibid.,* pp. 60, 120.

19. *Ibid.,* pp. 152–53, 155, 69.

20. *Ibid.,* pp. 64–65.

21. *Ibid.,* p. 70.

22. G. W. Leibniz, "Considerations on Vital Principles and Plastic Natures, by the Author of the System of Pre–Established Harmony" (written 1705), *Philosophischen Schriften,* vol. 6, p. 539; trans. in Leroy E. Loemker, ed. and trans. *Philosophical Papers and Letters,* 2 vols. (Chicago: University of Chicago Press, 1956), vol. 2, p. 954; William B. Hunter, Jr., "The Seventeenth Century Doctrine of Plastic Natures," *Harvard Theological Review* 43 (1950): 212; G. W. Leibniz, *Opera Omnia,* ed. Ludovici Dutens (Geneva, 1768), vol. 5, p. 359.

23. Conway, pp. 148, 159, 165, 168, 143.

24. Leibniz, *Philosophischen Schriften,* vol. 3, p. 217; Leibniz, "The Monadology," in *Philosophischen Schriften,* vol. 6, pp. 607–23, sec. 66, 67; Conway, p. 20

25. Leibniz, "The Monadology," sec. 69, 54; Conway, pp. 93, 144, 96–97.

26. Leibniz, "The Monadology," secs. 84–86, 1, 2, 70; Conway, pp. 122, 123.

27. Conway, pp. 132, 132, 36, 136.

28. Ritter, vol. 12, pp. 26, 27, 30.

29. On the mechanist–vitalist debates, see Hilda Hein, "Mechanism and Vitalism as Theoretical Commitments," *The Philosophical Forum* 1, no. 1, n.s. (Fall 1968): 185–205; Hilda Hein, "The Endurance of the Mechanism–Vitalism Controversy," *The Journal of the History of Biology* 5, no. 1 (Spring 1972): 159–88; L. R. Wheeler, *Vitalism: Its History and Validity* (London: Witherby, 1939).

30. Leibniz, *Opera Omnia*, vol. 6, p. 70; Leibniz, *Philosophischen Schriften*, vol. 3, p. 427; Leibniz, *Philosophical Papers*, vol. 2, p. 1027; *New Essays*, p. 67. Leibniz had visited Knorr von Rosenroth in 1688. He referred to Helmont's friendship with him in a note: "Ce fut M. Knorr de Sulzbach, qui donna la Cabbala denudata au public et quoyque Monsieur Helmont l'y ait porté et encouragé neanmoins ce sont proprement les sentimens de Monsieur Knorr, dont M. Helmont ne demeure pas toujours d'accord, comme il me l'a dit luy meme." Joachimus Feller, *Otium Hanoveranum*, 2nd ed. (Leipzig, 1719), p. 217, no. 163. See also, Politella, pp. 13–19; Stein, pp. 194, 206. On Leibniz's concept of individual substance, see Ian Hacking, "Individual Substance," in *Leibniz, A Collection of Critical Essays*, ed. Harry G. Frankfurt (Garden City, N.Y.: Doubleday, 1972), pp. 139–53.

31. On the scholarship surrounding the Bruno thesis and Stein's refutation, see Stein, pp. 198, 201, 204, 206. Leibniz referred to Bruno's book *De Monade* as follows: "Jordani Bruni, Nolani, de Monade Numero et figura liber, de minimo, magno, et mensura. Item de innumeralibus, immenso et insigurabili, seu de universo et mundis libri octo" (Feller, p. 142). In another instance Leibniz referred to Bruno as the native of Nola who had spent a long time in Germany and whose opinions on the pluralities of worlds and the indefinite extent of the universe closely approached those of Descartes (Feller, p. 142).

32. Conway, *Principles*, p. 28. Both Ritter and Stein attributed this passage to Van Helmont, who had only edited the Latin edition of Conway's *Principia Philosophiae* contained in the *Opuscula Philosophica* of 1690. The passage is quoted in full with the omission of the *Kab. Denud.* sources from the 1690 Latin edition in Ritter, vol. 12, p. 22 and in Stein, p. 212, note 1. In his text, Stein stated, "Sicherlich ist diese physische monade van Helmont's noch recht weit von der metaphysischen des Leibniz entfernt, wenn auch beide Denker gleicher weise die Bezeichnung der Substanz als mathematischen Punktes ablehnen."

33. Conway's references were to Francis Mercury Van Helmont in *Kabbala Denudata*, vol. 1, p. 310: "Dum materiam factam statuerem e coalitione *monadum* spiritualium torpentium" and Knorr von Rosenroth, "*Adumbratio Kabbalae Christianae*," *Kabbala denudata,* Tom. 2, last tract, [pub. in vol. 3], p. 28, Numb. 4: "... (*id est naturas has, quae facta sunt* monades *materiales, è quibus deinde combinatis facta est mundi materialis creatio:*). . . ." Numb. 5: *"Sicut autem de vasis illis delapsis dicitur, quod prolapsa sint in lucem sibi propriam, qua intelligerent et amarent se ipsas . . . hinc patet etiam* monadibus *istis materialibus remansisse, partim lucem aliquam propriam (quae si excitetur certo modo suos iterum posset emittere radios, ad quam pertinent formae materialis et seminales tam inanimatorum quam plantarum et brutorum) partim aliquam ad minimum ad istam eradiatiorem tendentiam." Ibid.,* p. 29 reads, *"Deinde dicatur materia in eo consistere quod singulae* monades *puncta saltem sint, motu proprio destituta, sed ad eundem prona; lucisque et eradiationis capacia."* Emphasis added.

34. Helmont, *Cabbalistical Dialogue,* quotations on pp. 4, 9, 13.

35. Stein, *ibid.,* p. 209. Foucher de Careil, ed. *Nouvelles Lettres et Opuscules Inédits de Leibniz* (Paris, 1857), p. 328. "*Mihi omnis substantia operationum mire fertilis videtur. Sed a substantia (praeterquam infinita) substantiam, id est monada, produci non arbitror."* See also Leibniz, "Letter to Fardella," in Feller, p. 104 (also in *Opera Omnia*, vol. 2, p. 234): "*De natura monadum substantiarum quod porro quaeris, putem facile satisfieri posse, si speciatim indices quid in ea re explicari velis. De origine earum puto me iam dixisse, omnes sine dubio perpetuas esse nec nisi creatione*

NOTES

oriri ac nonnisi annihilatione interire posse, id est, naturaliter nec oriri, quod tantum est aggregatorum. Vellum videre antea liceret, quae de meis sententiis dices in tuo, quod moliris, Augustiniano opere."

36. Leibniz, *Philosophischen Schriften*, vol. 4, p. 510; translation is from Loemker, *Philosophical Papers*, vol. 2, p. 817.

37. Leibniz, trans. Loemker, *Philosophical Papers*, vol. 2, pp. 819–20.

38. Leibniz, "On the Principles of Nature and of Grace," *Philosophischen Schriften*, vol. 6; pp. 598–606; "The Monadology," 607–623, quotations from secs. 69, 66, 67.

39. Conway, *Principles*, p. 20

40. Leibniz, "The Monadology," secs. 20, 24, 21.

41. On the historical development of Leibniz's monadology, see the work of Selver, Wendt, Auerbach, and Stein, cited earlier. For a more recent discussion, see Politella. On the role of Chinese thought in Leibniz's philosophy, see J. Needham, *Science and Civilization in China* (Cambridge, Mass.: Harvard University Press, 1956), vol. 2, pp. 291–343, 496–505. On the Cabala and Maimonides, see A. Foucher de Careil, *Leibniz, la Philosophie juivre et la Cabale: Trois lectures . . . avec manuscrits inédits de Leibniz* (Paris, 1861).

42. On Princess Caroline of Wales, pupil of Leibniz at Hanover, see Leibniz, "The Controversy Between Leibniz and Clarke," *Philosophical Papers*, vol. 2, pp. 1095–1169; Leibniz, *Philosophischen Schriften*, vol. 7, pp. 345–440. Leibniz's correspondence with Lady Masham is collected in Leibniz, *Philosophischen Schriften*, vol. 3, pp. 336–75. On Gabrielle Émelie du Châtelet as an exponent of Leibnizian thought, see Carolyn [Merchant] Iltis, "Madame du Châtelet's Metaphysics and Mechanics," *Studies in History and Philosophy of Science* 8 (1977): 29–48, and W. H. Barber, "Mme. du Châtelet and Leibnizianism: The Genesis of the *Institutions de Physique*," in W. H. Barber [and others] ed., *The Age of the Enlightenment: Studies Presented to Theodore Besterman* (Edinburgh and London: Oliver & Boyd, 1967), pp. 200–222.

43. Hannah Wooley, *The Gentlewomen's Companion* (London, 1673; first published, 1655); [Bathsua Makin], *An Essay to Re-*

vive the Antient Education of Gentlewomen, in Religion, Manners, Arts, and Tongues (London, 1673); Mary Astell, *A Serious Proposal to the Ladies for the Advancement of Their True and Greatest Interest . . .* (London, 1694). On seventeenth-century feminist ideas concerning women's education, see Hilda Smith, "Reason's Disciples: Seventeenth Century English Feminists, " doctoral dissertation, University of Chicago, 1975, pp. 167–79. On women's learning see Myra Reynolds, *The Learned Lady in England, 1650–1760* (Boston: Houghton Mifflin, 1920).

44. Feminist books reprinted in England included Henry Cornelius Agrippa, *De Nobilitate et Praecellentia foeminei sexus* (1525). English translations: H. C. Agrippa, *The Glory of Women: or a Lookingglasse for Ladies* (London, 1652); H. C. Agrippa, *Female Pre-eminence; or the Dignity and Excellency of that Sex, Above the Male* (London, 1670). François Poulain de la Barre, *The Women as Good as the Man; or the Equality of Both Sexes,* trans. A. L. (London, 1677; first published 1673). On Poulain de la Barre see Michael A. Seidel, "The Woman as Good as the Man," *Journal of the History of Ideas* 35 (July-Sept. 1974): 499–508.

45. [Margaret Cavendish] *The Philosophical and Physical Opinions* (London, Martin and Allestrye, 1655), preface, "To the Two Universities." Discussions of Margaret Cavendish's feminism and scientific work include Hilda Smith, "Reason's Disciples," pp. 106–14; Douglass Grant, *Margaret the First: A Biography of Margaret Cavendish, Duchess of Newcastle, 1623–1673* (London: Hart-Davis, 1957); Gerald Dennis Meyer, *The Scientific Lady in England* (Berkeley and Los Angeles: University of California Press, 1955), pp. 1–15; R. H. Kargon, *Atomism in England from Hariot to Newton* (Oxford, England: Clarendon Press, 1966), pp. 73–76.

46. [Margaret Cavendish,] *Poems and Fancies* (London: Martin and Allestrye, 1653), "An Epistle to Mistris Toppe," p. A4.

47. [Cavendish,] *Poems and Fancies,* Preface, "To All Writing Ladies."

48. [Cavendish,] *Poems and Fancies,* Pre-

334

face, "To All Noble and Writing Ladies," p. A3. Poetry quotations on pp. 3, 5.

49. Margaret Cavendish, *Grounds of Natural Philosophy* (London, 1668), pp. 1–3; Smith, "Reason's Disciples," pp. 109–11.

50. Meyer, pp. 10–11; Cavendish, *The Description of a New World Called the Blazing-World* (London, 1668), preface, "To the Reader," and pp. 4, 15, and passim.

51. Bernard de Fontenelle, *Entretiens sur la Pluralité des Mondes* (Paris, 1686); English trans.: B. de Fontenelle, *Week's Conversation on the Plurality of Worlds,* trans. William Gardiner (London, 1737), pp. iv–v, 16, xi–xii. Aphra Behn's translation was B. de Fontenelle, *"A Discovery of New Worlds,* from the French, made English by Mrs. Aphra Behn . . . Wholly New" (London, 1688). For a discussion see Meyer, pp. 21–22.

52. Meyer, pp. 49–70.

53. Francesco Algarotti, *Il Newtonianismo per le Dame* (Naples, 1737). For a discussion, see Meyer, pp. 29–32. Elizabeth Carter's translation was F. Algarotti, *Si, Isaac Newton's Philosophy Explain'd For the Use of the Ladies. In Six Dialogues on Light and Colours. From the Italian of Sig. Algarotti* (London, 1739).

54. Meyer, pp. 36–48.

CHAPTER 12:
NEWTON AND LEIBNIZ

1. Isaac Newton, *Optice sive de Reflexionibus, Refractionibus, Inflexionibus Lucis* (London, 1706). Citations refer to I. Newton, *Opticks*, 4th ed., 1730 (reprinted New York: Dover, 1952; first published 1704), see p. 403.

2. Newton, University Library, Cambridge, England, Additional Manuscripts 3970, folio 619r.

3. Isaac Newton, *Philosophiae Naturalis Principia Mathematica* (London, 1687). Citations refer to I. Newton, *Mathematical Principles of Natural Philosophy*, trans. A. Motte, 1729, rev. Florian Cajori (Berkeley: University of California Press, 1934), Definition III, p. 2. For a discussion of Newton's modification of the Cartesian ontology, see Ernan McMullin, *Newton on Matter and Activity* (South Bend, Ind.: University of Notre Dame Press, 1978), pp.

33–43. On Newton's transformation of Cartesianism, see Richard Westfall, *Force in Newton's Physics* (London: Macdonald, 1971); Alexandre Koyré, "The Significance of the Newtonian Synthesis," in *Newtonian Studies* (Chicago: University of Chicago Press, 1965), pp. 3–24. Alan Gabbey, "Force and Inertia in Seventeenth Century Dynamics," *Studies in History and Philosophy of Science* 2 (May 1971): 1–67.

4. Newton, *Opticks*, 4th ed., 1730, query 23 (31), p. 400. The Latin edition of 1706 contained twenty-three queries. To the 1717 edition, eight new queries were inserted and numbered 17–24. Query 23 of the 1706 edition appears as Query 31 in the 1717 edition. For a discussion of Newton's changing views on the "cause of gravitation" and on atoms and the void, see J. E. McGuire, "Force, Active Principles and Newton's Invisible Realm," *Ambix* 15 (1968): 154–208, see pp. 155, 174–75.

5. Arnold Thackray, *Atoms and Powers* (Cambridge, Mass.: Harvard University Press, 1970), Chap. 2.

6. On Newton's mechanics, see Westfall, *Force in Newton's Physics*; I. Bernard Cohen, "Newton's Second Law and the Concept of Force in the *Principia*," *Texas Quarterly* 10 (1967): 127–57; McMullin, *Newton on Matter and Activity*; Henry Guerlac, *Essays and Papers in the History of Modern Science* (Baltimore, Md.: Johns Hopkins University Press, 1977), pp. 69–242. On eighteenth-century chemistry, see Thackray, *Atoms and Powers*; on electricity, see I. B. Cohen, *Franklin and Newton* (Cambridge, Mass.: Harvard University Press, 1966). On the political and social context of Newtonianism, see David Kubrin, "How Sir Isaac Newton Helped Restore Law 'n' Order to the West" (San Francisco: Kubrin, 1972); Margaret C. Jacob, *The Newtonians and the English Revolution,* 1689–1720 (Ithaca, N.Y.: Cornell University Press, 1976); On the relationships between civil law and the laws of nature after the Restoration, see Barbara Shapiro, "Law and Science in Seventeenth-Century England," *Stanford Law Review* 21 (1969): 727–63.

7. Gottfried Wilhelm von Leibniz, "Brevis demonstratio erroris memorabilis Cartesii et aliorum circa legem naturalem, secundum quam volunt a Deo eandem semper quantitatem motus conservari; quo et in re mechanica abutuntur," *Acta Eruditorum* (1686): 161–63. English translation: Gottfried Wilhelm Leibniz, *Philosophical Papers and Letters*, trans. Leroy E. Loemker, 2 vols. (Chicago: Univ. of Chicago Press, 1956), vol. 1, pp. 455–63. G. W. Leibniz, *Discours de metaphysique* (written 1686), in *Die Philosophischen Schriften von Gottfried Wilhelm Leibnitz*, ed. C. I. Gerhardt, 7 vols. (Berlin, 1875–1890), vol. 4, pp. 442, 443. English translation in Loemker, vol. 1, pp. 464–506. G. W. Leibniz, "Essay de dynamique sur les loix du mouvement, ou il est monstre, qu'il ne se conserve pas la même quantite de mouvement, mais la même force absolue, ou bien la même quantité de l'action motrice," *Mathematische Schriften*, pp. 215–31. English translation: G. W. Leibniz, *New Essays Concerning Human Understanding*, ed. and trans. A. G. Langley (La Salle: Open Court, 1949), appendix pp. 657–70. On the development of Leibniz's dynamics, see Carolyn [Merchant] Iltis, "Leibniz and the *Vis Viva* Controversy," *Isis* 62 (1970): 21–35, and Pierre Costabel, *Leibniz et la Dynamique: Les Textes de 1692* (Paris: Hermann, 1960).

8. G. W. Leibniz, "Specimen Dynamicum," *Mathematische Schriften*, vol. 4, pp. 234–54; *Philosophical Papers*, vol. 2, pp. 714–18. See also Gerd Buchdahl, *Metaphysics and the Philosophy of Science* (Oxford, England: Blackwell, 1969), pp. 393, 410, 414, 417, 420, 422–23; C. D. Broad, *Leibniz: An Introduction* (Cambridge, England: Cambridge University Press, 1975), Chap. 4, pp. 87–129; George Gale, "The Physical Theory of Leibniz," *Studia Leibnitiana*, vol. 2, no. 2 (1970): 114–27.

9. G. W. Leibniz, "On the Ultimate Origin of Things" (written 1697) in Phillip Wiener, ed., *Leibniz Selections* (New York: Scribner's, 1951), p. 354; see also pp. *xix–xxi*. Clarence Glacken, *Traces on the Rhodian Shore* (Berkeley: University of California Press, 1967), pp. 477–78. On Leibniz and capitalism see Jon Elster, *Leibniz et la Formation de l'Esprit Capitaliste* (Paris: Montaigne, 1975).

10. Samuel Clarke, *A Collection of Papers Which Passed Between the Late Learned Mr. Leibniz and Dr. Clarke* (London, 1717). For a discussion, see Carolyn [Merchant] Iltis, "The Leibnizian-Newtonian Debates: Natural Philosophy and Social Psychology," *The British Journal for the History of Science* 4 (December 1973): 343–77.

11. On the concept of imposed law and the voluntarist tradition in the rise of modern science, see Francis Oakley, "Christian Theology and the Newtonian Science: The Rise of the Concept of the Laws of Nature," *Church History* 30 (1961): 443–57, esp. 433–38. On the voluntarist background to Boyle's philosophy, see J. E. McGuire, "Boyle's Conception of Nature," *Journal of the History of Ideas* 33 (1972): 523–542. On the Greek view of nature as an organism, see R. G. Collingwood, *The Idea of Nature* (Oxford, England: Clarendon Press, 1945), pp. 3–9. On internal versus external relations, see Alfred North Whitehead, *Adventures of Ideas* (New York: Macmillan, 1933), pp. 142–47.

12. G. W. Leibniz, "De ipsa natura sive de vi insita actionibusque Creaturum, pro Dynamicis suis confirmandis illustrandisque" (written 1698) in *Philosophischen Schriften*, pp. 504–16, quotations on pp. 512, 506–7. Trans. from L. E. Loemker in G. W. Leibniz, *Philosophical Papers*, vol. 2, pp. 819, 812–13, Italics added to quotations.

13. Leibniz, "The Monadology," *Philosophischen Schriften*, vol. 6, pp. 607–23, see p. 607, sec. 7; *Philosophical Papers*, vol. 2, pp. 1044–1061.

14. G. W. Leibniz, "De Ipsa Natura," *Philosophischen Schriften*, vol. 4, p. 508: "*Quodsi quis defensor philosophiae nova, inertiam rerum et torporem introducentis, eo usque progrediatur*"; G. W. Leibniz, *Discourse on Metaphysics, Correspondence with Arnauld, and Monadology*, trans. George R. Montgomery (LaSalle, Ill. Open Court, 1957), p. 233.

15. Leibniz, "The Monadology," pp. 598–606.

16. Newton, *Opticks*, p. 397.

17. Newton, University Library, Cam-

18. I. Newton, "Of Nature's Obvious Laws and Processes in Vegetation," Smithsonian Institution, Washington, D.C., Burndy Ms. 16. Quotations from fols. 4^r, 3^v, 4^r, 3^v, 5^r, 5^v, 6^r, 4^v. For an edition of and commentary on this important manuscript, see Betty Jo Teter Dobbs, "Newton on the Vegetation of Metals and Other Subjects," in Z. Bechler, ed., *Contemporary Newtonian Research* (1978), vol. 1. Other discussions of fermentation by Newton in this period include the "Letter to Boyle," Feb. 28, 1678–79, and the "Letter to Oldenburg," Jan. 25, 1675–76 in I. B. Cohen, ed., *Isaac Newton's Papers and Letters on Natural Philosophy* (Cambridge, Mass.: Harvard University Press, 1958), pp. 253, 254. On Newton's alchemy see B. J. T. Dobbs, *The Foundations of Newton's Alchemy* (Cambridge, England: Cambridge University Press, 1975).

19. James Murry and others, *Oxford English Dictionary* (Oxford: Clarendon Press, 1933), vol. 4, p. 163; quotation from John Dryden, *Absalom and Achitophel* (Dublin[?], 1681), p. 140. On the historical significance of fermentation, see Aristotle, *De Generatione Animalium*, trans. Arthur Platt (Oxford, England: Clarendon Press, 1910); Walter Pagel, *William Harvey's Biological Ideas* (Basel: Karger, 1967), pp. 267, 272; Audrey Davis, *Circulation Physiology and Medical Chemistry in England, 1650–1680* (Lawrence, Kans.: Coronado Press, 1973), pp. 102–5, 113, 208; Everett Mendelsohn, *Heat and Life* (Cambridge, Mass.: Harvard University Press, 1964), pp. 27–66. William Harvey, *Exercitationes de Generatione Animalium*, in *Works*, trans. R. Willis (reprint ed., New York: Johnson Reprint Corporation, 1965; first published, 1847), p. 375; W. Pagel, *New Light on William Harvey* (Basel: Karger, 1976), p. 137.

20. I. Newton, *Opticks*, pp. 379–80. Subsequent quotations on pp. 399, 400.

21. I. Newton, University Library, Cambridge, England, Add. ms. 3970 fol. 620^v and 619^r.

22. I. Newton, University Library, Cambridge, England, Add. 3970 fol. 619^r.

23. Newton, *Opticks*, p. 399; italics added. On gravitation see J. E. McGuire, "Force, Active Principles and Newton's Invisible Realm;" and P. M. Heimann, "'Nature is a Perpetual Worker': Newton's Aether and Eighteenth-Century Natural Philosophy," *Ambix* 20 (March 1973): 1–25; P. M. Heimann and J. E. McGuire, "Newtonian Forces and Lockean Powers: Concepts of Matter in Eighteenth-Century Thought," *Historical Studies in the Physical Sciences* 3 (1971): 233–306; David Kubrin, "Newton and the Cyclical Cosmos: Providence and the Mechanical Philosophy, *Journal of the History of Ideas* 28 (July-September 1967): 325–46.

24. Newton, *Opticks*, pp. 399–400, subsequent quotations in order on pp. 380, 401, 403. For more on the vegetative spirit and fermentation see I. Newton, University Library, Cambridge, England, Add. 3970 fol. 237^r "De Vita et Morte Vegetabili."

25. On the transformation in cosmology see Koyré, "Newtonian Synthesis," in *Newtonian Studies*, and Koyré, *From the Closed World to the Infinite Universe* (New York: Harper & Row, 1958). On the transformation in economic values from those of Aristotle to those of Adam Smith and the endowment of money with organic properties see Michael Taussig, "The Genesis of Capitalism Amongst a South American Peasantry: Devil's Labor and the Baptism of Money," *Comparative Studies in Society and History* 19 (April 1977): 130–53.

26. The impact of Newton and Leibniz on subsequent history is complex. On Newtonianism in the eighteenth century, see Thackray, *Atoms and Powers*; Heimann and McGuire, "Newtonian Forces and Lockean Powers"; Robert Schofield, *Mechanism and Materialism: British Natural Philosophy in an Age of Reason* (Princeton, N.J.: Princeton University Press, 1970); Guerlac, *Essays and Papers*. On Leibniz, see W. H. Barber, *Leibniz in France from Arnauld to Voltaire: A Study in French Reactions to Leibnizianism, 1670–1760* (Oxford, 1955); Carolyn [Merchant] Iltis, "The Decline of Cartesianism in Mechanics: The Leibnizian-Cartesian Debates," *Isis* 64 (1973): 356–73; C. [M.] Iltis, "The Leibnizian–Newtonian Debates," C. [M.] Iltis, "Madame du Châtelet's Metaphysics and Mechanics," *Studies* *in History and Philosophy of Science* 8 (1977): 29–40; C. [M.] Iltis "D'Alembert and the *Vis Viva* Controversy," *Studies in History and Philosophy of Science* 1 (1970): 134–35; Joseph Needham, *Science and Civilization in China* (Cambridge, Mass.: Harvard University Press, 1959), vol. 2, pp. 291–343, 496–505. On the "bootstrap model" in particle physics see Fritjof Capra, *The Tao of Physics* (Berkeley, Cal.: Shambala, 1975). On the incommensurability of the mechanical and organic world views and its implication for developmental psychology, see Willis F. Overton and Hayne W. Reese, "Models of Development: Methodological Implications" in John R. Nesselroade and Hayne W. Reese, eds., *Life Span Developmental Psychology* (New York: Academic Press, 1973). On organismic perspectives on nature see L. R. Wheeler, *Vitalism: Its History and Validity* (London: Witherby, 1939); Alfred North Whitehead, *Process and Reality* (Cambridge, England: Cambridge University Press, 1929); Dorothy Emmet, *Whitehead's Philosophy of Organism*, 2nd ed. (London: Macmillan, 1966); Howard L. Parsons, ed., *Marx and Engels on Ecology* (Westport, Conn.: Greenwood Press, 1977); Wilhelm Reich, *The Discovery of the Orgone* (New York: Orgone Institute Press, 1942); for a critique, see D. C. Phillips, *Holistic Thought in Social Science* (Stanford, Cal.: Stanford University Press, 1976).

EPILOGUE

1. Jacques Ellul, *The Technological Society*, trans. J. Wilkinson (New York: Random House, 1964), pp. 163 ff., quotation on p. 163; B. F. Skinner, *Beyond Freedom and Dignity* (New York: Random House, 1971). See also B. F. Skinner, *Walden Two* (New York: Macmillan, 1972; first published 1948). For a philosophical critique of systems theory, see Hubert Dreyfus, *What Computers Can't Do* (New York: Harper & Row, 1972), especially p. 170.

2. J. C. Smuts, *Holism and Evolution* (New York: Macmillan, 1926), pp. 86, 87. On holism in the biological sciences, see Arthur Koestler, "Beyond Holism and Reductionism: The Concept of the Holon," in *Beyond Reductionism: New Perspectives in the Life Sciences,* ed. A. Koestler and J. R. Smythies (Boston: Beacon Press, 1969).

3. On ecological cycles, see Barry Commoner, *The Closing Circle: Nature, Man, and Technology* (New York; Bantam Books, 1971), Chap. 2.

4. Samuel P. Hays, *Conservation and the Gospel of Efficiency: The Progressive Conservation Movement, 1890–1920* (Cambridge, Mass.: Harvard University Press, 1959), pp. 142–43.

5. Murray Bookchin, "Ecology and Revolutionary Thought," in *Post-Scarcity Anarchism* (San Francisco: Ramparts Press, 1971), pp. 57–82, and M. Bookchin, "Toward an Ecological Solution" (Berkeley, Cal.: Ecology Center Reprint, n. d.). See also Victor Ferkiss, *Technological Man* (New York: New American Library, 1969), Chap. 9, pp. 205–11; Theodore Roszak, *Where the Wasteland Ends* (Garden City, N.Y.: Doubleday, 1973), pp. 367–71; Paul Goodman and Percival Goodman, *Communitas,* 2nd ed., rev. (New York: Vintage, 1960); Paul Goodman, *People or Personnel* (New York: Random House, 1965); E. F. Schumacher, *Small Is Beautiful: Economics as if People Mattered* (New York: Harper & Row, 1973); Ernest Callenbach, *Ecotopia* (Berkeley, Cal.: Banyan Tree Books, 1976).

Index

339